# Polymer Biocatalysis and Biomaterials

ACS SYMPOSIUM SERIES **900**

# Polymer Biocatalysis and Biomaterials

**H. N. Cheng,** Editor
*Hercules Incorporated Research Center*

**Richard A. Gross,** Editor
*Polytechnic University*

Sponsored by the
**ACS Divisions of Polymer Chemistry, Inc. and Biological Chemistry**

American Chemical Society, Washington, DC

**Library of Congress Cataloging-in-Publication Data**

Polymer biocatalysis and biomaterials / H. N. Cheng, editor, Richard A. Gross, editor,; sponsored by the ACS Divisions of Polymer Chemistry, Inc. and Biological Chemistry.

    p. cm.—(ACS symposium series ; 900)

    Includes bibliographical references and index.

    ISBN 0-8412-3917-7 (alk. paper)

    1. Enzymes—Biotechnology—Congresses. 2. Polymers—Biotechnology—Congresses. 3. Biomedical materials—Congresses.

    I. Cheng, H. N. II. Gross, Richard A., 1957- III. American Chemical Society. Division of Polymer Chemistry, Inc. IV. American Chemical Society. Division of Biological Chemistry. V. American Chemical Society. Meeting (226$^{th}$ : 2003 : New York, N.Y.) VI. Series.

TP248.65.E59P64    2005
660.6´3—dc22                                            2004062661

The paper used in this publication meets the minimum requirements of American National Standard for Information Sciences—Permanence of Paper for Printed Library Materials, ANSI Z39.48–1984.

Copyright © 2005 American Chemical Society

Distributed by Oxford University Press

All Rights Reserved. Reprographic copying beyond that permitted by Sections 107 or 108 of the U.S. Copyright Act is allowed for internal use only, provided that a per-chapter fee of $30.00 plus $0.75 per page is paid to the Copyright Clearance Center, Inc., 222 Rosewood Drive, Danvers, MA 01923, USA. Republication or reproduction for sale of pages in this book is permitted only under license from ACS. Direct these and other permission requests to ACS Copyright Office, Publications Division, 1155 16th Street, N.W., Washington, DC 20036.

The citation of trade names and/or names of manufacturers in this publication is not to be construed as an endorsement or as approval by ACS of the commercial products or services referenced herein; nor should the mere reference herein to any drawing, specification, chemical process, or other data be regarded as a license or as a conveyance of any right or permission to the holder, reader, or any other person or corporation, to manufacture, reproduce, use, or sell any patented invention or copyrighted work that may in any way be related thereto. Registered names, trademarks, etc., used in this publication, even without specific indication thereof, are not to be considered unprotected by law.

PRINTED IN THE UNITED STATES OF AMERICA

# Foreword

The ACS Symposium Series was first published in 1974 to provide a mechanism for publishing symposia quickly in book form. The purpose of the series is to publish timely, comprehensive books developed from ACS sponsored symposia based on current scientific research. Occasionally, books are developed from symposia sponsored by other organizations when the topic is of keen interest to the chemistry audience.

Before agreeing to publish a book, the proposed table of contents is reviewed for appropriate and comprehensive coverage and for interest to the audience. Some papers may be excluded to better focus the book; others may be added to provide comprehensiveness. When appropriate, overview or introductory chapters are added. Drafts of chapters are peer-reviewed prior to final acceptance or rejection, and manuscripts are prepared in camera-ready format.

As a rule, only original research papers and original review papers are included in the volumes. Verbatim reproductions of previously published papers are not accepted.

**ACS Books Department**

# Contents

Preface..............................................................................................xi

1. Polymer Biocatalysis and Biomaterials.....................................................1
   H. N. Cheng and Richard A. Gross

## New Enzyme Methodologies

2. Biotechnology: Key to Developing Sustainable Technology
   for the 21$^{st}$ Century: Illustrated in Three Case Studies........................14
   Lori A. Henderson

3. Empirical Biocatalyst Engineering: Escaping the Tyranny
   of High-Throughput Screening................................................................37
   Jon E. Ness, Tony Cox, Sridhar Govindarajan, Claes Gustafsson,
   Richard A. Gross, and Jeremy Minshull

4. $^1$H NMR for High-Throughput Screening to Identify Novel
   Enzyme Activity.......................................................................................51
   Claire B. Conboy and Kai Li

5. Enzyme Immobilization onto Ultrahigh Specific Surface
   Cellulose Fibers via Amphiphilic (PEG) Spacers and
   Electrolyte (PAA) Grafts........................................................................63
   You-Lo Hsieh, Yuhong Wang, and Hong Chen

6. Nondestructive Regioselective Modification of Laccase
   by Linear-Dendritic Copolymers: Enhanced Oxidation
   of Benzo-α-Pyrene in Water..................................................................80
   Ivan Gitsov, Kevin Lambrych, Peng Lu, James Nakas,
   Joseph Ryan, and Stuart Tanenbaum

## Novel Biomaterials

7. Functional Hydrogel–Biomineral Composites Inspired by Natural Bone................................................................................................96
   Jie Song and Carolyn R. Bertozzi

8. Biomimetic Approach to Biomaterials: Amino Acid-Residue-Specific Enzymes for Protein Grafting and Cross-Linking................107
   Fianhong Chen, David A. Small, Martin K. McDermott, William E. Bentley, and Gregory F. Payne

9. Biodegradable Films from Pectin–Starch and Pectin–Poly(vinyl alcohol)..............................................................................119
   Marshall L. Fishman and David R. Coffin

10. Synthesis of Zein Derivatives and Their Mechanical Properties........141
    Atanu Biswas, David J. Sessa, Sherald H. Gordon, John W. Lawton, and J. L. Willett

## Silicon-Containing Materials

11. Protein Mediated Bioinspired Mineralization........................................150
    Siddharth V. Patwardhan, Kiyotaka Shiba, Christina Raab, Nicola Huesing, and Stephen J. Clarson

12. Biocatalysis of Siloxane Bonds................................................................164
    Alan R. Bassindale, Kurt F. Brandstadt, Thomas H. Lane, and Peter G. Taylor

13. "Sweet Silicones": Biocatalytic Reactions to Form Organosilicon Carbohydrate Macromers............................................182
    Bishwabhusan Sahoo, Kurt F. Brandstadt, Thomas H. Lane, and Richard A. Gross

## Polysaccharides

14. Enzymatic Synthesis of Complex Bacterial Carbohydrate Polymers....................................................................................................192
    Hanfen Li, Hesheng Zhang, Wen Yi, Jun Shao, and Peng George Wang

15. Enzymatic Polymerization: In Vitro Synthesis of
    Glycosaminoglycans and Their Derivatives..................217
    Shiro Kobayashi, Shun-ichi Fujikawa, Ryosuke Itoh,
    Hidekazu Morii, Hirofumi Ochiai, Tomonori Mori,
    and Masashi Ohmae

16. Sugar Polymer Engineering with Glycosaminoglycan
    Synthase Enzymes: 5 to 5,000 Sugars and a Dozen Flavors.............232
    Paul L. DeAngelis

17. Enzyme-Catalyzed Regioselective Modification of Starch
    Nanoparticles..................246
    Soma Chakraborty, Bishwabhusan Sahoo, Iwao Teraoka,
    Lisa M. Miller, and Richard A. Gross

18. Reactions of Enzymes with Non-Substrate Polymers..................267
    H. N. Cheng, Qu-Ming Gu, and Lei Qiao

# Condensation Polymers: Whole Cell and Related Approaches

19. Preparation, Properties, and Utilization of Biobased
    Biodegradable *Nodax*™ Copolymers..................280
    Isao Noda, Eric B. Bond, Phillip R. Green, David H. Melik,
    Karunakaran Narasimhan, Lee A. Schechtman,
    and Michael M. Satkowski

20. Novel Synthesis Routes forPolyhydroxyalkanoic Acids
    with Unique Properties..................292
    Bo-Zhang, R. Carlson, E. N. Pederson, B. Witholt, and F. Srienc

21. Kymene® G3-X Wet-Strength Resin: Enzymatic Treatment
    during Microbial Dehalogenation..................302
    Richard J. Riehle

# Condensation Polymers: Enzymatic Approaches

22. Lipase Catalyzed Polyesterifications..................318
    Anil Mahapatro, Ajay Kumar, Bhanu Kalra, and
    Richard A. Gross

23. Versatile Route to Polyol Polyesters by Lipase Catalysis ........ 327
    Ankur S. Kulshrestha, Ajay Kumar, Wei Gao,
    and Richard A. Gross

24. Biocompatibility of Sorbitol-Containing Polyesters:
    Synthesis, Surface Analysis, and Cell Response In Vitro ........ 343
    Ying Mei, Ajay Kumar, Wei Gao, Richard Gross,
    Scott B. Kennedy, Newell R. Washburn, Eric J. Amis,
    and John T. Elliott

25. Enzyme-Catalyzed Synthesis of Hyperbranched Aliphatic
    Polyesters ........ 354
    Ingo T. Neuner, Mihaela Ursu, and Holger Frey

26. Enantioenriched Substituted Polycaprolactones by Enzyme
    Catalysis ........ 366
    Kirpal S. Bisht, Leelakrishna Kondaveti, and Jon D. Stewart

27. Enzymatic Ring Opening Polymerization of ε-Caprolactone
    in Supercritical $CO_2$ ........ 393
    Takahiko Nakaoki, Makoto Kitoh, and Richard A. Gross

28. Synergies between Lipase and Chemical Polymerization
    Catalyst ........ 405
    Bhanu Kalra, Irene Lai, and Richard A. Gross

## Other Examples of Biocatalysis

29. Polymers from Sugars: Chemoenzymatic Synthesis and
    Polymerization of Vinylethylglucoside ........ 420
    Bhanu Kalra, Mania Bankova, and Richard A. Gross

30. Enzyme-Catalyzed Condensation Reactions for Polymer
    Modifications ........ 427
    Qu-Ming Gu and H. N. Cheng

Epilogue: A Rhyme on Enzymes ........ 437
    H. N. Cheng

Author Index ........ 439

Subject Index ........ 441

# Preface

Biocatalysis involves the use of enzymes, microbes, and higher organisms to carry out chemical reactions. These biocatalysts are known to operate under mild conditions, with impressive selectivity, on a diverse range of natural and non-natural substrates. They furnish exciting opportunities to manipulate polymer structures, to discover new reaction pathways, and to devise environmentally friendly processes. Recently, the rapid pace of developments in biotechnology has made possible more, better, and cheaper biocatalysts. These and other technical advances have expanded the scope and the capabilities of this technology and set the stage for real progress in reaching commercial targets.

Another dynamic, and related, field is that of biomaterials. Indeed, biomaterials found in nature are mostly made in vivo via biocatalytic reactions. There is a lot of current research activity in this area, and much creativity is being displayed in producing these biomaterials in vitro via chemical and biocatalytic means.

It is of interest to note that these fields are highly multidisciplinary. A successful research endeavor often requires the collaborative efforts of people with different backgrounds (e.g., organic chemistry, polymer chemistry, material science, chemical engineering, biochemistry, molecular biology, microbiology, and enzymology). It is useful for the researchers to communicate with one another and to share information through meetings, symposia, and books.

In view of the rapid progress and increasing interest in these fields, we embarked upon this book endeavor. The purpose of this book is to publish, in one volume, the latest findings of the leading researchers in polymer biocatalysis and biomaterials. A total of 30 chapters are included, covering the following topics:
1. novel syntheses of monomers and polymers with enzymes or microorganisms.
2. biocatalyzed polymer modification reactions

3. biodegradation and bioconversion of materials
4. new biocatalytic methodologies (e.g., protein engineering and directed evolution, metabolic engineering, in vitro or whole cell strategies, and immobilization methods)
5. new and improved biomaterials

This book is based on a successful symposium organized by the editors and held at the American Chemical Society (ACS) National Meeting in New York in September 2003. The symposium comprised 56 technical papers and posters, and attracted a sizeable and enthusiastic audience. Many exciting new techniques, new materials, new reactions, and new processes were reported. Hopefully this book serves to transmit the latest information to the readers and also captures some of the excitement of the symposium in the process.

We thank the authors for their contributions and their patience while the manuscripts were being reviewed and revised. Thanks are also due to the ACS Division of Polymer Chemistry, Inc. for sponsoring the symposium and the ACS Division of Biological Chemistry for cosponsorship. We also acknowledge the generous funding from the ACS Division of Polymer Chemistry, Inc. and ACS Corporation Associates.

## H. N. Cheng
Hercules Incorporated Research Center
500 Hercules Road
Wilmington, DE 19808–1599

## Richard A. Gross
NSF I/UCRC for Biocatalysis and Bioprocessing of Macromolecules
Polytechnic University
6 Metrotech Center
Brooklyn, NY 11201

# Chapter 1

# Polymer Biocatalysis and Biomaterials

## H. N. Cheng[1] and Richard A. Gross[2]

[1]Hercules Incorporated Research Center, 500 Hercules Road, Wilmington, DE 19808–1599
[2]NSF I/UCRC for Biocatalysis and Bioprocessing of Macromolecules, Polytechnic University, 6 Metrotech Center, Brooklyn, NY 11201

This overview briefly surveys the use of enzymatic and whole-cell approaches in polymers. Three types of reactions are covered: polymer syntheses, polymer modifications, and polymer hydrolyses. Thus far, most of the enzyme-related R&D activities involve hydrolases, oxidoreductases, and transferases, with occasional use of lyases and isomerases. Whole-cell methods continue to be valuable in both academic and industrial labs. All these research areas display continued vitality and creativity, as evidenced by the large number of publications. Advances in biotechnology have provided new and improved enzymes and additional tools. Also included in this overview is the related topic of biomaterials.

The use of enzymes and whole-cell approaches in polymers is now fairly widespread. A large number of reactions and processes has been developed, and new developments continue to appear in both the open and the patent literatures. Several excellent books *(1,2)* and reviews *(3)* are available.

Biomaterials comprise an equally exciting field of research that finds many applications in dental, surgical, and medical areas *(4)*. Both fields are highly interdisciplinary, requiring (at various times) knowledge and expertise in organic and polymer chemistry, material science, biochemistry, molecular biology, and chemical engineering.

This overview covers some of the recent developments in these fields, with a particular emphasis on the articles included in this book (5-33).

## Enzyme Biocatalysis

Enzymes are commonly classified, via a system of Enzyme Commission (EC) numbers, into six divisions: oxidoreductases, transferases, hydrolases, lyases, isomerases, and ligases (34). In this work, we are concerned with three types of polymer reactions: polymer syntheses, polymer modifications, and polymer degradation and hydrolyses. For these reactions, only hydrolases, oxidoreductases, and transferases are being used extensively in polymers and biomaterials. A summary is given in Table 1.

Table 1. The Use of Biocatalysts in Polymer Reactions

| Biocatalyst Type | EC | Polymer Syntheses | Polymer Modifications | Polymer Hydrolyses |
|---|---|---|---|---|
| Oxidoreductase | 1 | X | X | X |
| Transferase | 2 | X | X | |
| Hydrolase | 3 | X | X | X |
| Lyase | 4 | | | X |
| Isomerase | 5 | | X | |
| Ligase | 6 | | | |
| Whole-cell | - | X | X | X |

### Hydrolases (EC 3)

As a group, hydrolases are used more often in polymers than all other enzymes. Many hydrolases can accept different substrates and have utility for a variety of reactions. Their popularity is assisted by the commercial availability of many hydrolases and their relatively lower prices.

*Hydrolase-Catalyzed Polymerizations*

A very active and fruitful area of research is the use of lipases for the synthesis of polyesters, polylactones, and polycarbonates (35). Many creative reactions have been devised. In this book, lipase-catalyzed polycondensation of diols/diacids was reported separately by Mahapatro et al (25), Kulshrestha et al

*(26)*, and Mei et al *(27)*. The synthesis of several chiral substituted poly(ε-caprolactone) was achieved by Bisht et al *(29)*. Nakaoki et al *(30)* described the enzymatic polymerization of poly(ε-caprolactone) in supercritical $CO_2$. Frey et al *(28)* reported the synthesis of hyperbranched poly(ε-caprolactone) using concurrent ring-opening polymerization and polycondensation. Kalra et al *(31)* copolymerized L-lactide and ω-pentadecalactone using lipase and an organometallic catalyst.

Another class of polymers where hydrolases are used successfully comprises polysaccharides. This approach has been pioneered by Kobayashi et al *(3c,36)* and is exemplified by their article *(18)* where hyaluronan and chondroitin were prepared via hyaluronidase-catalyzed polymerizations. A slightly different approach is to use glycosidases to prepare oligosaccharides *(17,37)*.

A recent development is silicon bioscience *(38)*. An example in this book is the paper by Bassindale et al *(15)* where several lipases and proteases were engaged to catalyze the condensation of alkoxysilanes.

It may be noted that lipases are also commonly used for the synthesis of organic compounds, including monomers and reactive oligomers. For example, Kalra et al *(32)* prepared vinylethylglucoside, and Gu et al *(33)* prepared substituted acrylic monomers, both with lipases. These monomers were subsequently polymerized by conventional methods.

*Hydrolase-Catalyzed Polymer Modifications*

Instead of polymer synthesis, hydrolases can be used to add, change, or remove functional groups on existing polymers. Several examples are given in this book, including protease-catalyzed acylation of polysaccharides *(21)*, papain-catalyzed amidation of pectin *(21)*, hydrolase-catalyzed amidation of carboxymethylcellulose *(33)*, and lipase-catalyzed syntheses of fatty acid diester of poly(ethylene glycol) *(33)*, fatty acid ester of cationic guar *(33)*, modified starch *(26)*, and glycosilicone conjugates *(16)*.

*Hydrolysis with Hydrolases*

As their name implies, hydrolysis is the preferred reaction for hydrolases in water under optimal conditions. Some examples in the polysaccharide area include cellulolytic enzymes for biomass conversion *(5)*, proteases for the degradation of guar gum *(21)*, cellulase for viscosity reduction of xanthan gum *(21)*, and β-D-galactosidase for pectin hydrolysis *(21)*.

Two examples in the non-polysaccharide area are the use of Alcalase® protease for the hydrolysis of end-terminated esters in polyamide *(24)* and nitrilase for the bioconversion of nitriles to carboxylic acids *(7)*.

## Oxidoreductases (EC 1)

Many enzymes in this family require cofactors, which are inconvenient and often expensive in commercial applications. Not surprisingly, most of the reactions reported thus far in polymer science deal with enzymes that require no cofactors.

*Oxidoreductases in Polymer Synthesis*

Thus far, oxidoreductases are mostly engaged in polymer synthesis in two major areas. First, they are employed for the polymerization of phenols and anilines. In the former case, the products include polyphenols *(39)* and poly(phenylene oxide) *(40)*. In the latter case, water-soluble polyaniline polymers can be made *(41)*. Secondly, oxidoreductases are employed for the free-radical polymerization of vinyl monomers *(42)*.

*Oxidoreductases in Polymer Modifications*

Oxidoreductases tend to be specific with respect to their substrates. Whereas this feature limits the scope of the reactions, these enzymes can be used in suitable cases for specific functionalization. An example is galactose oxidase which can specifically oxidize the C6 alcohol on galactose in guar to an aldehyde *(43)*. Another oxidoreductase is tyrosinase, which has been shown to functionalize chitosan *(44)*. In this book, Payne et al *(11)* grafted two proteins onto chitosan with this enzyme.

Although strictly speaking a hydrolase, a suitable lipase when combined with $H_2O_2$ and a carboxylic acid can carry out oxidation reaction through the formation of a peracid *(45)*. This reaction has been used for polymer modification, e.g., epoxidation of polybutadiene *(46a)* and oxidation of hydroxyethylcellulose *(46b)*.

*Oxidoreductases in Polymer Hydrolysis*

One type of reactions that has attracted a lot of attention is the enzymatic removal of lignin in wood pulp (also known as "biobleaching") *(47)*. The enzymes involved include laccase (in combination with oxygen and a mediator), lignin peroxidase (in conjunction with $H_2O_2$), and manganese peroxidase (also with $H_2O_2$). These enzymatic reactions modify the lignin molecules, which can then be removed in a subsequent alkaline washing step.

### Transferases (EC 2)

A lot of creative work has been done with transferases, primarily for polymer synthesis and modification reactions.

*Transferases in Polymer Synthesis*

A major application is the use of glycosyltransferase for the synthesis of oligosaccharides and polysaccharides *(48)*. Elegant methods have been developed to recycle the cofactors and render the processes economical. In this book, Wang et al *(17)* reviewed several approaches relating to glycosyltransferases. DeAngelis *(19)* described two approaches using recombinant *Pasteurella multocida* synthase to generate glycosaminoglycans.

Another transferase reaction involves dextransucrase, which catalyzes the formation of dextran and some oligosaccharides *(49)*. Yet another transferase reaction entails the use of potato starch phosphorylase in the synthesis of low-molecular-weight amylose *(50)*.

*Transferases in Polymer Modifications*

Glycosyltransferases can also carry out glycan chain modifications, especially at outer or terminal positions *(17)*.

Transglutaminases are acyl transfer enzymes that catalyzed the condensation of glutamine and lysine residues of proteins *(51)*. They have been utilized for food processing. In this book a calcium-independent microbial transglutaminase has been used by Payne et al *(11)* to crosslink the protein in gelatin-chitosan blends

### Lyases (EC 4)

Lyases are effective enzymes for the degradation of polysaccharides. Examples are pectin lyase, pectate lyase, xanthan lyase, alginate lyase, hyaluronate lyase, and heparin lyase. BioPreparation™, which was developed by NOVOZYMES A/S for the removal of polymeric materials from the cotton surface, contains the pectate lyase *(5)*.

### Isomerases (EC 5)

Isomerases are involved in polymer science only in specific occasions. An example is the recent use of epimerase to convert mannuronate to guluronate in alginates *(52)*, oxidized konjac glucomannan *(53a)*, and oxidized galactomannan *(53b)*.

**Ligases (EC 6)**

This class of enzymes has not found much application in conventional polymeric materials thus far.

## Whole-Cell Biocatalysis

Whole-cell biotransformations typically utilize the metabolic pathways of microorganisms (or plant or animal cells) to produce desirable products *(1)*. These are among the oldest technologies used for food products. Some common items include alcohol, cheese, soy sauce and tofu, among others *(54)*.

Microbial biocatalysis is often employed for polymer synthesis. A successful application area is the poly(hydroxyalkanoate) (PHA). Thus, Noda et al *(22)* reported the development of Nodax™ family of copolymers, which were shown to have versatile physical properties. Srienc et al *(23)* engineered yeast to produce PHAs comprising 6-13 carbon monomers. Henderson *(5)* included whole-cell approaches and metabolic engineering in the biomass conversion program.

The synthesis of oligosaccharides and polysaccharides can also be achieved via microbial biocatalysis. For example, Wang et al *(17)* devised the "Superbug" method that could produce oligosaccharides of less than 4 sugar units efficiently.

A successful industrial process was described by Riehle *(24)*, using a microbial process to convert the waste materials to innocuous compounds

## New Enzyme Methodologies

One of the reasons for the increasing popularity of biocatalysis is the rapid pace of progress in biotechnology, which has a synergistic effect on biocatalysis. Bioprocess engineering has also helped to improve the product yields and the process economics. Given below is a sampling of the methodologies:

*Enzyme discovery and improvement:* high-throughput screening, biodiversity, directed evolution, gene shuffling, combinatorial chemistry, computational chemistry, bioinformatics, DNA microarray technology

*Enzyme applications:* cofactor recycling techniques, enzyme immobilization, enzyme solubilization, enzyme recycling, multi-enzyme clusters, mini- and micro-scale application tests, new enzyme analytical and separation techniques.

*Bioengineering:* enzyme production (including gene expression, enzyme purification, and formulation), fermentation, membrane technology, process systems engineering.

*Microbial methodologies:* culture enrichment and microbial strain improvement, metabolic engineering, integration of multi-pathways in cells, cell entrapment, functional genomics

These advances are reflected in a number of articles in this book. For example, Henderson (5) mentioned several programs at Novozymes A/S where some of these techniques were involved. Conboy et al (7) used $^1$H NMR as a tool for high throughput screening to identify new enzymatic activity. An ingenious alternative to high throughput assays was proposed by Minshull (6).

Genetically modified enzymes were involved in the articles by DeAngelis (19), Wang et al (17), and Henderson (5). A genetically modified protein was prepared by Patwardhan et al (14). Wang et al (17) devised multi-enzyme clusters (Superbeads) and cells containing multiple pathways (Superbugs).

Immobilization is a useful tool for biocatalysis. Several methods are often used: 1) Binding to a carrier, e.g., through covalent bonding, physical adsorption, electrostatic interaction, or biospecific binding. 2) Crosslinking with bifunctional or multifunctional reagents, or derivatization with a reactive group and subsequent polymerization. 3) Entrapment in a gel, microcapsules, liposomes, hollow fibers, or ultrafiltration membranes. 4) Combination of the above methods *(3a)*. Nevertheless, new and improved methods are desirable. In this book, Hsieh et al (8) described the successful binding of lipase onto cellulose. Gitsov et al (9) trapped laccase in an amphiphilic linear-dendritic block copolymer. DeAngelis *(19)* used an immobilized enzyme reactor in one of his approaches.

A popular and effective lipase is Novozym® 435, an immobilized *Candida antartica* Lipase B (CALB), available commercially from Novozymes A/S. This enzyme was invoked in at least 11 of the articles in this book *(16,20,25-33)*.

## Novel Bio-Related Materials

Biomaterials are synthetic or natural materials that are in contact with biological tissues or fluids and may enhance or replace tissues, bones, organs, or body functions *(4)*. They include metals, alloys, glasses, ceramics, natural or synthetic polymers, biomimetics, and composites. Typical biomaterials may be used in artificial skin, tissues, and bones, dental fillings, wire plates and pins for bone repair, artificial hips and joints, implantable drug delivery systems, and other dental, surgical, and medical devices. New and improved biomaterials continue to be sought.

In this book Song et al *(10)* described a novel nucleation and mineral growth process to produce a bone-like biomineral composite. The crosslinked gelatin-chitosan blend made by Payne et al *(11)* may perhaps be used as biomimetic soft tissue or for bioencapsulation. The sorbitol-based polyesters synthesized by Mei at al *(27)* and Kulshrestha et al *(26)* may possibly find applications in tissue engineering. Biswas et al *(13)* described the preparation and the mechanical properties of modified zein. Fishman et al *(12)* made pectin-starch and pectin-poly(vinyl alcohol) blends and found them to be strong, flexible films.

Several papers in this book dealt with silicon-containing materials. Patwardhan et al *(14)* carried out bioinspired mineralization of silica in the presence of a genetically engineered protein. Bassindale et al *(15)* screened several enzymes to look at the condensation of silanes. Sahoo et al *(16)* prepared glycosilicone conjugates.

## References

1. General books on biocatalysis include: (a) *Biocatalysis and Biodegradation*; Wackett, L. P.; Hershberger, C. D.; ASM Press, Washington, DC, 2001. (b) *Biotransformations in Organic Chemistry*, 3$^{rd}$ Ed.; Faber, K.; Springer, Berlin, Germany, 1997. (c) *Introduction to Biocatalysis Using Enzymes and Micro-organisms*; Roberts, S. M., Turner, N. J.; Willetts, A. J.; Turner, M. K.; Cambridge Univ. Press, Cambridge, UK, 1995. (d) *Applied Biocatalysis*; Cabral, J. M. S.; Best, D.; Boross, L.; Tramper, J., Eds.; Harwood, Char, Switzerland, 1994. (e) *Enzymes in Synthetic Organic Chemistry*; Wong, C. H.; Whiteside, G. M.; Elsevier, Oxford, UK, 1994.
2. Some recent books on polymer biocatalysis include: (a) *Biocatalysis in Polymer Science*; Gross, R. A.; Cheng, H. N., Eds.; Amer. Chem. Soc., Washington, DC, 2003. (b) *Enzymes in Polymer Synthesis*; Gross, R. A.; Kaplan, D. L.; Swift, G., Eds.; Amer. Chem. Soc., Washington, DC, 1998.
3. Some reviews on polymer biocatalysis include: (a) Cheng, H. N.; Gross, R. A. *ACS Symp. Ser.* **2002**, 840, 1. (b) Gross, R. A.; Kumar, A.; Kalra, B. *Chem. Rev.* **2001**, 101, 2097. (c) Kobayashi, S.; Uyama, H.; Kimura, S. *Chem. Rev.* **2001**, 101, 3793.
4. Some books on biomaterials include: (a) *Biomaterials: Principles and Applications*; Park, J. B.; Bronzino, J. D.; CRC Press, 2002. (b) *Biomaterials Science*; Ratner, B. D.; Hoffman, A. S.; Schoen, F. J.; Lemons, J. E. Eds.; Academic Press, Orlando, 1996. (c) *Biomaterials Science and Biocompatibility*; Silver, F. H.; Christiansen, D. L.; Springer-Verlag, New York, 1999.
5. Henderson, L. A. *ACS Symp. Ser.* (this volume), Chapter 2.
6. Ness, J. E.; Cox, T.; Govindarajan, S.; Gustafsson, C.; Gross, R.A.; Minshull, J. *ACS Symp. Ser.* (this volume), Chapter 3.

7. Conboy C. B.; Li, K. *ACS Symp. Ser.* (this volume), Chapter 4.
8. Hsieh, Y.-L.; Wang, Y.; Chen, H. *ACS Symp. Ser.* (this volume), Chapter 5.
9. Gitsov, I.; Lambrych, K.; Lu, P.; Nakas, J.; Ryan, J.; Tanenbaum, S. *ACS Symp. Ser.* (this volume), Chapter 6.
10. Song, J.; Bertozzi, C. R. *ACS Symp. Ser.* (this volume), Chapter 7.
11. Chen, T.; Small, D. A.; McDermott, M. K.; Bentley, W. E.; Payne, G. F. *ACS Symp. Ser.* (this volume), Chapter 8.
12. Fishman, M. L.; Coffin, D. R. *ACS Symp. Ser.* (this volume), Chapter 9.
13. Biswas, A.; Sessa, D. J.; Gordon, S. H.; Lawton, J. W.; Willett, J. L. *ACS Symp. Ser.* (this volume), Chapter 10.
14. Patwardhan, S. V.; Shiba, K.; Raab, C.; Huesing, N.; Clarson, S. J. *ACS Symp. Ser.* (this volume), Chapter 11.
15. Bassindale, A. R.; Brandstadt, K. F.; Lane, T. H.; Taylor, P. G. *ACS Symp. Ser.* (this volume), Chapter 12.
16. Sahoo, B.; Brandstadt, K. F.; Lane, T. H.; Gross, R. A. *ACS Symp. Ser.* (this volume), Chapter 13.
17. Li, H.; Zhang, H.; Yi, W.; Shao, J.; Wang, P. G. *ACS Symp. Ser.* (this volume), Chapter 14.
18. Kobayashi, S.; Fujikawa, S.; Itoh, R.; Morii, H.; Ochiai, H.; Mori, T.; Ohmae, M. *ACS Symp. Ser.* (this volume), Chapter 15.
19. DeAngelis, P. L. *ACS Symp. Ser.* (this volume), Chapter 16.
20. Chakraborty, S.; Sahoo, B.; Teraoka, I.; Miller, L. M.; Gross, R. A. *ACS Symp. Ser.* (this volume), Chapter 17.
21. Cheng, H.N.; Gu, Q.-M.; Qiao, L. *ACS Symp. Ser.* (this volume), Chapter 18.
22. Noda, I.; Bond, E. B.; Green, P. R.; Melik, D. H.; Narasimhan, K.; Schechtman, L. A.; Satkowski, M. A. *ACS Symp. Ser.* (this volume), Chapter 19.
23. Zhang, B.; Carlson, R.; Pederson, E. N.; Witholt, B.; Srienc, F. *ACS Symp. Ser.* (this volume), Chapter 20.
24. Riehle, R. J. *ACS Symp. Ser.* (this volume), Chapter 21.
25. Mahapatro, A.; Kumar, A.; Kalra, B.; Gross, R. A. *ACS Symp. Ser.* (this volume), Chapter 22.
26. Kulshrestha, A. S.; Kumar, A.; Gao, W.; Gross, R. A. *ACS Symp. Ser.* (this volume), Chapter 23.
27. Mei, Y.; Kumar, A.; Gao, W.; Gross, R. A.; Kennedy, S. B.; Washburn, N. R.; Amis, E. J.; Elliott, J. T. *ACS Symp. Ser.* (this volume), Chapter 24.
28. Neuner, I. T.; Ursu, M.; Frey, H. *ACS Symp. Ser.* (this volume), Chapter 25.
29. Bisht, K. S.; Kondaveti, L.; Stewart, J. D. *ACS Symp. Ser.* (this volume), Chapter 26.
30. Nakaoki, T.; Kitoh, M.; Gross, R. A. *ACS Symp. Ser.* (this volume), Chapter 27.

31. Kalra, B.; Lai, I.; Gross, R. A. *ACS Symp. Ser.* (this volume), Chapter 28.
32. Kalra, B.; Bankova, M.; Gross, R. A. *ACS Symp. Ser.* (this volume), Chapter 29.
33. Gu, Q.-M.; Cheng, H. N. *ACS Symp. Ser.* (this volume), Chapter 30.
34. For example, *Source Book of Enzymes*; White, J. S.; White, D.C.; CRC Press, Boca Raton, 1997.
35. For example, (a) Mahapatro, A.; Kumar, A.; Gross, R. A. *Biomacromolecules* **2004**, *5*, 62. (b) Mahapatro, A.; Kumar, A.; Kalra, B.; Gross, R. A. *Macromolecules* **2004**, *37*, 35.(c) Divakar, S. *J. Macromol. Sci., Pure Appl. Chem.* **2004**, *A41*, 537. (d) Uyama, H.; Mai, K.; Takashi, T.; Kobayashi, S. *Biomacromolecules* **2003**, *4*, 211. (e) Kikuchi, H.; Uyama, H.; Kobayashi, S. *Polym. J.* **2002**, *34*, 835. (f) Kim, D.-Y.; Dordick, J. S. *Biotechnol. Bioeng.* **2001**, *76*, 200. (g) Tsujimoto, T.; Uyama, H.; Kobayashi, S. *Biomacromolecules* **2001**, *2*, 29. (h) Mesiano, A. J.; Beckman, E. J.; Russell, A. J. *Biotechnol. Prog.* **2000**, *16*, 64.
36. For example: (a) Shoda, S.; Kobayashi, S. *Trends Polym. Sci.* **1997**, *5*, 109. (b) Kobayashi, S.; Sakamoto, J.; Kimura, S. *Prog. Polym. Sci.* **2001**, *26*, 1525.
37. For example: (a) Jahn, M.; Withers, S. G. *Biocatalysis Biotransformation*, **2003**, *21*, 159. (b) Perugino, G.; Trincone, A.; Rossi, M.; Moracci, M. *Trends Biotechnol.* **2004**, *22*, 31, and references therein.
38. For example, (a) Morse, D. E. *Trends Biotechnol.* **1999**, *17*, 230. (b) Tacke, R. *Angew. Chem. Int. Ed.* **1999**, *38*, 3015. (c) Cha, J. N.; Stucky, G. D.; Morse, D. E.; Deming, T. J. *Nature* **2000**, *403*, 289. (d) Kroger, N.; Deutzmann, R.; Sumper, M. *J. Biol. Chem.* **2001**, *276*, 26066. (e) Kroger, N.; Lorenz, S.; Brunner, E.; Sumper, M. *Science* **2002**, *298*, 584. (f) Naik, R. R.; Brott, L. L.; Clarson, S. J.; Stone, M. O. *J. Nanosci. Nanotech.* **2002**, *2*, 95.
39. For example: (a) Kadota, J.; Fukuoka, T.; Uyama, H.; Kasegawa, Kobayashi, S. *Macromol. Rapid Commun.* **2004**, *25*, 441. (b) Shutava, T.; Zheng, Z.; John, V.; Lvov, Y. *Biomacromolecules* **2004**, *5*, 914. (c) Naruyoshi, M.; Shin-ichiro, T.; Hiroshi, U.; Kobayashi, S. *Macromol. Biosci.* **2003**, *3*, 253. (d) Uyama, H.; Maruichi, N.; Tonami, H.; Kobayashi, S. *Biomacromolecules* **2002**, *3*, 187. (e) Bruno, F.; Nagarajan, R.; Stenhouse, P.; Yang, K.; Kumar, J.; Tripathy, S. K.; Samuelson, L. A. *J. Macromol. Sci.* **2001**, *A39*, 1417.
40. For example: (a) Ikeda, R.; Sugihara, J.; Uyama, H.; Kobayashi, S. *Macromolecules* **1996**, 29, 8702. (b) Ikeda, R.; Uyama, H.; Kobayashi, S. *Macromolecules*. **1996**, 29, 3053.
41. For example: (a) Kumar, J.; Tripathy, S.; Senecal, K.J.; Samuelson, L. *J. Am. Chem. Soc.* **1999**, *121*, 71. (b) Tripathy, S. *Chem. Eng. News.* **1999**, *77*,

68. (c) Sahoo, S.; Nagarajan, R.; Samuelson, L.; Kumar, J.; Cholli, A. L.; Tripathy, S. K. *J. Macromol. Sci.* **2001**, *A38*, 1315.
42. For example: (a) Teixeira, D.; Lalot, T.; Brigodiot, M.; Marechal, E. *Macromolecules* **1999**, *32*, 70. (b) Kalra, B.; Gross, R.A. *Biomacromolecules,* **2000**, *1*, 501. (c) Singh, A.; Roy, S.; Samuelson, L.; Bruno, F.; Nagarajan, R.; Kumar, J.; John, V.; Kaplan, D. L. *J. Macromol. Sci.* **2001**, A38, 1219. (d) Tsujimoto, T.; Uyama, H.; Kobayashi, S. *Macromol. Biosci.* **2001**, *1*, 228. (e) Kadokawa, J.; Kokubo, A.; Tagaya, H. *Macromol. Biosci.* **2002**, *2*, 257. (f) Kalra, B.; Gross, R. A. *ACS Symp. Ser.* **2002**, *840*, 297.
43. For example: (a) Brady, R. L. ; Leibfried, R. T.; Nguyen, T. T. *U. S. Patent* 6,124,124, 9/26/2000. (b) Chiu, C.-W.; Jeffcoat, R.; Henley, M.; Peek, L. *U. S. Patent* 5,554,745, 9/10/1996. (c) Frollini, E.; Reed, W. F.; Milas, M.; Rinaudo, M. *Carbohydrate Polym.* **1995**, *27*, 129, and references therein.
44. For example: (a) Govar, C. J.; Chen, T.; Liu, N.-C.; Harris, M. T.; Payne, G. F. *ACS Symp. Ser.* **2002**, *840*, 231. (b) Chen, T.; Kumar, G.; Harris, M. T.; Smith, P. J.; Payne, G. F. *Biotechnol. Bioeng.* **2000**, *70*, 564.
45. (a) Bjorkling, F.; Godtfredsen, S. E.; Kirk, O. *J. Chem. Soc., Chem. Commun.* **1990**, 1301-1303. (b) Bjorkling, F.; Frykman, H.; Godtfredsen, S. E.; Kirk, O. *Tetrahedron* **1992**, 4587-4592.
46. (a) Jarvie, A. W. P.; Overton, N.; St. Pourcain, C. B. *J. Chem. Soc., Perkin Trans.1* **1999**, 2171. (b) Hu, S.; Gao, W.; Kumar, R.; Gross, R. A.; Qu, Q.-M.; Cheng, H. N. *ACS Symp. Ser.* **2002**, *840*, 253.
47. For example, (a) Bourbonnais, R.; Paice, M. G. *FEBS Lett.* **1990**, *267*, 99. (b) Bourbonnais, R.; Paice, M. G.; Reid, I. D.; Lanthier, P.; Yaguchi, M. *Appl. Environ. Microbiol.* **1995**, *61*, 1876. (c) Call, H. P.; Mucke, I. *J. Biotechnol.* **1997**, *53*, 163.
48. For example, (a) Wang, P. G.; Fitz, W.; Wong, C.-H. *Chemtech* **1995** (April), 22. (b) Guo, Z.; Wang, P. G. *Appl. Biochem. Biotechnol.* **1997**, *68*, 1. (c) Palcic, M. *Curr. Opin. Biotechnol.* **1999**, *10*, 616, and refs. therein. (d) Chen, X.; Kowal, P.; Wang, P. G. *Current Opin. Drug Disc. Dev.* **2000**, *3*, 756. (e) Sears, P.; Wong, C.-H. *Science* **2001**, *291*, 2344.
49. For example: (a) Kubik, C.; Sikora, B.; Bielecki, S. *Enzyme Microb. Technol.* **2004**, *34*, 555. (b) Kim, D.; Robyt, J. F.; Lee, S. Y.; Lee, J. H.; Kim, Y. M. *Carbohydr. Res.* **2003**, *338*, 1183. (c) Kim, Y. M.; Park, J. P.; Sinha, J.; Lim, K. H.; Yun, J. W. *Biotechnol. Lett.* **2001**, *23*, 13. (d) Demuth, B.; Jordening, H. J.; Buchholz, K. *Biotechnol. Bioeng.* **1999**, *62*, 583.
50. For example: (a) Pfannmueller, B. *Naturwissenschaften* **1975**, *62*, 231. (b) Ziegast, G.; Pfannmueller, B. Carbohydr. Res. **1987**, *160*, 185.

51. For example: (a) Kamiya, N.; Takazawa, T.; Tanaka, T.; Ueda, H.; Nagamune, T. *Enzyme Microb. Technol.* **2003**, *33*, 492. (b) Dickinson, E. *Trends Food Sci. Technol.* **1997**, *8*, 334.
52. For example, (a) Ertesvag, H.; Doseth, B.; Larsen, B.; Skjak-Braek, G.; Valla, S. J. Bacteriol. **1994**, *176*, 2846. (b) Franklin, M. J.; Chitnis, C. E.; Gacesa, P.; Sonesson, A.; White, D. C.; Ohman, D. E. *J. Bacteriol.* **1994**, *176*, 1821. (c) Harmann, M.; Holm, O. B.; Johansen, G.-A. B.; Skjak-Braek, G.; Stokke, B. T. *Biopolymers* **2002**, *63*, 77. (d) Harmann, M.; Dunn, A. S.; Markussen, S.; Grasdalen, H.; Valla, S.; Skjak-Braek, G. *Biochim. Biophys. Acta* **2002**, *1570*, 104.
53. (a) Crescenzi, V.; Skjak-Braek, G.; Dentini, M.; Maci, G.; Bernalda, M. S.; Risica, D.; Capitani, D.; Mannina, L.; Segre, A. L. *Biomacromolecules* **2002**, *3*, 1343. (b) Crescenzi, V.; Dentini, M.; Risica, D.; Spadoni, S.; Skjak-Braek, G.; Capitani, D.; Mannina, L.; Viele, S. *Biomacromolecules* **2004**, *5*, 537.
54. For example, *Wiley Encyclopedia of Food Science and Technology*; Francis, F. J., Ed.; Wiley, 1999.

# New Enzyme Methodologies

Chapter 2

# Biotechnology: Key to Developing Sustainable Technology for the 21st Century: Illustrated in Three Case Studies

Lori A. Henderson

Novozymes North America, Inc., 77 Perry Chapel Church Road, Franklinton, NC 27525

The maturation of technology from scientific breakthroughs to commercial applications is driven by the symbiotic relationship that exists between society, academia, industry and government. With vast improvements and increasing availability of new bioengineering techniques today, the future in developing enzyme technologies to meet the demands of many industrial applications holds great promise. From textiles to ethanol production, the unique advantage of enzymes serves as the impetus for increasing R&D projects and their commercialization. Herein, describes the emergence of three technologies related to BioPreparation™, Bioethanol and Biocatalysis, whose commercial success is contingent on the discovery and bioprocess engineering of enzymes for industrial applications. The research activities and strategies in meeting the needs of all 3 industries is discussed via 3 case studies - *I. Discovery and Exploratory of BioPreparation™, II. Development of Low Cost Technology for Biomass to Ethanol Production and, III. Discovery of Next Generation Biocatalysts for Chemical Synthesis.*

# Case Study 1: The Discovery of Biopreparation

**Introduction**

In textiles, one of the most negative impacts on the environment originates from traditional processes used to prepare cotton fiber, yarn and fabric. Fabric preparation consists of a series of various treatments and rinsing steps critical to obtaining good results in subsequent textile finishing processes. These water-intensive wet processing steps contribute large volumes of wastes, particularly from alkaline scouring and continuous/batch dyeing. Such treatments generate large amounts of salts, acids and alkali. Scouring is a cleaning process that removes impurities from cotton substrates during textile processing. In view, of the 40 billion pounds of cotton fiber that are prepared annually on an international level, it becomes clear that the preparation process is a major source of environmentally harsh chemical contribution to the effluent, with the major offender being sodium hydroxide and its salts. Conventional chemical preparation processes involve treatment of the cotton substrate with hot solutions of sodium hydroxide, chelating agents and surface active agents, often followed by neutralization step with acetic acid. The scouring process is designed to break down or release natural waxes, oils and contaminants and emulsify or suspend these impurities in the scouring bath. Typically, scouring wastes contribute high BOD loads during cotton textile preparation. According to the EPA (*1*), ~50% of the total BOD in preparing knitted fabrics is due to scouring chemicals.

Cotton preparation in the textile mill is a sequence of events that define all of the industrial steps leading from fiber to fabric. Cloth is created by weaving or knitting mostly raw cotton yarn that is then processes for dyeing. In this case, greige fabric is converted to a fully dyeable yarn or fabric using different wet processing steps. Many of the processes represent aggressive chemical treatments that incorporate high concentrations of harsh, corrosive chemicals like sodium hydroxide, hydrogen peroxide, lubricants, sequestrants, etc. Moreover, the textile processing industries are driven by time-consuming production processes, which consume much energy and resources. Thus, Novozymes initiated an innovative research project with the intent of developing biological alternatives to chemical preparative routes in cotton textile preparation. The goal of this project was to develop a technology that would significantly reduce pollution and resources without increasing capital expenses. This technology referred to as BioPreparation™ has been reduced to practice via several field and industrial trials conducted across the globe (*2*).

The idea originated from a technology review of plant cell wall physiology, cotton fiber morphology and its' composition to determine the different structure-property relationships that exists within cotton fibers. Pioneering studies were subsequently pursued in the laboratory to develop test methods and protocols for evaluating the effect of various enzymes on woven cotton fabrics. Several exhaustive empirical screening tests were carried out in a labomat to simulate the scouring of fabric. In summary, BioPreparation™ is an enzymatic process for treating cotton textiles that meets the performance characteristics of alkaline scour systems while reducing chemical and effluent load.

**Results and Discussion**

The goal of cotton preparation is primarily to improve the wettability of cotton fiber. A cotton fiber is the seed hair of plants of the genus *Gossypium* where each fiber is generated from a single cell. A mature fiber is composed of three main layers: a primary cell wall, secondary wall and lumen. Cotton wax, located in the outer layer of cotton fibers called the cuticle, is regarded as a major obstacle influencing the quality of dyeing as well as the "dynamics" of wetting in wet textile preparation of fabric. It was believed for quite some time that these hydrophobic substances (long chain fatty acids, esters, alcohols) and other impurities formed a *distinct* protective layer that surrounds the primary cell wall based on the morphology of cotton fibers (2, 3). The composition of the primary cell wall consists of mainly cellulose and xylogucan located within the fibrils. Part of the primary cell wall also contains a small percentage of impurities like pectin, protein and waxes. The secondary wall constitutes the bulk of a mature fiber in fibril from spirally arranged around the fibril axis. The fibrils consist of bundles of cellulosic microfibrils that are 0.025 μm thick and at least 10 μm long. The lumen is the central canal and contains residual proteins. As the mysteries in this project started to unravel, it was realized that only the primary cell plays a vital role in textile preparation.

In this investigation, it was postulated that the cotton waxes are oriented in a 3-dimensional structure within the primary cell wall architecture and can be removed by degradation of polymeric materials. The objective was to examine the *physco-chemical* nature of the interlinked networks within a cotton cell wall. As shown in Figure 1, the architecture of a plant cell wall consist of a series of polymer networks that when superimposed upon one another gives rise to a very complex structure. The model is a simplified view of the specific interactions between three networks: 1 – cellulose-hemicellulose characterized by the hydrogen bonding of xyloglucan to the cellulose, 2 – the pectic polymers held in the matrix through ionic bonding of the egg-box type and 3 – glycoprotein (extensin) that are structurally independent of the polysaccharides. Moreover, the cellulose/xyloglycan network described above is embedded by the outer pectin netting. The junction zone of pectin network is held together by ionic, hydrophobic or hydrogen bonds depending on the degree of methylation for the different pectic polymers.

This research project investigated the interactions between polymers and substrates like waxes by experimenting with enzymes and studying their effect

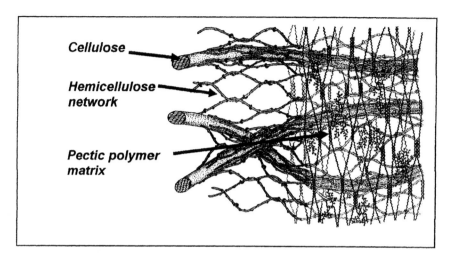

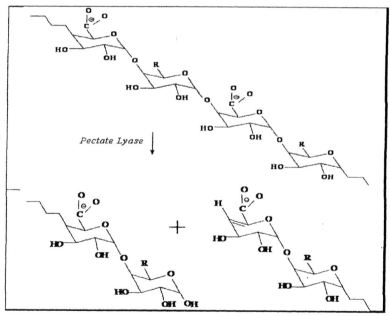

*Figure 1. The top drawing is a proposed model simplified to demonstrate the various interpenetrating networks within the primary cell wall of cotton[3]. The reaction scheme belowt illustrates the products formed by the enzymatic hydrolysis of pectic polymers via a β-elimination reaction. Note the complexity between the interactions-degradation of the polysaccharides.*

on cotton textiles. Enzymes such as glucanases, cellulases, hemicellulases, and pectinases were screened in simulated scouring tests and the absorbency measured. The results consistently proved that enzymes, referred to as pectinases, would degrade polymeric substrates like pectin from the surface of cotton and produce a highly absorbent material. This success led to cloning, purifying and isolating a bacterial pectate lyase at high yields with one proteolytic activity. The results also support the proposed mechanism of action.

The mechanism of polymer degradation (Figure 1) is a β-elimination reaction which cleaves the α-1,4 links between galacturonosyl residues to produce two products: an unsaturated oligosaccharide (C4-C5 bond) and a hydroxy-terminated chain end oligomer (referred to as the reducing chain). The conclusions from this research are consistent with the proposed substrate model and reaction mechanism postulated in Figure 1. The BioPreparation™ of cotton takes place by: *i)* cleaving polyglacturonic acid and methoxylated derivatives to break the outer pectin network and hence alter the morphology of the cell wall and *ii)* removing the waxes, proteins and other components within this matrix via solubility and emulsification with the aid of surfactants and chelants. The enzyme is also compatible with other enzymatic preparations (amylases, cellulases) used to improve the performance properties of cotton fabrics (*4*).

The process also decreases both effluent load and water usage to the extent that the new technology becomes an economically viable alternative. A reduced need for sodium hydroxide significantly reduces BOD and COD in the effluent, as was determined by analyzing spent scouring baths from numerous field trials with cotton knits and yarn. When similar process auxiliaries (e.g. surfactants) were used, the BioPreparation™ process decreased BOD and COD load by 25 and 40%, respectively, relative to conventional sodium hydroxide treatments. To appreciate the significance of these reductions, consider the cost savings to a woven processing mill (desize-dyeing) that produces ~2.3 million lbs of goods per month. Assuming the charges for BOD and COD is $0.20 and $0.122/lb for amounts over the permissible limits[1], the annual cost to the mill for water and waste charges is $637,000 per year when calculated with a water consumption of 12 gal/lb. fabric. If the waste values are decreased according to the above field data, the cost is $424, 980/yr which represents ~33% savings to the mill. Similar cost savings were documented at a few targeted mills. Additional field trials indicated a 30% reduction in water use by elimination of both the acetic acid neutralization step and several rinse cycles. Even further opportunities to reduce effluent load and water use were realized when BioPreparation™ was used to combine treatment steps. Results from a 1-step scour-dyeing process combined in a single bath showed a 20 and 50% reduction in BOD and COD, respectively, compared to the conventional 2-step alkaline treatment. Assuming the same BOD/COD charges as before, such reductions would allow the mill to save 30% of their costs.

## Conclusions

Application studies on both a pilot and industrial scale were explored using BioPrep 3000L as the main component in scouring knit and woven textiles. Variations in process variables such as enzyme concentration, pH and temperature were performed in pilot scale equipment designed for exhaust and pad operations. The "treated" fabrics were evaluated against the conventional alkaline processes based on standardize test methods including performance characteristics like dyeing (5). Successful full-size industrial trials in preparing yarn and knitted textiles were conducted and confirmed the processing advantages described herein. These advantages relative to chemical preparative routes are as follows: (*i*) ease of operation (no modification of existing equipment), (*ii*) improve environmental effluents based on BOD, COD, and TDS, (*iii*) selective degradation of components that enhance properties with minimal weight loss (maintaining the quality/integrity of the cotton fiber) and (*iv*) consumes significantly less water, time and energy. In conclusion, this biocompatible process provides an economical & an environmentally friendly alternative to alkaline scour systems or any combinations thereof, currently used in the textile industry today. The technology also emphasizes a novel strategy in the industry focusing on pollution prevention rather than innovative treatments.

## Case Study 2: The Future of Bioethanol Production

### Introduction

The US needs alternatives to foreign oil for transportation to wean the country from its dependency on imported oil. Using biomass as a feedstock for ethanol production could expand the domestic ethanol market, improve national security, create jobs, dispose of burdensome biomass waste and produce a clean transportation fuel. From a biomass energy standpoint, unhealthy forests are only one of many sources that could eventually support a biomass energy industry. DOE's new technology for biomass conversion to ethanol could increase production efficiencies up to about 4:1. Using biofuels such as ethanol provides measurable air quality benefit by reducing vehicle emissions and abating field burning of some agriculture residues. It will reduce air pollution and the greenhouse gases that are implicated in the problems of global climate change. Estimates are that 40% of today's smog, 33% of annual $CO_2$ emissions and 67% of CO production come from automobiles and other forms of transportation.

The market opportunity in the production of ethanol is expanding rapidly due to the elimination or significant reduction in the use of MTBE, an oxygenate that is traditionally blended with fuel. The potential for a new process technology referred to as bioethanol is receiving significant attention with vast number of government sponsored research programs emerging. It has the potential to match the features of petroleum at a low price. Bioethanol can be produced from domestically abundant sources of biomass including agricultural and forestry residues, wastepaper and other municipal solid waste and ultimately woody and herbaceous crops grown on underutilized land. Because the fossil fuel inputs in growing such materials and converting to ethanol is low, a high ratio of energy production is achieved and the net release of $CO_2$ that contributes to global climate changes is zero. When ethanol is added to gasoline it improves fuel combustion, thereby reducing tailpipe emissions of CO and unburned hydrocarbons that form smog. By applying the rapidly advancing sources of biotechnology to bioethanol production, the technology can be improved to make bioethanol competitive for blending with gasoline with that from corn in the US.

This innovative technology is based on integrating chemoenzymatic routes and microbial conversion to the production of ethanol from renewable resources. Novozymes has two research projects related to the development of this process technology using biotechnology tools/applications to: i) Find more efficient enzymes through a combination of diversity mining, protein engineering and DNA discovery tools, and ii) Conduct lab-scale application tests for proof-of-concept in biomass-ethanol production.

**Biomass Treatment Technologies**

The U.S. Department of Energy (DOE) and the National Renewable Energy Laboratories (NREL) have been working closely with state agencies, academic institutions and a wide range of industrial partners to accelerate the advancement of "new bioethanol technology". Researchers are working to demonstrate biochemical conversion processes in real-world applications with emphasis on improving the efficiency and economics of the process technologies by focusing their efforts on the most challenging steps in the process. Through the US government subsidized programs, the major thrust of the advanced R&D on biochemical conversion technologies is on pretreatment, cellulase enzymes and catalyst development for products from sugars (see Figure 2). More recent sponsorships are also directed at demonstrating process integration at a pilot plant scale to produce bioethanol and other chemicals – steps towards establishing a true "biorefinery" with value-added coproducts.

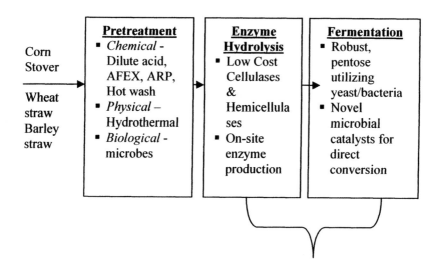

*Figure 2: An illustration of the various research activities underway to develop the next generation bioethanol plant. The 3 key areas of R&D are in pretreatments, enzyme hydrolysis and fermentations. Significant efforts focus on process engineering and plant designs with the following at the core: SHF - separate hydrolysis and fermentation, SSF - a simultaneous saccharification-fermentation, or SSCF - a simultaneous saccharification-cofermentation.*

## Pretreatment Technologies for Lignocellulosic Conversion (Step 1)

In the utilization of enzymes for the hydrolysis of the cellulose, a pretreatment of the lignocellulosic material is necessary to break up the barrier made by both lignin and hemicelluloses. The lignin is one of the major obstacles in enzymatic hydrolysis because it binds a large part of the enzymes. In some studies, only 40-50% of the total protein added can be recovered in solution. The remaining portion is irreversibly bound to lignin (6). Furthermore, the physical and/or chemical pretreatment may increase the accessible surface area and change the crystallinity of the cellulose, which results in increased digestibility of the cellulose. The efficiency of these pretreatments is normally evaluated by measuring the sugar release during the combined pretreatment-enzyme hydrolysis. Thus, an effective pretreatment technique renders the cellulose more digestible, avoids degradation of hemicellulose sugars and removes lignin. The barriers to develop a robust pretreatment process include a lack of fundamental understanding related to the chemistry at work in pretreatment of biomass and the hydrolysis of hemicellulose, reactor design fundamentals, equipment reliability, and materials of construction.

During the past decade, the science of pretreatments has received increasing attention because of its significance in the first stage of processing biomass. Among the variety of available pretreatment technologies reported in the literature, there are at least 5 general classes: base-catalyzed, acid-catalyzed, non-catalyzed, solvent and chemical based systems. From the developmental work by NREL Advance Pretreatment Project (7,8), and the Biomass Refining Consortium of Applied Fundamentals and Innovations Team (CAFI) (7,8) a strong fundamental knowledge base of biomass pretreatment chemistries, kinetics and process economics is underway. The Advance Pretreatment Projects focus on clean fractionation, hemicellulases and accessory enzymes, biomass compositional analysis and the application of NREL's Bioethanol Pilot Development Unit. Together, their goal is to expand ongoing efforts in understanding the impact of reactor design and configuration on thermochemical cellulose hydrolysis and include a broader range of biomass pretreatments and fractionation approaches. This partnership continues to investigate 5 leading technologies based on performance and process economics using corn stover as the model substrate. The aim is to identify the strongest possibilities of achieving broad commercial applicability in an advanced technology sugar/lignin platform. Upon conclusion of the CAFI program *at least* four leading, potentially viable pretreatment processes were analyzed in detail: dilute acid, ammonia fiber explosion (AFEX), ammonia recycle percolation (modified ARP) and Pressurized Hot Wash treatment (PHW) (8-13). No single pretreatment is currently a clear front-runner, however, for meeting the *advanced* technology targets. All have some process performance, economic or complexity issues that have not been adequately addressed. The different pretreatment methods solubilize primarily the hemicellulose in different lignocellulosic material – with marginal impact on lignin and cellulose and hemicellulose. The following is a brief description of those methods having the most commercial potential *(8-13)*.

- *Dilute Sulfuric Acid* - Depending on acid content and residence time, it can achieve near complete solubilization of hemicellulose resulting in high yields of xylose. The dilute process has a little to no impact on solubilizing lignin and cellulose crystallinity. Dilute acid (<1%) demands the use of high temperatures in order to achieve acceptable reaction rates for the hydrolysis of cellulose and to obtain high glucose yields. The steam pretreatment of $H_2SO_4$-Impregnated biomass treatment enhances digestibility. For example, at 10% dry matter, 200°C, 5 min, 2% $H_2SO_4$, 50% of the hemicellulose and 20% of the lignin is solubilized while 80% of the cellulose remains in the solid fraction. However, an acid catalyzed steam explosion process does remove ~80% of lignin and results in lower enzyme loading for same degree of conversion. Details of this advance process technology is available in the NREL report TP-510-32438 *(8, 10, 11, 14)*.

- *AFEX* – The ammonia fiber explosion treats lignocellulosic plant materials with liquid ammonia at moderate temperatures (60-100°C) for short times. By fine-tuning the AFEX process conditions, it may be possible to achieve high conversion of cellulose and hemicelulose to sugars. The physical and chemical effects increase conversion of cellulose and hemicellulose to fermentable sugars at low enzyme loadings. The AFEX process does not require neutralization of biomass nor produce inhibitory compounds. Preliminary economic analysis indicates that this process could produce ethanol at a substantially lower cost than dilute acid hydrolysis. However, recent data indicates that complete cellulose degradation is not yet possible at reasonable doses of cellulases (15 to 7 FPU/g cellulose). It is also believed that a hemicellulase is needed to further enhance the enzymatic degradation of the cellulose present in the prehydrolyzates (increase costs- Bruce Dale) *(8, 12, 14)*.

- *ARP* – The ammonia recycle percolation process is being developed by Dr. Y.Y. Lee at Auburn University. It results in near complete fractionation of biomass into the 3 major constititents – xylan, cellulose, lignin. The solids have a high glucan - low lignin content with short chain cellulose materials *(8, 14)*.

- *PHW* - This pretreatment has been found to be effective on a variety of biomass feedstocks. The addition of a pressurized hot wash step immediately following thermochemical pretreatment (e.g., dilute acid) significantly improves the overall process. It utilizes water in the liquid state even under high pressure and reaction temperatures (160°C). The process utilizes a flowthrough or percolation strategy in which hot water is preheated and passed through a packed bed of biomass. This results in high yields of xylose (primarily in oligomeric form) and highly digestible solids. It does not affect the cellulose or lignin content and requires an enzyme cocktail for complete hydrolysis. The process is very water intensive and is difficult to process (*8, 13, 14*).

In summary, the dilute acid pretreatment is the most widely studied with proof-of-concept well established (pilot scale) and is a commercially viable technology. It was selected for Novozymes's "process integration" project based on a detailed techno-economic analysis (TEA) and industry acceptance (*15*). While a well-coordinated knowledge gathering process development and evaluation efforts are underway (CAFI group), much work is still needed to explore the effect of recent enzyme developments for baseline feasibility, technical hurdles and defining success criteria. This, too, was accomplished via Novozymes project by taking a closer look at the chemistry of plant cell walls – overcoming the barriers to crystalline cellulose and lignin adsorption.

## Enzyme Hydrolysis of Lignocellulosics (Step 2)

*Fundamentals.* The prevalent understanding of the cellulolytic system divides the cellulolytic enzymes into three classes: exo-1,4- β-D-glucanases or cellobiohydrolases (CBH, EC 3.2.1.91), which cleave off cellobiose units from the ends of cellulose chains, endo-1,4- β-D-glucanases (EG, EC 3.2.1.4), which hydrolyzes internal β-1,4-glucosidic bonds randomly in the cellulose chain, and 1,4- β-D-glucosidases (EC 3.2.1.21), which hydrolyzes cellobiose to glucose and also cleaves glucose from cellooligosaccharides. Within each enzyme class, most fungi are capable of producing multiple enzymes with apparently similar enzymatic properties with distinct physical properties (molecular mass, isoelectric point). The molecular architecture of the EG and CBH's is important for catalytic activity of these enzymes. For example, the exo and endo enzymes (Family 7) shown in Figure 3 shows similar topography but with different binding domains – a tunnel versus a cleft leading to selective hydrolysis. Moreover, the removal of the cellulose-binding domain from CBHI from *Trichoderma reesei* has been shown to reduce their adsorption and activitiy on crystalline cellulose significantly without affecting their activity on amophorous material (*16*).

*Practical.* The hydrolysis of lignocellulosic material by enzymes at commercial scale is confronted by a number of difficulties. The enzymes are

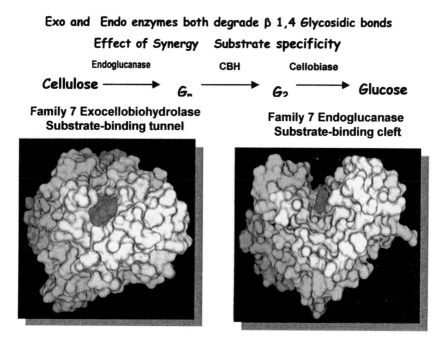

*Figure 3: The above topography illustrates the difference in the active site of two Family 7 cellulases. The cellulose binds to the tunnel of exocellobiohydrolase to produce cellobiose from the chain ends whereas the endoglucanase hydrolyzes internal bonds of the chain. All three activities are important for ethanol production.*

costly and the loading should consequently be reduced but then requires longer reaction times to achieve near complete hydrolysis. Operating at high substrate concentrations increases the problem of product inhibition, which results in a less effective enzyme. The presence of lignin, which shields the cellulose chains and irreversibly adsorbs the enzymes, also affects hydrolysis efficiency (*17, 18*). The accumulation of cellobiose strongly inhibits the activity of the cellobiohydrolases and even low amount of cellobiose has a significant impact on the performance. The addition of sufficient β-glucosidase activity to remove the formed cellobiose is therefore of uttermost importance, particularly in SHF. The high glucose concentration in the last part of the hydrolysis will inhibit the action of β-glucosidases resulting in the build up of cellobiose and thereby decreasing the hydrolysis rate. Even in SSF mode, sufficient β-glucosidase is needed for the rate with little impact on sugar yield. Therefore, to achieve a more cost effective bioconversion process, the NREL funded two leading enzyme companies, Novozymes BioTech and Genencor, Int. to engineer novel cellulases and/or their production to lower the current cost of cellulases. The NREL Subcontract No. ZCO-1-30017-02 with Novozymes BioTech, for example, has reach the goal of a 10X cost reduction by applying advance molecular biology tools. With the aid of proteomics, DNA microarray, bioinformatics and protein engineering techniques, novel cellulases from *T. reesei* were discovered and improved to include the three key activities. Of course, the molecular engineering, and protein discovery research is an iterative process as shown in Figure 4. In addition to identifying novel secreted proteins and mRNA by growing them on corn stover, the new strains were expressed, fermented and characterized for activity, protein content and thermostability. Directed evolution was also used to improve enzyme performance and properties. Additional work focused on improving expression and protein yield. By integrating this R&D and utilizing high throughput assays and screening tests, a multitude of strains were examined in a short period of time. As a result, both Novozymes and Genencor were able to produce an enzyme prep exceeding 10X reduction in cost. In addition, new glycosyl hydrolases were identified based on DNA microarray and proteomics (see Figure 5).

**Biomass-to-Ethanol Production (Step 3)**

*Mode of operation.* In the bioethanol production plant, the hydrolysis can be performed in two modes: Separate Hydrolysis and Fermentation (SHF) or Simultaneous Saccharification and Fermentation (SSF). In SHF, the hydrolysis can be performed under conditions optimal for hydrolysis, (e.g. pH around 5 and temperatures from 45-50°C). The disadvantages are product inhibition of the enzymes by the high sugar concentrations in the final stage of hydrolysis, and that the hydrolysis and fermentation requires two steps thereby increasing the

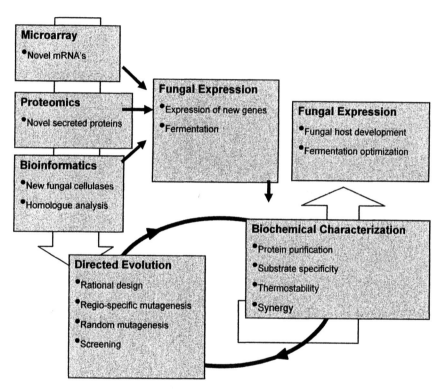

*Figure 4: The diagram illustrates the strategy used to discover more cost efficient cellulases for the hydrolysis of biomass substrates (dilute acid pretreated corn stover). It includes the discovery and improvement of new fungal strians.*

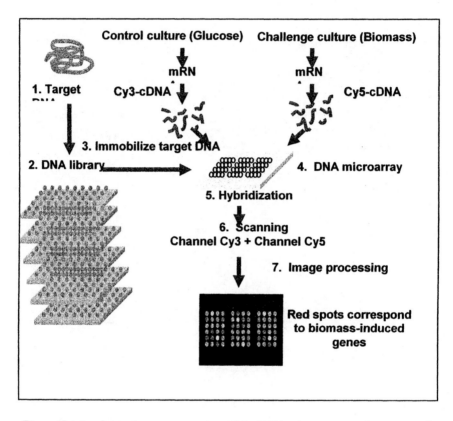

*Figure 5: A schematic representation of the DNA microarray technique used to identify genes induced during biomass growth. The numbers show the different steps in isolating-imaging the genes.*

capital costs and process time (*19*). The advantages of SSF are that the whole process is performed in one reactor and the released glucose is quickly fermented into ethanol, which is less inhibiting than glucose (*20*). One disadvantage of employing SSF is that the conditions have to be a compromise between optimum conditions for hydrolysis and fermentation. Most fermenting microorganisms, such as *Saccharomyces cerevisiae*, have optimum performance at temperatures 10-15°C lower than the optimum for hydrolysis. The hydrolysis is normally performed using a substrate concentration below 15% dry weight (w/v) due to the problems of mixing more concentrated slurries. The use of the whole slurry obtained after pretreatment of the lignocellulosic material can significantly reduce the performance of the enzyme due to end-product inhibition and other degradation products present in the pretreatment. A better performance is therefore obtained by washing the solid before hydrolysis, however, at the expense of increasing operation costs. Presently, the SSF process is viewed as being the most attractive mode to operate (*15,21*). According to literature data, the cellulose conversion under SSF is ~80-90% and SHF is 70-90% with enzyme loadings just below 15 FPU/g cellulose (*22*). The cost to production ethanol under such process conditions is not competitive with current grain processing. For an industrial process, cellulose conversion of biomass is expected to be 80-90% using either SHF or SSF preferably at a cost of $0.05/gallon of ethanol produced (economic analysis using corn stover) (*23*). In the Novozymes DOE funded project, efforts are underway to test both processes utilizing the low cost cellulases.

*Metabolic Engineering of Organisms.* In the establishment of novel bioprocesses for production of chemicals, it is usually desirable to develop generic cell factories that ensure a fast and efficient conversion of the raw material to the product of interest. The yeast, *S. cerevisiae,* is an attractive cell factory due to its widespread use in biotechnology and the robustness of those applications. Much of the research in the past decade, was dedicated to overcoming the obstacles in fermentations: ethanol tolerance, prehydrolzyate tolerance, temperature, sugar concentrations, etc. Extensive R&D into the metabolic engineering of both yeasts and bacteria targeted improvements in the assimilation of xylose and other five carbon sugars in fermentations (*24, 25*). The aim is to construct an efficient xylose fermenting *S. cerevisiae* strain with increased ethanol yield and volumetric productivity that could be used in bioethanol production from raw materials such as lignocellulosic hydrolyzates. Several strategies (*24, 25*) are often employed: Engineering the xylose pathway consisting of the xylose reductase and xylitol dehydrogenase from *P. stipitis* and the endogenous xylulokinase, metabolic engineering of the central metabolism via the construction of glucose-depressed strains or the redox balance to modify the cofactors in the cell. Many researchers like Professors Nancy Ho (Purdue), Merja Pentilla (VTT), Lisbeth Olsson (DTU) and Lonnie Ingram (Florida University) are just a few leading scientist in this field. In the end, the biomass-to-ethanol production plant requires progress in all 3 technologies to compete with grain.

## Case Study 3: Next Generation of Biocatalysts (Chemical Industry)

*Opportunities and Needs.* The chemical industry is both diverse and complex. It encompasses the field of pharmaceuticals, fine chemicals, and commodities. Interestingly, whether producing fine chemicals or commodities, the research needs are similar with a recent focus on chemical biotechnology. "Chemical Biotechnology" is the application of biotechnology to chemical production. It is utilized today in developing new products, manufacturing methods, and improved economics with lesser demands for energy and deleterious impact on the environment. Examples include the use of renewable feedstock in making ethanol, 1,3 propanediol and polylactic acid. Its integration into chemical production has taken the lead in pharmaceuticals (fermentations), food chemicals (high fructose corn syrup), and commodities (ethanol) employing microbes or enzymes. Within the research field, just identifying the needs in biocatalysis (biotransformation) alone is an important goal for chemistry-biotech-related industries. The following passage describes the biotech tools/research needs to develop the next generation enzyme catalysts.

Biocatalysis is a research area that can lead to unprecedented opportunities in this industry but has not been fully exploited. Significant work has been done, however, in the field of polymers. Several enzyme-catalyzed transformations were clearly demonstrated with unique advantages (see Table I). Such enzymatic and/or chemoenzymatic routes to synthesis ranged from the preparation of monomers/precursors, synthesis of polymers, to processing/modification of polymers. A few of the pathways whereby functionality is introduced into a molecule by hydroxylation, halogenation, cycloaddition, etc., are listed in Table I. For a detailed review of this work, this author suggests searching the research carried out by the "Biocatalysis and Bioprocessing of Macromolecules" center at Polytechnic University (note an NSF-Industry sponsored center) (*26*). The next generation catalysts for industrial use, however, are not well defined or identified. Quite often, the emergence of new applied technology is based on establishing a "Biocatalytic-based Technology Platform". The creation of such a platform stems from identifying the contents of a biocatalytic *developmental* toolbox and should reflect the breadth of diversity needed for the chemical industry. Based on feedback from the industry, enzyme producers, and academia, several focus areas have been identified. Examples for the biotech researchers include: achieving higher activity of enzymes in polar solvents, discovering/developing new catalysts in *all* 6 classes of enzymes, lowering cost of enzyme production,

**Table I. Examples of Enzyme-catalyzed Reactions.**

| Type of Reaction | Alteration of Target Molecule |
|---|---|
| Hydroxylation<br>Halogenation<br>Halohydrin formation<br>Cycloadditions | Introduction of functional groups |
| Lactonization<br>Isomerization<br>Oxidation of alcohols to aldehydes & ketones<br>Reduction of aldehydes and ketones to alcohols<br>Oxidation of amino groups to nitro groups<br>Hydrolysis of nitriles to amides and carboxylic acids | Functional group alteration |
| Esterification<br>Carbonate formation<br>Glycosylation<br>Amidation<br>Phosphorylation | Addition onto functional groups |

developing economic approaches to bypass cofactors and developing biocatalysts that carry out multi-step reactions. Although, areas for improvement can be postulated, the actual enzyme molecule to target is based on industrial application. The industry is far more ambiguous in commercializing new enzymes.

*The Paradox.* The creation of a toolbox is an expensive, labor intensive, long-term task with lower than desired probability of success. Many research projects are market driven based on net present values for a given application. Thus a paradox is formed – enzymes are needed for testing, application tests are performed, and enzymes modified for enhanced properties. Moreover, the enzyme discovery-development process requires years of R&D to reach commercialization. Therefore, having knowledge of the performance characteristics and desired properties in the early-stages of research is crucial for sharpening enzyme developmental efforts. When obtained early in the new lead phase, this information can increase hit rates to market and speed in innovation. Interestingly, there are advance biotech tools that help to manage the costs, time and diversity in developing new technologies.

The application of high through-put screening techniques (HTS) in combinatorial catalysis and miniature bioprocessing research has proven advantageous as a new platform for the discovery and *development* of new enzymes. This information can be applied to help prioritize leads that favor desirable processing characteristics and identify candidates for molecular biology research. Moreover, the coupling of this platform with DNA microarray technology incorporates metabolic information directly into the "creating diversity" and "discovery phase" of a new lead. The research conducted at Novozymes has clearly shown the benefits of enzyme screening, selection and application testing. Such examples include the NREL subcontract work with cellulases, and the DOE-Abengoa biomass-to-ethanol project (see Case Study 2), and the drug discovery process in health-care businesses. These techniques were used to screening many *T. reesei* transformants for improved activity and thermostability. The R&D efforts in Case Study 2 resulted in identifying interesting genetically engineered strains (3 cellulolytic activities), new structural motifs for directed evolution studies and auxillaries that boost activities. The resulting candidates were tested in miniature, enzyme application tests that hydrolyze and ferment biomass (corn stover). As a result, new leads that result in both desirable catalytic and metabolic activity were identified for further genetic manipulation/application testing. New structural motifs (enzyme catalytic binding sites) were also recognized with the aid of other miniature application tests (saccharification, fermentation). This resulted in a new library of catalysts.

The enzyme discovery-development process is analogous to the R&D facing the pharmaceutical industry. In the health care business, combinatorial catalysts is extremely important in the drug discovery phase providing a multitude of new molecules for testing. Examples of target molecules undergoing further chemical modification to produce a vast array of new leads for screening are shown in Figure 6. See reference 27 for details (27). In summary, these tools can be adapted to build the "biocatalytic-based technology platform" needed for next generation technology. The emergence of new technologies with such combinatorial approaches to discovery can significantly impact the pathway to commercialization.

## Conclusion

The rate at which new biocatalysts are introduced into the market is accelerated with the use of advanced biotech tools. The first wave technology in applied biotechnology for the chemical industry consists of high through-put screening methods, combinatorial catalysis/synthesis and DNA microarray. Recent advances in applied research utilizing the same tools have also demonstrated an increase in the hit rate and speed in innovation. Examples in our laboratories include HTS of new cellulases, the automation of microscale and miniscale hydrolysis tests and fermentations. With the aid of fast, reliable assays, enzymes with specific performance characteristics can be identified earlier in the research phase. In speculating future technology waves, one might envision enhancing the potential of HTS processes (heterogeneous, chips, arrays) such that they are consistent with the breadth of biodiversity available.

## Acknowledgement

I would like to acknowledge my fellow scientists in Novozymes A/S, Novozymes Biotech and Novozymes North America, Inc. for their significant contributions to these projects.

## References

1. *Development Document for Effluent Limitations Guidelines and Standards for the Textile Mills;* U.S. EPA, Document EPA 440/1-79-0226; October 1979.
2. Lange, N.; Liu; J.; Husain, P.; Condon, B. BioPreparation of Cotton, AATCC International Conference & Exhibition, 1998.
3. Brett, C.; Waldron, K. *Physiology & Biochemistry of Plant Cell Walls;* Chapman & Hall: NY, 1996; $2^{nd}$ edition, pp. 20-80.
4. Liu, J.; Condon, B. WO Patent 9932708 A1, 1998.

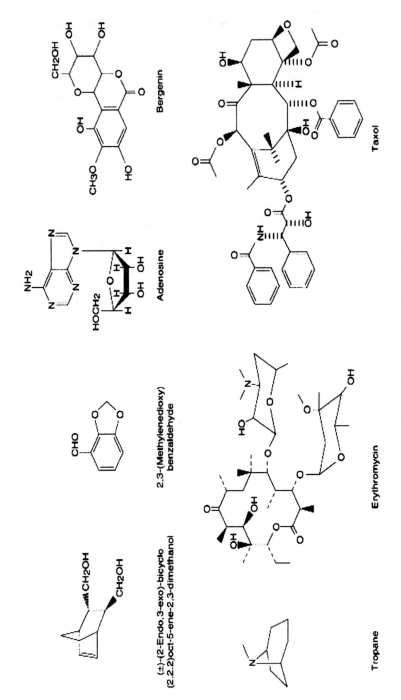

*Figure 6: Examples of New lead molecules in the discovery of drugs.*

5. Etters, N.; Henderson, L.A.; Liu, J.; Sarkar, K. *AATCC Review.* 2001, 22-24.
6. Lu, Y.; Yang B.; Gregg, D.; Saddler, J.N.; Mansfield, S.D. *Applied Biochemistry and Biotechnology.* Davison, B.H.; Lee, J.W.; McMillan, J.D.; Finkelstein, M.; Eds.; Humana Press: NJ, 2002; 98-100, pp. 641-654.
7. Dale, B. E.; Elander, R.T.; Holtzapple, M.; Ladisch, M.R.; Lee, Y.Y.; Eggemen, T.; Wyman, C.E. Comparative Data from Application of Leading Pretreatment Technologies to Corn Stover. 25$^{th}$ Annual Meeting of the AICHE, November 2003.
8. Schell, D.J.; Farmer, J.; Newman, M.; McMillian, J.D. *Applied Biochemistry and Biotechnology.* Walt, D.R.; Davison, B.H.; Lee, J.W.; McMillan, J.D.; Finkelstein, M.; Eds.; Humana Press: NJ, 2003; 105-108, pp. 69-85. ,
9. Negro, M.J.; Manzanares, P.; Ballesteros, I.; Oliva, J.M.; Cabanas, A.; Ballesteros, M. *Applied Biochemistry and Biotechnology.* Walt, D.R.; Davison, B.H.; Lee, J.W.; McMillan, J.D.; Finkelstein, M., Eds.; Humana Press: NJ, 2003; 105-108, pp. 87-100.
10. Tucker, M. P.; Kim, K. H.; Newman, M.; Nguyen, Q.A. *Applied Biochemistry and Biotechnology.* Walt, D.R.; Davison, B.H.; Lee, J.W.; McMillan, J.D.; Finkelstein, M.; Eds.; Humana Press: NJ, 2003; 105-108, pp. 165-178.
11. Varga, E.; Reczey K.; Zacchi, G. Optimisation of Steam Pre-Treatment of H2SO4 impregnated Corn Stover for Enhance Enzymatic Digestibility, Poster Presentation, 3-50, 24$^{TH}$ Symposium on Biotechnology for Fuels and Chemicals, Spring 2002.
12. Hasan, A.; Teymouri, F.; Laureano-Perez; Dale, B.E. Ammonia Fiber Explosion Process (AFEX): A Rapid and Flexible Laboratory Scale Unit. 24$^{TH}$ Symposium on Biotechnology for Fuels and Chemicals, Spring 2002.
13. Nagle, N.; Elander, R.; Newman, M.; Rohrback, B.; Ruiz, R.; Torget, R. *Biotechnology Progress.* **2002**, 18, 734-738.
14. *Handbook on Bioethanol: Production and Utilization, Applied Energy Technology Series;* Wyman, C.E., Ed.; Taylor & Francis: Washington, DC., 1996, pp. 213-303.
15. McAloon, A.; Taylor, F.; Yee, W.; Ibsen, K.; Wooley, R. Determining the Cost of Producing Ethanol from Corn Starch and Lignocellulosic Feedstocks NREL/TP-580-28893, October 2000, 30 pages. http://www.nrel.gov.
16. Tomme, P.; Van Tilbeurgh, H.; Petterson, G.; Van Damme, J.; Vandekerckhove, J.; Knowles, J.; Teeri, T.T.; Clacyssens, M. *European Journal of Biochemistry.* **1988**. 170, 575-581.
17. Kim, T.H.; Lee, Y.Y. Effects of Residual and Soluble Lignin on Enzymatic Hydrolysis of Cellulose, 24$^{TH}$ Symposium on Biotechnology for Fuels and Chemicals, Spring 2002.

18. Mansfield, S.D.; Mooney, C.; Saddler, J.N. *Biotechnology Progress.* **1999**. 15, 804-816.
19. Galbe, M.; Zaachi, G. *Appl. Microbiol. Biotechnol* **2002,** 59, 618-628.
20. Holtzapple, M.; Shu, Y.; Hendrickson, C. *Biotechnol. Bioeng.* **1990,** 36, 275-287.
21. Wooley, R.; Ruth, M.; Glassner, D.; Sheehan, J. *Biotechnol, Prog.* **1999,** 15, 794-803.
22. Alfani, F.; Gallifuoco, A.; Saporosi, A.; Spera, A.; Cantarella, M. *Microbiol. Biotechnol.* .**2000,** 25,184-192.
23. Hatzis, C.; Riley, C.; Philippidis, G.P. Detailed Material Balance and Ethanol Yield Calculations for the Biomass-to-Ethanol Conversion Process Applied Biochemistry and Biotechnology. Proceedings of the 17$^{th}$ Symposium on Biotechnology for Fuels and Chemicals, Spring 1996, 57-58, pp. 443-459.
24. Jin, Y.; Jeffries, T. *Applied Biochemistry and Biotechnology.* Walt, D.R.; Davison, B.H.; Lee, J.W.; McMillan, J.D.; Finkelstein, M.; Eds.; Humana Press: NJ, 2003, pp. 277-286.
25. Lawford, H.G.; Rousseau, J.D. *Applied Biochemistry and Biotechnology.* Walt, D.R.; Davison, B.H.; Lee, J.W.; McMillan, J.D.; Finkelstein, M.; Eds.; Humana Press: NJ, 2003, pp. 457-469.
26. Gross, Richard A.; Kumar, Ajay; Kalra, Bhanu. *Chemical Reviews.* **2001**. 101(7), 2097-2124.
27. Michels, Peter C.; Khmelnitsky, Yuri L.; Dordick, Jonathan S.; Clark, Douglas S. Combinatorial biocatalysis for drug discovery and optimization. EnzyMed, Inc., Iowa City, IA, USA. Book of Abstracts, 213th ACS National Meeting, San Francisco, April 13-17 (1997).

## Chapter 3

# Empirical Biocatalyst Engineering: Escaping the Tyranny of High-Throughput Screening

Jon E. Ness[1], Tony Cox[1], Sridhar Govindarajan[1], Claes Gustafsson[1], Richard A. Gross[2], and Jeremy Minshull[1,*]

[1]DNA 2.0, 1455 Adams Drive, Menlo Park, CA 94025
[2]NSF I/UCRC for Biocatalysis and Bioprocessing of Macromolecules, Polytechnic University, 6 Metrotech Center, Brooklyn, NY 11201

High throughput assays are a commonly used tool in empirical protein engineering. Such screens often do not accurately predict the behavior of proteins, including biocatalysts, under commercially relevant conditions. We are developing an alternative approach based on mathematical data-mining tools that allow us to determine sequence-activity relationships for proteins. Sequence-activity relationships can be used in the design of proteins with specified properties. The availability of methods to quickly synthesize designed sets of proteins make this a cost-effective technique for biocatalyst engineering. Because only small numbers of variants (<100) need to be tested, proteins engineered using sequence-activity relationships can be tested directly under final application conditions. This avoids the expenses and errors of developing and implementing high throughput surrogate screens.

Protein engineering has to date been tackled according to two quite different philosophies. The first of these can broadly be termed mechanism-based design. This involves describing protein function as completely as possible so that desired changes can be effected by calculation from first principles. Typically the protein mechanic first asks what three dimensional structure an amino acid sequence will adopt, then what function will be performed by such a structure and finally how alterations to the structure may cause the function to change. Practitioners of the opposing methodology, empiricists, generally eschew the confines of structural logic. Instead they construct many protein variants and test them, imitating natural selection by choosing the best performing amino acid sequences to incorporate into a subsequent set of protein variants.

The advantages and disadvantages of these two approaches to protein modification have been extensively discussed in the literature (1, 2). The lines between them are becoming blurred as empiricists incorporate variations identified by structural studies into their designs while structural biologists are creating small libraries for functional analysis. A key limitation for any library-based protein engineering, however, is in developing a good assay for protein function. The assay must measure protein properties that are relevant to the final application, but must also be capable of testing a sufficient number of variants to identify the small fraction that are improved. The difficulty of creating such an assay is particularly relevant when optimizing proteins for the synthesis or degradation of polymers, since reaction conditions during polymer modification must usually be controlled more precisely than can be achieved reliably in high throughput. An effective methodology for engineering proteins useful in polymer modification should therefore be capable of obtaining improved variants whilst testing only small numbers (<100) of variant proteins.

Protein engineering based on sequence-activity relationships is attractive because of the small numbers of variants that need to be tested in order to obtain improvement. In principle this frees the investigator from the tyranny of high throughput screening. It also allows one to design and create the most informative set of variants, rather than relying upon a stochastically constructed library (3-6).

This paper describes the application of sequence-activity modeling techniques to identify amino acid residues within one protein (proteinase K) that contribute to specific functions. We show that combining amino acids identified as beneficial by mathematical modeling results in a protein with improved activity. We also show that different amino acids are important for different functions, underscoring the importance of using a screen that accurately reflects the commercially desired protein function.

## Sequence-Activity based Protein Engineering Strategies

Protein optimization is conceptually identical to many other complex system and process optimization problems in which there are many variables and an incomplete understanding of the system. Advances in several areas are now increasing the acceptance of protein engineering methods based on modern engineering principles (7, 8). The cost of DNA sequencing is decreasing continuously, synthesizing specific genes or gene sets is becoming increasingly feasible and cost effective (9, 10), and a number of techniques developed to model complex systems have been used effectively in deriving sequence activity relationships for proteins. In this way protein engineers are embarking upon a similar course to that charted successfully by medicinal chemists who use quantitative structure-activity relationships (QSAR) to optimize the pharmaceutical properties of small molecules.

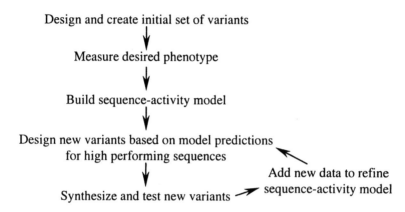

*Figure 1. Protein engineering process using sequence activity relationships.*

One basic procedure used to optimize molecules using sequence-function modeling is shown in Figure 1. This iterative synthesis /testing/modeling method was used to create a 250-fold improved analog of the neuropeptide substance P in just 3 cycles testing a total of 45 variants (*11*). In this case the modeling was performed using partial least squares regression, a technique that, with its close relatives, has also been successfully applied to modeling and optimizing protein function using test sets of less than 50 variants (*12, 13*).

## Selection of Initial Amino Acid Substitutions

For this study we chose to modify the activity, pH optimum and thermostability of the enzyme proteinase K from *Tritirachium album*. Hydrolytic enzymes such as proteases and lipases are finding many applications in polymer synthesis and degradation (*14-21*). The activity and physical properties of these enzymes determine whether or not they will make suitable biocatalysts for polymer modification.

Cloning of the proteinase K from *Tritirachium album Limber* and expression of the active protease has been described (*22*). We did not attempt to obtain a sample of the cloned gene: this path is often fraught with difficulties and delays and can even result in significant work being expended on a construct that ultimately reveals itself to be an imposter. Instead we decided to synthesize the gene *de novo*. Gene synthesis provides a simple and convenient route from database to physical gene. In this case it also allowed us to alter the codon bias from that found in the fungal source organism to one more favourable to high levels of expression in *E coli* genes, change the secretion signal to improve periplasmic expression in *E coli* and remove an intron. When we expressed the synthesized gene in *E coli* we found that it hydrolyzed casein, as had been previously reported for the cloned *Tritirachium album* gene expressed from *E coli* (22).

| | |
|---|---|
| N95C | Homolog properties from published literature |
| P97S | Structural considerations |
| S107D | Homolog sequence alignment analysis |
| S123A | Thermostable homolog sequence alignment analysis |
| I132V | Substitution matrix-derived change |
| E138A | Homolog properties from published literature |
| M145F | Homolog properties from published literature |
| Y151A | Homolog properties from published literature |
| V167I | Substitution matrix-derived change |
| L180I | Substitution matrix-derived change |
| Y194S | Variation observed in highly active clone |
| A199S | Substitution matrix-derived change |
| K208H | PCA modelling of thermostable homologs collected from GenBank. |
| A236V | PCA modelling of thermostable homologs collected from GenBank. |
| R237N | Homolog properties from published literature |
| P265S | Structural considerations |
| V267I | Substitution matrix-derived change |
| S273T | Homolog sequence alignment analysis |
| G293A | Thermostable homolog sequence alignment analysis |
| L299C | Homolog properties from published literature |
| I310K | Structural considerations |
| K332R | Thermostable homolog sequence alignment analysis |
| S337N | Thermostable homolog sequence alignment analysis |
| P355S | Structural considerations |

*Figure 2. Initial set of 24 amino acid variations to alter substrate specificity and physical properties of proteinase K.*

To create a set of variants we identified variable amino acid residues by comparing proteinase K with homologous sequences from the serine protease family using standard phylogenetic analysis, structural comparisons and principal component analysis. Principal component analysis can be used to reduce the high dimensional complexity of sequence variations by creating new composite dimensions which account for the majority of the differences within a set of genes (*23*). Gene families cluster in some of these new dimensions according to function. The amino acids responsible for clustering, and by extension the positions responsible for specific functions can thus be identified (*24*). Once we had identified sites that appeared to vary with protein function and phylogeny, we selected amino acid changes that are likely to be acceptable at those sites using an assortment of substitution matrices. These matrices are derived empirically from analyses of the rules governing amino acid changes in naturally occurring homologous and functional proteins (*25*). The full set of initial variations are shown in Figure 2.

## Creation and Testing of Initial Variants

We first designed 24 variants for synthesis. We chose to make each variant with 6 of the 24 changes from the wild type sequence shown in Figure 2. We ensured that the variant set included a statistically representative distribution of amino acid changes, incorporating each change 6 times in the set of 24 variants. We confirmed the sequence of each designed variant and then tested it for the ability to hydrolyze casein: this was a published measurable activity and was also the initial test that we used for the wild type protein. Of the first set of 24 variants that we synthesized, only 3 showed activity towards casein. These 3 variants contained a total of 15 of the 24 changes. We then synthesized 2 additional sets of variants. The first of these consisted of 9 genes, each containing only 1 of the 9 changes not seen in any of the active variants from the first set, 3 of these were active and 6 were not. The second consisted of 15 new variants containing between 3 and 5 changes. This made a total of 48 individually designed synthetic variants.

One of our aims in this work was to test whether we could engineer enzyme properties by testing small numbers of variants and modeling their sequence activity relationship. We were also interested in determining the extent to which different assays are predictive of one another, since this is the basis of any high throughput approach to protein engineering. We therefore selected two different assays for measuring the activities and physical properties of our proteinase K variants.

We measured the activity of the variants towards casein, which is a large polymeric protease substrate whose hydrolysis has previously been used as a high-throughput screen to engineer proteases for use in laundry detergents (*3, 26*). We also measured variant activity towards a modified tetrapeptide, *N*-succinyl-Ala-Ala-Pro-Leu-*p*-nitroanilide (AAPL-*p*-NA) which undergoes a colorimetric change upon protease-mediated hydrolysis, and which has also been used in high throughput protease screening (*27*). Using AAPL-*p*-NA we measured the activity of the variants at 3 different pH values (7, 5.5 and 4.5). We also used AAPL-*p*-NA to measure the activity of variants following a 5 minute heat treatment at 65°C.

## Modeling the Proteinase K Sequence Activity Relationship for Heat Tolerance

For each of the proteinase K activities tested, we used partial least squares regression (PLSR) to model the relationship between amino acid variation and the variations in proteinase activities (the sequence-activity relationship). The application of these methods to nucleic acids, peptides and proteins has been described previously (7, 11, 12, 28-30).

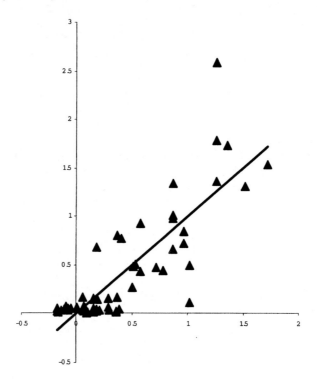

*Figure 3. Modeling of proteinase K variant activity towards AAPL-p-NA following a 5 minute 65°C heat treatment. Measured activities on the y-axis are compared with those predicted by the model on the x-axis. All activities were measured at 37°C and pH 7.0 using purified protein and are normalized to the activity of wild type proteinase K.*

We used our PLSR-based sequence activity relationship to assign a regression coefficient to each varied amino acid. We then calculated the predicted activity for a proteinase K variant by summing the regression coefficients for amino acid variations that are present in that variant. In this case we did not include terms to account for interactions between the varied amino acids, though this can also be done (*13*). Figure 3 shows a good correlation between the predictions of our calculated sequence-activity relationship and the measured ability of heat-treated proteinase K variants to hydrolyze AAPL-*p*-NA.

## Design, Synthesis and Testing of an Improved Proteinase K

The utility of a sequence-activity model lies in its ability to predict the activity of variants that have not been measured, or to identify amino acid variations that contribute positively to a specific protein property and that can then be experimentally combined. To test our sequence activity model for heat-tolerant hydrolyzers of AAPL-*p*-NA, we analyzed the regression coefficients from the model, as shown in Figure 4.

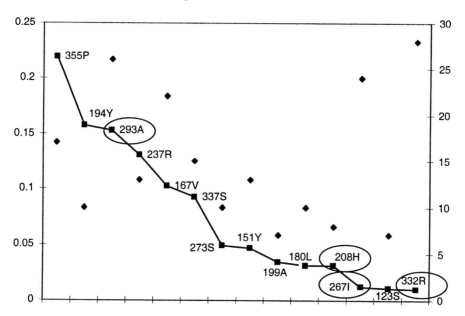

*Figure 4. Regression coefficients (squares, left axis) of variant amino acids derived from the PLSR sequence-activity model. Number of occurrences (diamonds, right axis) are also shown. Changes from the wild type sequence are circled.*

Four of the amino acid changes that we had incorporated into the library were predicted to have a positive effect on the activity of proteinase K after heating. These were K208H, V267I, G293A and K332R. Among the variants synthesized in the initial set of 48, one (NS40) contained 3 of these changes (V267I, G293A and K332R) and one (NS19) contained the other (K208H). To test the predictive power of our model we therefore synthesized a variant (NS56) containing all four of these changes and compared its activity with that of NS19 and NS40.

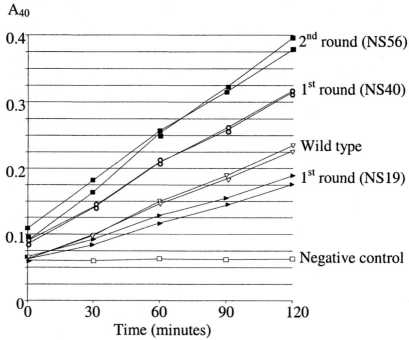

*Figure 5. Activity of designed 1$^{st}$ and 2$^{nd}$ round variants activity towards AAPL-p-NA following a 5 minute 65°C heat treatment. Purified proteins were heated to 65°C then incubated with AAPL-p-NA at pH7.0. The reaction was followed by measuring the absorbance at 405 nm. Alterations from wild type sequence are: NS19, K208H (filled triangles); NS40, V267I, G293A, K332R (open circles); NS56, K208H, V267I, G293A, K332R (filled squares).*

As shown in Figure 5, combining the four changes identified by our PLSR model produced a variant with greater post-heat treatment activity towards AAPL-*p*-NA than the single or triple changes. We are extremely encouraging that by synthesizing and measuring the activities of only 48 variants we were able to design a new variant that was further improved for activity following heat treatment. This suggests that tests of protein function that are low-throughput but faithfully replicate the final desired application may be combined with mathematical determination of protein sequence- function relationships to make commercially relevant changes to protein function.

## Different Amino Acid Variations are Beneficial for Different Functions

The current paradigm for empirical protein engineering is to employ high throughput screens to test libraries of thousands of variants (*31*). In general, high throughput screens do not measure all of the properties that are important for the final application. One common way of overcoming this discrepancy is the use of tiered screens, in which high throughput screens that measure only one or two of the properties of interest are followed by lower throughput screens that more accurately reflect the desired protein characteristics (*32*). This technique relies on the assumption that the high throughput primary screen will identify the amino acid variations that are important for the final function but will also select some false positives. False positives do not actually contribute to the final function and are eliminated by subsequent screens. The alternative possibility, that amino acids that would be beneficial for the final application may be missed by the initial high throughput screen (false negatives) is seldom considered. By prematurely discarding variations that would be beneficial for the desired function, the protein engineering process may be unnecessarily prolonged or even fail.

Having measured several properties of our proteinase K variants and validated the predictive power of our sequence-activity modeling for one of these properties, we wished to explore the validity of the high throughput screening approach in more depth. Although we performed no high throughput screening ourselves, all of our assays could easily be adapted for use as high throughput primary screens. Hydrolysis of casein incorporated into media plates has been used as a primary screen for protease libraries (*3, 26*), as has hydrolysis of AAPL-*p*-NA (*27*). Testing AAPL-*p*-NA hydrolysis at lowered pH (5.5 or 4.5) might be considered an appropriate surrogate for the low pH tolerance that will be required by an enzyme that is modifying a polymer at low pH. Similarly testing AAPL-*p*-NA hydrolysis following heat treatment may be thought to measure the stability that will be required for an enzyme that must resist the thermal stresses of commercial process conditions. We expressed thermostability in three ways: as the absolute level of activity remaining following heat treatment, as the activity remaining relative to the activity prior to heat treatment, and as the product of these two values. Having obtained values for each of these proteinase K properties, we wished to examine the correlation between the properties, and to compare the amino acid variations that would be selected by each screen, as if it had been used as a high throughput primary screen.

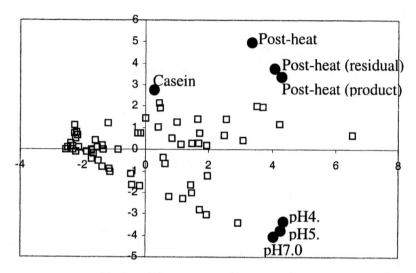

*Figure 6. Principal component analysis of proteinase K variant activities. Seven measured proteinase K properties are shown compressed into two principal components. PC1 (horizontal axis) capturing 60% of variance, PC2 (vertical axis) capturing 22% of variance. Open squares represent each proteinase K variant, filled circles show the positions of each property in principal component space. Properties are: activity towards casein, activity towards AAPL-p-NA at pH7.0, pH5.5 and pH4.5, activity towards AAPL-p-NA after heating at 65° for 5 minutes (absolute), fraction of original activity towards AAPL-p-NA remaining after heat treatment (residual), product of absolute and residual activities (product).*

Figure 6 shows the distribution of 7 proteinase K properties following calculation of the principal components of the 7-dimensional activity data (33). Eighty two percent of the variation in the data could be captured in just two dimensions. The activities of variants towards AAPL-$p$-NA at different pHs was highly correlated, as were the different methods of measuring heat tolerance. Substrate specificity of the proteinase K variants is the most distinguishing trait within the dataset, accounting for 60% of all variance present. This variance is captured in principal component 1 (horizontal axis). Thermostability is the second most differentiating trait in the dataset; captured by principal component 2 it accounts for 22% of the variance in the functional data.

The principal component analysis also clusters related functional measures. We selected three representative activities based on the functional clustering shown in Figure 6 for further analysis: activity towards AAPL-*p*-NA at pH 7.0, absolute activity towards AAPL-*p*-NA following 5 minutes at 65°C and activity towards casein. For each of these activities we constructed PLSR models similar to that shown in Figure 3, and calculated the regression coefficients for each amino acid variation as shown for thermal tolerance in Figure 4. The changes calculated to contribute positively to each property are shown in Table 1.

| Variation position | Casein hydrolysis | Thermal tolerance | AAPL-p-NA pH7.0 |
|---|---|---|---|
| 107 | <u>D</u> | S | S |
| 123 | <u>A</u> | S | S |
| 151 | <u>A</u> | Y | <u>A</u> |
| 167 | <u>I</u> | V | V |
| 180 | <u>I</u> | L | <u>I</u> |
| 194 | <u>S</u> | Y | Y |
| 199 | <u>S</u> | A | A |
| 208 | K | <u>H</u> | K |
| 267 | V | <u>I</u> | V |
| 273 | <u>T</u> | S | S |
| 293 | G | <u>A</u> | <u>A</u> |
| 332 | <u>R</u> | <u>R</u> | <u>R</u> |

*Table 1. Beneficial amino acid variations calculated by PLSR for 3 different proteinase K properties. Changes from the wild type sequence are underlined.*

The amino acids that we predict to be beneficial based on the three different representative assays are strikingly different. This may explain the frequent failure of high throughput screens to produce proteins that are useful for commercial applications. Use of any one of our three representative assays as the primary screen would have selected some amino acid changes that we predict would not be important for the others. These would be false positives, for example 6 beneficial changes identified using casein hydrolysis as a primary screen (S107D, S123A, V167I, Y194S, A199S and S273T) are all predicted to have a negative effect on activity towards AAPL-*p*-NA, with or without heating. Perhaps even more importantly, using casein as a primary screen falsely attributes a negative value to 3 of the 4 changes we found to be important for thermal tolerance (K208H, V267I and G293A).

This failure of a tiered screening strategy is not simply a result of selecting an inappropriate surrogate substrate. Similar results would have been seen had activity towards AAPL-*p*-NA been used as a primary screen followed by a test for thermal tolerance. In this case half of the beneficial changes would still have been discarded as false negatives (K208H and V267I). Half of the 24 initial changes that we incorporated into our design are predicted by our sequence-activity relationships to have a positive effect on at least one of the functions that we measured, but only one (K332R) is consistently beneficial in all dimensions. From this analysis it appears that measuring properties that are different to those of the final application may result both in incorporation of sequence changes that do not contribute to the desired phenotype, as well as omission of those that do.

## Conclusions

We have described a method for engineering proteins based on design, synthesis and testing of small numbers of individual variants followed by mathematical modeling to determine a sequence-activity relationship. We have also shown that sequence-activity models can be used predictively to design improved variants.

By incorporating the principles of experimental design, individual design and synthesis of sequence variants allows a more efficient search of sequence space than a library approach (*34*). Another extremely important advantage of the modeling approach is that it facilitates empirical protein engineering but requires only very low numbers of variants to be tested. This means that the need for high throughput screens are obviated. The value of this cannot be overstated. As well as many anecdotal reports from throughout the protein engineering community, our analysis suggests that high throughput and tiered screening may be fundamentally flawed strategies for protein engineering. Both conserved reaction conditions and use of the same substrate appear susceptible to selection of false positives and rejection of false negatives. The performance of high throughput screens will be further compromised when the primary screen is selected on the basis of throughput rather than faithful replication of the final application.

## Acknowledgements

We thank DARPA for funding part of this work.

## References

1. Ryu, D.D.; Nam, D.H. *Biotechnol Prog* **2000**, *16*, 2-16.
2. van Regenmortel, M.H. *J Mol Recognit* **2000**, *13*, 1-4.
3. Ness, J.E.; Kim, S.; Gottman, A.; Pak, R.; Krebber, A.; Borchert, T.V.; Govindarajan, S.; Mundorff, E.C.; Minshull, J. *Nat Biotechnol* **2002**, *20*, 1251-5.
4. Coco, W.M.; Encell, L.P.; Levinson, W.E.; Crist, M.J.; Loomis, A.K.; Licato, L.L.; Arensdorf, J.J.; Sica, N.; Pienkos, P.T.; Monticello, D.J. *Nat Biotechnol* **2002**, *20*, 1246-50.
5. Zha, D.; Eipper, A.; Reetz, M.T. *Chembiochem* **2003**, *4*, 34-9.
6. Stoop, A.A.; Craik, C.S. *Nat Biotechnol* **2003**, *21*, 1063-8.
7. Geladi, P.; Kowalski, B.R. *Analytica Chimica Acta* **1986**, *186*, 1-17.
8. Gustafsson, C.; Govindarajan, S.; Minshull, J. *Curr Opin Biotechnol* **2003**, *14*, 366-370.
9. Beattie, K.L.; Logsdon, N.J.; Anderson, R.S.; Espinosa-Lara, J.M.; Maldonado-Rodriguez, R.; Frost, J.D.r. *Biotechnol Appl Biochem* **1988**, *10*, 510-521.
10. Cello, J.; Paul, A.V.; Wimmer, E. *Science* **2002**, *297*, 1016-1018.
11. Norinder, U.; Rivera, C.; Unden, A. *J Pept Res* **1997**, *49*, 155-62.
12. Bucht, G.; Wikstrom, P.; Hjalmarsson, K. *Biochim Biophys Acta* **1999**, *1431*, 471-82.
13. Aita, T.; Hamamatsu, N.; Nomiya, Y.; Uchiyama, H.; Shibanaka, Y.; Husimi, Y. *Biopolymers* **2002**, *64*, 95-105.
14. Gross, R.A.; Kumar, A.; Kalra, B. *Chem Rev* **2001**, *101*, 2097-124.
15. Mahapatro, A.; Kalra, B.; Kumar, A.; Gross, R.A. *Biomacromolecules* **2003**, *4*, 544-51.
16. Uyama, H.; Fukuoka, T.; Komatsu, I.; Watanabe, T.; Kobayashi, S. *Biomacromolecules* **2002**, *3*, 318-23.
17. Park, H.G.; Chang, H.N.; Dordick, J.S. *Biotechnol Bioeng* **2001**, *72*, 541-7.
18. Ulijn, R.V.; Baragana, B.; Halling, P.J.; Flitsch, S.L. *J Am Chem Soc* **2002**, *124*, 10988-9.
19. Kumar, A.; Gross, R.A. *Biomacromolecules* **2000**, *1*, 133-8.
20. Deng, F.; Gross, R.A. *Int J Biol Macromol* **1999**, *25*, 153-9.
21. Park, O.J.; Kim, D.Y.; Dordick, J.S. *Biotechnol Bioeng* **2000**, *70*, 208-16.
22. Gunkel, F.A.; Gassen, H.G. *Eur J Biochem* **1989**, *179*, 185-194.
23. Casari, G.; Sander, C.; Valencia, A. *Nat Struct Biol* **1995**, *2*, 171-178.
24. Gogos, A.; Jantz, D.; Senturker, S.; Richardson, D.; Dizdaroglu, M.; Clarke, N.D. *Proteins* **2000**, *40*, 98-105.
25. Benner, S.A.; Cohen, M.A.; Gonnet, G.H. *Protein Eng* **1994**, *7*, 1323-1332.

26. Ness, J.E.; Welch, M.; Giver, L.; Bueno, M.; Cherry, J.R.; Borchert, T.V.; Stemmer, W.P.; Minshull, J. *Nat Biotechnol* **1999,** *17*, 893-896.
27. Sroga, G.E.; Dordick, J.S. *Biotechnol Bioeng* **2002,** *78*, 761-9.
28. Hellberg, S.; Sjostrom, M.; Skagerberg, B.; Wold, S. *J Med Chem* **1987,** *30*, 1126-35.
29. Eriksson, L.; Jonsson, J.; Hellberg, S.; Lindgren, F.; Skagerberg, B.; Sjostrom, M.; Wold, S. *Acta Chem Scand* **1990,** *44*, 50-55.
30. Jonsson, J.; Norberg, T.; Carlsson, L.; Gustafsson, C.; Wold, S. *Nucleic Acids Res.* **1993,** *21*, 733-739.
31. Lin, H.; Cornish, V.W. *Angew Chem Int Ed Engl* **2002,** *41*, 4402-25.
32. Ness, J.E.; del Cardayre, S.B.; Minshull, J.; Stemmer, W.P. *Adv Protein Chem* **2000,** *55*, 261-292.
33. O'Connel, M.J. *Comp. Phys. Comm.* **1974,** *8*, 49.
34. Hellberg, S.; Eriksson, L.; Jonsson, J.; Lindgren, F.; Sjöström, M.; Skagerberg, B.; Wold, S.; Andrews, P. *Int J Pept Protein Res* **1991,** *37*, 414-424.

Chapter 4

# $^1$H NMR for High-Throughput Screening to Identify Novel Enzyme Activity

### Claire B. Conboy and Kai Li

The Dow Chemical Company, Midland, MI 48686

$^1$H NMR spectroscopy provides an unbiased and universal detector which can be applied to the discovery of novel biocatalytic activity. A high throughput NMR sampling system has been modified to facilitate throughput requirements of genomic screening. To demonstrate the utility of this technique, high throughput NMR was used to screen a genomic library of a *Pseudomonas* strain for nitrilase activity. A novel hydrolyzing enzyme was found. Based on assays of whole cell and cell lysate of the recombinant gene cloned in *E. coli*, this enzyme was found to have both nitrilase and nitrile hydratase activity when hydrocinnamonitrile was used as the substrate. Elucidation of the activities was accomplished by using NMR to detect the corresponding amide and acid products of the substrate.

Biocatalysts are becoming increasingly important in the production of chiral molecules, pharmaceutical intermediates, and medium to high value chemicals and monomers for biomaterials, from alternative and petrochemical feed stocks (1-4). Discovery and screening of biocatalysts for specific chemical transformations are extremely important component in the development of industrial biotechnology. In general, biocatalyst discovery consists of three independent but closely related aspects: gene discovery, microbe discovery and enzyme discovery. In gene discovery, the goal is to isolate and identify genes that have desired catalytic functions from gene libraries. Such libraries can be either microbial gene libraries or environmental DNA (eDNA) libraries (4).

The importance of the identifying specific enzyme activity has prompted the development of high throughput screening methods that are applicable to the large number of genomic variants that must be analyzed. Screening methods for biocatalysts are usually designed for a specific biotransformation. Screening techniques include the of chromogenic substrates, colorimetric pH indication and chromatographic methods (5-8). Hybridization techniques are also used for the screening of novel bioactivities (9). These techniques are based on homology of polynucleotides. The methodology does not specifically determine function, and also suffers from high rates of false-positives (10). These techniques provide the throughput required for screening expression libraries for a predetermined chemical specificity. However, these assay methods are case-dependent. Finding a suitable assay for a specific reaction requires development and may not always possible. NMR spectroscopy provides an unbiased and a "universal" detection method that is not restricted to the presence of a chromophore, does not require standard calibration, and requires minimal analytical method development.

Considerations of protein isolation, solubility and stability are also important criteria for screening. One approach to circumvent these issues is to analyze whole cell cultures (7). This further eliminates the necessity for protein purification which requires considerable resources. $^1$H NMR spectroscopy provides an alternate approach to the detection of biocatalytic activity that is applicable to whole cell cultures, cell lysates and membrane bound proteins. $^1$H NMR spectroscopy can provide quantitative information without the requirement of primary standards or an a priori knowledge of the analyte(s). The later of these attributes is essential for the discovery of "new" biocatalytic activity. $^1$H NMR spectroscopy provides a function based assay without the development time required for chromatographic or wet chemical methods. When the spectroscopy is practiced at sufficient field strength, the resulting spectral dispersion is sufficient to allow for the simultaneous detection of multiple substrates and products. Another important consideration is the sensitivity of the detection method. At a field strength corresponding to 600 MHz for $^1$H NMR, detection limits on the order of μM are achievable using a 20-second per sample acquisition time.

High throughput analysis is essential for a successful application of any detection method for functional screening of DNA libraries. Until recently, high throughput applications of NMR were not available. Driven by the need to accommodate the increasing demands of combinatorial chemistry, "direct-injection" (DI) - NMR techniques have been developed (11). Current commercial sampling systems allow for analysis of approximately 400 samples per day. Screening "typical" cDNA/genomic libraries from microbial hosts using inserts on the order of 2kb corresponds to a library size on the order of 20,000 for genomic coverage. To satisfy the requirement of rapid high throughput to analyze a library of this size, the existing hardware has been modified to allow for a sample throughput of 2400 sample per day (12).

Nitrilases catalyze the conversion of organonitriles directly to the corresponding carboxylic acids. Synthetic hydrolysis of nitriles into the corresponding amides and carboxylic acids requires severe reaction conditions. A typical synthetic approach would require the use of 70% $H_2SO_4$ and heat (13). Such a reaction condition is not compatible when selectivity and the conservation of other hydrolysable functional groups in a substrate are desired. Biotransformation of nitriles can be accomplished under mild conditions, in an aqueous environment (13). Additionally, enantioselectivity of the biocatalytic conversion of nitriles to chiral acids has been demonstrated (14-16). Therefore, nitrilases provide an alternative route for synthetic processes that require conversion of nitriles to corresponding acids.

Nitrilases are emerging as an important tool in industrial synthesis of bulk chemicals and the production of pharmaceuticals and pharmaceutical intermediates. Acrylamide (17) and nicotinamide (18) are produced with the help of a nitrile hydratase (19). Furthermore the organinc solvent 1,5-dimethyl-2-piperidone is produced using a nitrilase (20). High selectivity and enatiomerically pure products are particularly attractive features of Biocatalysis (21). This is demonstrated by the enatioselectivty of a nitrilase from *Arabiposis thaliana* (22). This enzyme has been used to produce enatiopure (E)-cis-3-(2-Cyanocyclohex-3-enyl)propionic acid from a aryl dinitrile substrate, Figure 1. This enzyme has also been used for the enatioselective hydrolysis of fluoroarlynitriles. Organofluoro compounds are important in the applications of medicinal diagnostics including magnetic resonance imaging (MRI) and positron emission tomography (PET) (14). A specific example of the resolution of racemic 2-fluoroarylnitirles is shown in figure 2.

To demonstrate the utility of $^1$H NMR high throughput screening of enzymes, nitrilase activity was the target biocatalyst for a library screening. A library of random 1.5-2.5 kb fragments of genomic DNA from a strain of *Psuedomonas* was constructed using the pET21(+) vector. This library was subsequently screened for the specific transformation of hydrocinnemonitrile (3-phenyl propionitrile) to hydrocinnemic acid (3-phenyl propionic acid), shown in Figure 3. The modified high throughput sampling conditions described else where (12) were used to analyze the genomic library.

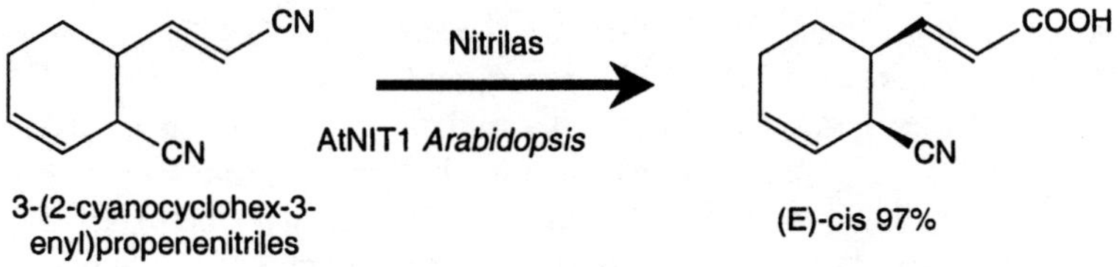

*Figure 1. AtNIT1 exhibits both diastereoselectivity and regioselectivity (22).*

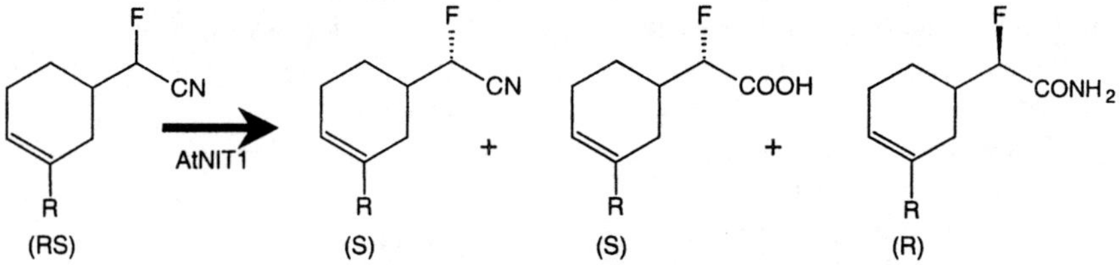

*Figure 2. Resolution of racemic 2-fluoroarylnitriles using nitrilase from Arabidopsis thaliana (14).*

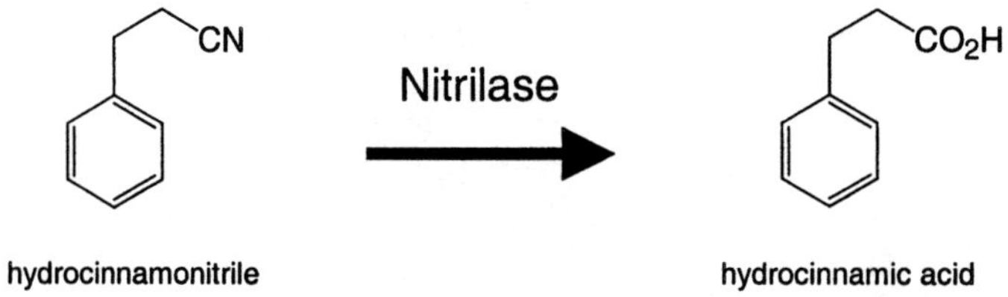

*Figure 3. Biotransformation of nitrile to corresponding acid via hydrolysis. Screening target for $^1H$ NMR high throughput application.*

# Experimental

## Cultivation and Biocatalysis

For screening the genomic library of a species of *Pseudomonas*, random 1.5-2.5 kb fragments of genomic DNA were constructed in the pET21(+) vector. This DNA was subsequently transformed by electroporation into E. coli strain Electro-Ten-Blue. The cells were plated out on LB medium containing ampicillin and grown overnight at 37°C. This DNA preparation representing the genomic DNA library of the species of Pseudomonas was subsequently used to transform *E. coli* BL21(DE3) to analyze individual clones. Cultures were grown in 96-well titer plate format and approximately 20,000 randomly selected clones were screened.

Using the high throughput NMR screening method, a colony of recombinant *E. coli* expressing a putative nitrilase gene in the genome of the species of Pseudomonas was identified. The identified gene was subcloned into an expression vector in *E. coli* JM109 to achieve optimal expression. The resulting strain was subsequently inoculated into 6 mL of LB medium containing the appropriate antibiotics and grown at 36 °C with agitation at 250 rpm until an $OD_{600}$ value of 0.5 to 0.8 was achieved. Expression of the recombinant protein was induced and cultivation continued overnight under the same conditions. The culture was centrifuged and the pellet was rinsed. The pellet was then re-suspended in an equal volume of reaction medium with appropriate antibiotics. The biotransformation was initiated by the addition of substrates and subsequently analyzed using $^1$H NMR.

Enzyme activity was also studied using cell lysates. Cultures of the recombinant *E. coli* strain were prepared as described in preceding paragraph. After a 24 hour cultivation period, the cultures were centrifuged, the pellet re-suspended using minimal salt buffer and the cells sonicated. Following sonication, the cell debris was separated using centrifugation. Supernatants were aliquoted into solutions containing nitrile substrates and biocatalysis was performed at 37°C. The reaction mixtures were sampled at various times and the samples subsequently analyzed using $^1$H NMR.

## NMR Analysis

A Varian Inova 600AS (actively shielded) MHz NMR spectrometer equipped with the Versatile Automated Sample Transport (VAST) system was used for the high throughput screening (11). The Gilson 215 liquid handling system that is part of the VAST accessory had been modified to allow for increased sample throughput under screening conditions (12). A 120μL $^1$H{X} flow-probe was used to collect the $^1$H NMR spectra.

NMR samples from single cultures were prepared by mixing the supernatant with a solution containing the sodium salt of 3-(trimethylsilyl)propionic-2,2,3,3-$d_4$ acid (TSP) in $D_2O$. Concentrations of substrates, products and metabolites in the supernatant were determined by comparison of integrals corresponding to each compound with the integral corresponding to TSP. The $^1$H NMR spectra for these samples were recorded on a Varian 600AS MHz NMR using a 5mm pulsed field gradient (PFG) $^1$H{X} probe and using the appropriate water suppression technique.

## Results and Discussion

Screening of the genomic library of the species of *Pseudomonas* expressed in *E. coli* identified a clone that demonstrated the conversion of the hydrocinnamonitrile substrate. The expected bioconversion was to a single product, the hydrocinnamic acid. However, the $^1$H NMR spectrum shown in Figure 4 of the resulting culture supernatant of the reaction clearly indicates the presence of an additional product.

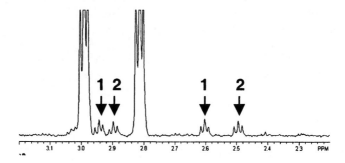

*Figure 4. $^1$H NMR spectrum of the hydrolysis of hydrocinnamonitrile catalyzed by a single E. coli clone from the genomic library of the species of Pseudomonas. Triplets at 2.992 ppm and 2.815 ppm correspond to the $CH_2$ protons of the hydrocinnamonitrile substrate. Two sets of triplets are observed, indicated by the inserted arrows.*

This clone was recovered and re-cultivated to verify the observed activity. Subsequently the gene was PCR amplified and expressed in *E. coli*. The recombinant *E. coli* enzyme was used for additional experiments to identify the observed products and to elucidate the hydrolyzing activity of this enzyme. The products were identified as hydrocinnamic acid and hydrocinnamide using mass spectrometry and NMR.

Experiments using whole cell cultures of the *E. coli* clone were performed to determine the mechanism of this recombinant enzyme. The $^1$H NMR spectra corresponding to a time course of the biotransformation is shown in Figure 5. At the 40-hour time point, only the corresponding hydrocinnamide is detected. This is due to the degradation of the hydrocinnamic acid by native *E. coli* enzymes. It has also been demonstrated that when the amide is used as the sole substrate, the corresponding acid was not produced by the *E. coli* clone carrying the Pseudomonas gene.

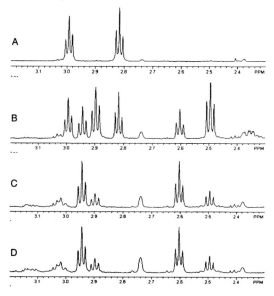

*Figure 5. NMR spectra obtained for supernatants of E. coli whole cell cultures expressing the Pseudomonas gene, catalyzing the hydrolysis of hydrocinnamonitrile. (A) Initial time point, only the nitrile substrate is observed. (B) At 6 hours, nitrile, acid and amide are observed. Triplets at 2.987 ppm and 2.497 ppm are assigned to hydrocinnamic acid; triplets at 2.944 ppm and 2.604 ppm are assigned to the hydrocinnamide. (C) and (D) At 18 and 40 hours, both the products are detected.*

Data collected from these experiments are shown in quantative form in Figure 6. From this data, the specific activity is ambiguous. Decrease in the acid concentration concurrent with a continued increase in the amide data suggests the possibility of amidase activity (23, 24). Additional whole cell experiments, using both the recombinant and wild type *E. coli* strains, were conducted using the amide and the acid as substrates for the bioconversion. These experiments indicate no bioconversion of the amide. However, the acid was catabolized by both the recombinant and wild type *E. coli* cells. This suggests that the enzyme activity produces two products, the acid and amide, from the single nitrile substrate.

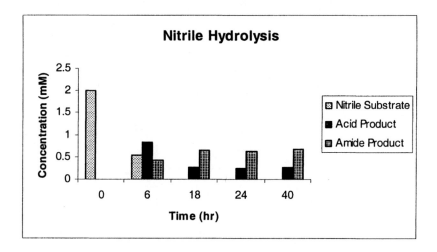

*Figure 6. Nitrile hydrolysis bioconversion of the putative Pseudomonas gene with hydrocinnemonitirle as the substrate. Concentrations were determined from NMR spectra shown in Figure 5.*

To further validate the mechanism of this activity, similar experiments were conducted using cell lysates of the recombinant strain. Cells were cultivated, isolated and sonicated to provide the crude lysate. The resulting $^1$H NMR spectra collected for the time course experiment are shown in Figure 7.

Compilation of the biotransformation information has led to the postulation that the cloned *Pseudomonas* gene has both nitrilase and nitrile hydratase activities. This novel nitrile hydrolyzing activity is summarized in Figure 8.

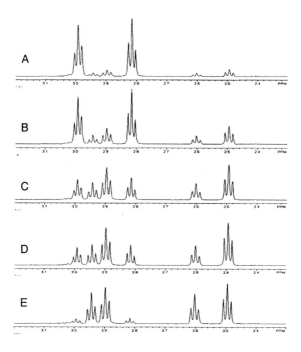

*Figure 7. NMR spectra obtained for supernatants of E. coli whole cell lysates expressing the Pseudomonas gene, catalyzing the hydrolysis of hydrocinnamonitrile. (A) At 2 hours, the nitrile and products are observed. Triplets at 2.992 ppm and 2.815 ppm are assigned to the aliphatic protons of hydrocinnamonitrile; triplets at 2.987 ppm and 2.497 ppm are assigned to hydrocinnamic acid; triplets at 2.944 ppm and 2.604 ppm are assigned to the hydrocinnamide. (C), (D) and (E) Spectra corresponding to 18, 24 and 72 hour sample points.*

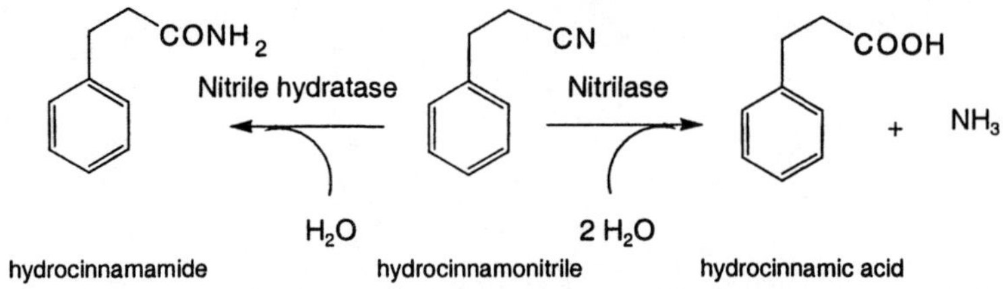

Figure 8. *Proposed scheme for the hydrolysis of hydrocinnamonitrile by the recombinant Pseudomonas protein expressed in E. coli.*

**Summary of Hydrolysis Activity**

Nitrilases are typically classified by substrate specificity (25). Three general classes are aromatic nitrilases, aliphatic nitrilases, and arylacetonitrilases. DSM has reported a nitrilase from *Pseudomonas fluorescens*, DSM 7155, that exclusively hydrolyzes arylacetonitrile substrates (26). The preferred substrate of the DSM enzyme is benzyl cyanide and the primary product the corresponding acid. The nitrilse isolated from the Dow *Pseudomonas* strain hydrolyzes arylacetonitrile substrates, aliphatic dinitrile substrates and exhibits limited activity toward aromatic nitrile substrates. In the case of the arylacetonitirle and aliphatic dinitrile substrates, both the corresponding acid and amide are produced. For the specific case of hydrocinnamonitrile substrate, 40% amide is produced. Nitrilases and nitrile hydratases from other sources, including another homologous enzyme from a different strain of *Pseudomonas* and *Arabidopsis thaliana* (22), were shown to produce only the corresponding acid or amide from hydrocinnamonitrile.

# Conclusions

This work has demonstrated the utility of high throughput $^1$H NMR for the discovery of novel enzymes from genomic libraries. The use of NMR to detect biotransformation products enabled the unexpected observation of multiple product formation for the nitrile hydrolyzing enzyme. This observation resulted in the identification and proposal of a novel mechanism. Additional NMR was used to elucidate the enzymatic activity. Furthermore, it was shown that hydrolysis of the substrate was not due to a combination of nitrile hydratase and amidase.

Future applications of biocatalysis will be related to discovery of novel activities and substrate relationships (21). As mentioned earlier in this paper, high throughput $^1$H NMR spectroscopy provides a mechanism to discover such new enzymatic activities and chemistry. To date, this technology has been applied to substrate bioconversions. However, the application can be extended to the discovery of biocatalysis applications to polymer science.

## Acknowledgements

The authors acknowledge the support and contributions of Hank Talbot, Barbara Miller, Chris Christenson and Mani Subramanian of the Dow Chemical Company.

## References

1. Gross, R. A; Kumar, A.; Kalra, B. *Chem. Rev.* **2001**, 101, 2097-2124.
2. Kumar, A.; Gross, R. A. *J. Am. Chem Soc.* **2001**, 124, 1850.
3. Singh, S. K.; Felse, A.P.; Nunez, A.; Foglia, T.A.; Gross, R.A. *J. Org. Chem.* **2003**, 68, 5466-5477.
4. Robertson, D.E.; Chaplin, J.A.; DeSantis, G.; Podar, M.; Madden, M.; Chi, E; Richardson, T.; Milan, A.; Miller, M.; Weiner, D.P.; Wong, K.; McQuaid, J.; Farwell, B.; Preston, L.A.; Tan, X.; Snead, M. A.; Keller, M.; Mathur, E.; Kretz, P.L., Burk, M.J.; Short, J.M. *Appl. Eviron. Microbiol.* **2004**, 70, 2429-2436.
5. Sarubbi, E; Nolli, M. L.; Andronico, F.; Stella, S.; Saddler, G.; Selva, E.; Siccardi, A.; Denaro, M. *FEBS Lett*, **1991**, *279(2),* 265.
6. Wahler, D.; Reymond, J. *Curr Opin. Biotech* **2001**, 12, 535.
7. Taylor, S.J.C.; Brown, R.C.; Keene, P.A.; Taylor, I.N.; *Bioorg & Med. Chem.* **1999**, 7, 2163-2168.
8. Hendrickx, H.; Duchateau, A.; Raemakers-Franken, P.; *J. Chromatogr. A* **2003**, 1020, 69.
9. Short, J.M.; US patent 6,030,779. 2/2000.
10. Kothapalli, R.; Yoder, S. J.; Mane, S.; Loughran, Jr. T. P. *BMC Bioinformatics* **2002**, 3:22.
11. Keifer, P. *Curr. Opin. Biotech* **1999**, 10, 34.
12. U.S. Patent Applications 60/465,699; 60/465,701; 60/465,700, **2004**.
13. Geresh, S.; Giron, Y.; Gilboa, Y.; Glaser, R. *Tetrahedron Letters* **1993**, 49, 10099.
14. Effenberger, F.; Oβwald, S. *Tetrahedron: Asymmetry* 2001, 12, 279.

15. Wang, M.X.; Feng, G.Q *J. Org. Chem.* 2003, 68, 621.
16. DeSantis, G.; Zhu, Z.; Greenberg, W.; Wong, K.; Chaplin, J.; Hanson, S.; Farwell, B.; Nicholson, L; Rand, C.; Weiner, D.; Robertson, D.; Burk, J. *J. Am. Chem. Soc.* 2002, 124, 9024.
17. Yamada, H.; Kobayashi, M. *Biosci. Biotech. Biochem.*, **1996**, 60, 1391.
18. Schmid, A.; et al. *Nature* **2001**, 409, 258.
19. Peterson, M.; Kiener, A. *Green Chem.* **1999**, 1, 99.
20. Thomas, S.M.; Dicosimo, R.; Nagrarajan, V. *Trends Biotechnol.* **2002**, 13, 548.
21. Schoemaker, H.E.; Mink, D.; Wubbolts, M.G. *Science* **2003**, 299 1694.
22. Effenberger, F.; Oβwald, S. *Tetrahedron: Asymmetry* **2001**, 12, 2581.
23. Klempier, N.; Raadt, A.; Faber, K.; Grieng, H. *Tetrahedron Letters* **1991**, 12(3), 344.
24. Geresh, S.; Giron, Y.; Gilboa, Y.; Glaser, R. *Tetrahedron* **1993**, 49(44), 10099.
25. Kobayashi, S.; Shimazu, M. *FEMS Microbiology Letters* **1994**, 120, 217.
26. Layh, N.; Parrant, J.; Willets, A. *J. Mol. Cat. B: Enzymatic* **1998**, 5, 467.

Chapter 5

# Enzyme Immobilization onto Ultrahigh Specific Surface Cellulose Fibers via Amphiphilic (PEG) Spacers and Electrolyte (PAA) Grafts

[1]You-Lo Hsieh, Yuhong Wang, and Hong Chen

Fiber and Polymer Science, University of California at Davis, Davis, CA 95616

Our group has investigated the incorporation of proteins in fibers by two approaches. One involves physical encapsulation of proteins in the bulk of ultrafine and porous fibers and the other bound proteins to fiber surfaces with reactive spacers and functional grafts. This paper reports two examples from the second approach. Both involve the addition of covalently bonded polymeric chains to ultra-fine cellulose fiber surfaces. One tethers enzyme proteins by covalently bonds via amphiphilic PEG spacers that carry reactive end groups. The other adds polyelectrolyte PAA grafts that are sufficiently polar to attract enzymes via secondary forces. Both surface polymer systems are compatible with aqueous and organic media. Lipase (E.C. 3.1.1.3, from *Candida rugosa*) enzyme was used. PEG was introduced by reacting cellulose with PEG diacylchloride followed by amide covalent bond formation between COOH of PEG and amine ($NH_2$) of the lipase. PAA (0.76-40.9 mmol of COOH per g cellulose) was grafted via ceric ion initiated polymerization. On the PEG attached cellulose, reactive COOH end groups ranging from 0.2 to 1.0 mmol per g cellulose. The adsorbed lipase on the PAA grafted cellulose fibers exhibited significantly higher immobilization efficiency, *i.e.*, up to 391 Unit/mmol COOH, than the covalently bound lipase via PEG spacers. The immobilized lipases via both methods possessed much superior retention of catalytic activity following exposure to hydrocarbons, including cyclohexane, toluene, and hexane, than the free lipase. The covalently bound lipase exhibited significantly higher catalytic activity retention levels at elevated temperatures than the adsorbed and free form. In addition, both adsorbed and covalently bound lipases can be repeatedly used for 4 cycles.

## Introduction

Enzymes are versatile biocatalysts, capable of catalyzing diverse and unique reactions that can be highly specific in mechanism, often stereo specific, enabling simplified steps toward structurally specific product formation and making them efficient and unique for targeted reactions. Advances in biotechnology, in recent years, have made more efficient generation of specific enzymes available, expanding their potential, practical use in large-scale conversion of chemicals and materials.

Immobilized enzymes provide many important advantages over the native enzymes in soluble form, namely, improved reusability, simple or no separation and purification, improved storage and operational stability, and continuous processing and versatile reactor design. Many approaches have been explored to incorporate enzymes to a great variety of organic and inorganic supports. The methods and supports employed for enzyme immobilization are chosen to ensure the retention, stability and durability of enzyme activity.

The major considerations for enzyme immobilization on solids are the mechanisms and the chemical and physical nature of the support materials. The mechanisms to immobilize enzymes include covalent binding, crosslinking, physical adsorption, or encapsulation. Each has its advantages and limitations. Chemical covalent bonding offers the strongest link and yields most stable enzyme-polymer bioconjugates. Encapsulation is a simple yet efficient method, where high enzyme loading and incorporation of more than one enzyme are possible. Physical adsorption is the easiest process.

The structure of the support materials, which may be porous or nonporous, has a great impact on the performance of the immobilized enzymes. Nonporous materials, to which enzymes are attached to the surfaces, have the advantage of posing minimal limitation to substrate diffusion. However, the enzyme loading per unit mass of support is limited to the available surface area. The improved loading and efficiency of immobilized enzymes are usually achieved with increased specific surface by reducing particle sizes from micrometer *(1,2)* to nanometer range *(3)*. Particles smaller than 100 nm have shown to have very high (up to 10 wt%) loading. However, other issues including dispersion of particles in the reaction media, high pressure drops during the reaction and the subsequent recovery for reuse remain. Alternatively, high enzyme loading, low pressure drop and easy separation can be achieved with porous membranes or films and gel matrices. The major drawback of porous materials is their limitation to diffusion. For instance, the rate of the heterogeneous reaction of α-chymotrypsin that was incorporated in a porous polymethylmethacrylate powders was only one-tenths of that form the homogeneous reaction *(4)*.

## Fibers for Enzyme Immobilization

Fibers are the ideal candidates for supporting catalysts, but are also the least studied materials in comparison to other solid forms. Fibrous materials with high intrinsic inter-fiber porosity typically have low pressure drop and negligible mass transfer resistance, thus are more suitable than non-fibrous materials, such as spheres and films. The varied geometrically integrated forms of fibrous materials also have the advantage of easy handling, self-sustaining physical integrity and shape-retention over loose particulates. The high specific areas improve the accessibilities of the active sites resulting in higher reaction rate and conversions. Furthermore, their superior mechanical and handling properties not only simplify but also diversify their applications.

Yang et al reported the immobilization of *Aspergillus oryzae* β-galactosidase on knitted cotton cloth via a tosylated intermediate by reacting cellulose with p-toluenesulfonyl chloride (tosyl-Cl) *(Scheme 1) (5)*. Enzyme loading was about 50 mg per gram of cotton. The bound enzyme showed no significant differences in its kinetics than the free enzyme. The same group studied the formation of polyethyleneimine(PEI)–enzyme aggregates *(6)* and their multilayer structure on cotton cloth to produce a much higher enzyme loading of 250 mg per gram of cotton *(5)*. As expected, high enzyme loading resulted in exceptionally high reactor productivity (6 kg/L/H) yet did not affect the kinetics of galacto-oligosaccharide formation comparing to the soluble form.

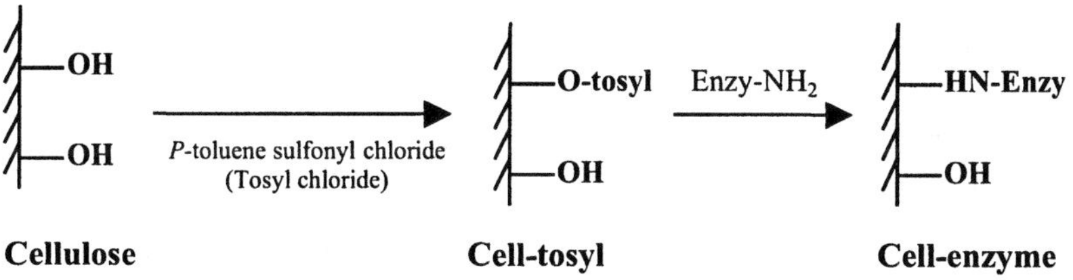

*Scheme 1. Enzyme immobilization mechanism on tosylated cotton*

A technique developed to restore damaged paper documents *(7,8)* covalently bound thermolysin to polyamide (PA) nonwovens by hydrolyzing PA followed by reactions with diamine or dihydrazide spacers *(7)*. The immobilized thermolysin showed improved stability with respect to prolonged exposure to an elevated temperature (2.5 hrs, 70 C), over broader range of pH (6 to 11), and up to 20% ethanol in water. The same group investigated the covalent bonding of trypsin to polyester fleece by multi-step reactions *(8)*. The polyester was first hydrolyzed, then activated with N-hydroxysuccinimide and finally reacted with varying spacers, including PEG-diamine, aldehyde dextran, and bovine serum albumin (BSA). The highest activity was observed with globular BSA spacer. Similarly improved stability against heat, pH, and the ethanol/water mixtures were also reported on the trypsin immobilized on polyester, as on bound thermolysin *(7)*.

Covalent binding and molecular recognition mechanisms were studied to bind urease, xanthine oxidase, uricase, and catalase to cellulose nonwovens for removing urea from wastes *(9,10)*. The enzymes were covalently bound to the aldehyde and isocyanate groups on the cellulose created by reactions with periodate and 1,4-phenylene-diisothiocyanate, respectively. The molecular recognition process involved reacting the cellulose with *N*-hydroxysuccinimide biotin, then impregnated the biotin-modified cellulose in avitin, and finally coupled with enzymes, also biotinylated with biotin-amidocaproate N-hydrocysuccinimide ester. The molecular recognition approach was proven to be much more efficient than covalent bonded enzymes with respect to both reaction rate and specific activity. The improved efficiency was attributed to the biotin-avitin molecular recognition, creating a biomimetic microenvironment favorable for the catalytic activity. Another possible reason was the relatively mild immobilization process in the molecular recognition approach.

A three-step process *(Scheme 2)* was carried out to covalently attach α-chymotrypsin to functionalized polystyrene (PS) fibers *(11)*. PS was functionalized by reacting with 4-nitro-phenyl chloroformate to yield nitrophenyl ending PS (PS-NPh) which was electrospun into fibers, then coupled with α-chymotrypsin. The fiber bound α-chymotrypsin retained 65% of the activity of free enzyme in aqueous solutions, and exhibited nearly three order of magnitude higher activity in transesterification of *n*-acetyl-*L*-phenylalanine ethyl ester in hexane and isooctane. Also, the half life of the bound α-chymotrypsin was 18 fold longer than that of free form in methanol.

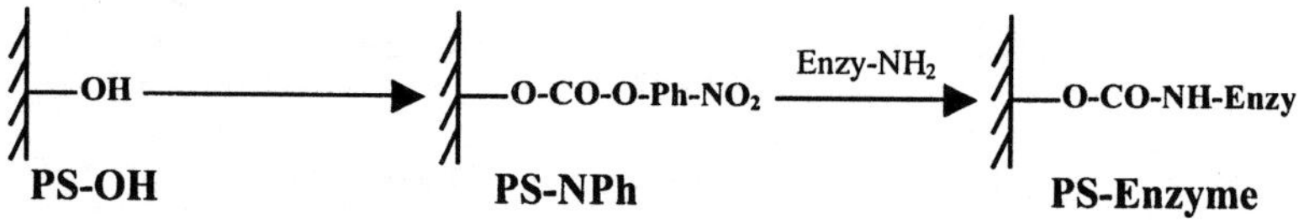

$$\text{PS-OH} \xrightarrow{} \text{PS-NPh (O-CO-O-Ph-NO}_2\text{)} \xrightarrow{\text{Enzy-NH}_2} \text{PS-Enzyme (O-CO-NH-Enzy)}$$

*Scheme 2. Functionalization of polystyrene and subsequent enzyme attachment*

## PEG Spacer

The basic function of "spacer" is to distance the tethered molecules, mostly bioactive compounds, from the surface of the solid support to minimize the steric hindrance and maximize the functions of linked molecules. Polyethylene glycol (PEG) is a neutral polyether which is soluble in water as well as most organic solvents. PEG is unusually effective in excluding other polymers from its presence, which translates into protein rejection, low cell adhesion, nonimmunogenicity, and nonantigenicity. In addition, PEG is nontoxic and doesn't harm active proteins or cells when it interacts with cell membranes *(12)*. Therefore, PEG-modified solid supports provide benefits far beyond those of the common spacers because of its unique properties and have been the focus of much interest in the biotechnological and biomedical applications.

The attachment of proteins and other biomolecules to PEG-grafted surfaces is also of interest for a number of applications. In solid-phase immunoassay and extracorporeal therapy, antibodies or other bioactive molecules are immobilized to a support that interacts with cells, blood, or plasma. Biocompatibility of implants and artificial organs can be improved by the attachment of growth factors to the surface via PEG spacers. These applications are all based on the regulating function of PEG in the interaction between a biomolecule, usually a protein, and another biomolecule or cell. More specifically, immobilization of the biologically active molecules to the free end of grafted PEG chains offers a way to minimize the interactions (deformation and nonspecific adsorption) of attached biomolecules with underlying surface, thus maximize the functions of immobilized biomolecules.

PEG has also been attached to polymers with the use of star-PEG and crosslinkers. To enhance antigen-antibody interaction and kinetics for immunoadsorption, immunoreactive molecules were tethered to the ends of the multiple flexible arms of hyperbranch (star) PEG that was covalently bonded to polymethymethacrylate (PMMA) or polysulfone membranes *(13)*. This star configuration produced high ligand densities, improved hydrophilicity of the surface, and spacing of reactive molecules away from the solid supports. However, neither antigen-antibody interaction nor immuno-adsorption kinetics was reported. While the structural integrity of PS was not altered, the activated PMMA fibers became brittle, potentially limiting its use. Another approach using spacers was to immobilize proteins to glass via thiol-terminal silane and heterobifunctional crosslinkers containing meleimide and succinimide to create biosensors *(14)*. However, no relationship was found between the length of the spacer or the surface density and the activity of the immobilized antibodies.

## Graft Copolymerization on Polymers and Enzyme Binding

Graft copolymerization on the polymer surfaces is usually achieved by the formation of highly reactive species such as trapped polymer radicals and peroxides, via γ-ray irradiation, electron-beam activation, ultraviolet (UV) light irradiation, plasma, or ozone gas treatment, followed by radical-initiated polymerization of monomers at elevated temperatures. Each method of initiation has its own characteristics. For instance, electron-beam irradiation is more relevant for modifying films and fibers in industry due to normal pressure treatment and high efficiency. Due to the inherent nature of radical polymerization, where highly reactive radicals undergo various adverse reactions such as coupling and/or disproportionation, control of above reactions is generally lacking. Ozone oxidation is not as effective with regard to the graft yield, but penetrates into the bulk as the ozone vapor can permeate into polymer interior.

Other than the chemical nature, the most significant aspects of grafted structures as related to protein binding capacity and activity of bound enzymes are graft length and density. Although other polymerization techniques such as living polymerization *(15)* have generated surface grafts with better controlled length and density for protein immobilization, graft copolymerization by free-radicals is by far the most easily applied and widely used approach. The need to determine the length and density of the polymer grafts have been met by estimation based on the mass of polymer grafts and the molecular weight (MW) of the homopolymer during grafting polymerization, assuming their MW to be the same.

Ulbricht and Riedel *(16)* studied the covalent binding of alkaline phosphatase (APh) on polyacrylic acid (PAA)-grafted polysulfone ultrafiltration porous membrane. The molecular weight and graft density of PAA were estimated by HP-GPC analyses of the homopolymer formed during the grafting. Enzyme binding was found to be proportional to the quantity of carboxylic acid. However, the activity of bound enzyme was much higher with surfaces containing few but long PAA chains than with those having dense layers of many long chains. The surfaces with few and short PAA chains, although immobilized less enzymes, exhibited relatively high activity due to lower diffusion limitation. In the immobilization of a variety of proteins and/or peptides on the surface of poly (D,L-lactide) grafted with PAA *(17)*, the average length of the PAA chains was estimated to range from 540 nm to 1000 nm with a DP from 1800 to 3000, assuming that about 10 grafted polymer chains may be accommodated on 10 $nm^2$.

## Grafted Fibers for Enzyme and Protein Binding

In comparison to the work on the non-fibrous solid supports, graft copolymerization of fibrous supports for enzyme *(18,19)* and protein *(20-23)* binding is much less reported.

PAA was grafted onto ultrafine denier poly(ethylene terephthalate) (PET) fibers *(18)*, having a specific surface area of 0.83 $m^2/g$, by ozone gas treatment at levels between 0.03 to 2.5 $\mu g/cm^2$. Trypsin was immobilized with or without the use of 1-ethyl-3-(3-dimethylaminopropyl) carbodiimide hydrochloride (EDC). The density of covalently bound trypsin increased with

PAA grafts linearly, and reached about 6.0 µg/cm² on 2.5 µg/cm² of PAA. In contrast, the quantity of adsorbed trypsin peaked at 0.5 µg/cm² on a low PAA graft level of 0.5 µg/cm². Enzyme activity of covalently bound trypsin was proportional to the PAA concentration up to 0.7 µg/cm², and reached a saturation value of 2.1 unit/cm². On the other hand, activity of adsorbed trypsin saturated at 0.2 unit/cm² on PET fibers containing 0.5 µg/cm² PAA. Using the MW of the homopolymer formed during the ozone-initiated graft polymerization, Kulik et al *(18)* reported that the MW of PAA chains on the surface of PET microfibers was in the range of $4 \times 10^4$ to $6 \times 10^6$, and the surface density of grafts was from $1 \times 10^{-12}$ to $4 \times 10^{-12}$ mol/cm².

Fontaine et al *(19)* introduced 2-vinyl-4,4-dimethylazlactone (VDM) onto polypropylene (PP) fabrics by an electron-beam activated polymerization. The electrophilic azlactone groups were reactive toward nucleophiles without by-product elimination. It was demonstrated that this VDM-modified PP fabric bond sericin by reacting with the nitrogen nucleophilic amino acid side groups of sericin such as lysine, arginine, and histidine, with up to 10% mass uptake.

Glycidyl methacrylate (GMA) was grafted onto a porous membrane of polyethylene hollow fibers by electron beam radiation-induced graft polymerization *(20)*. The epoxide groups were converted to diethylamino (DEA) groups by reacting with 50% (v/v) aqueous diethylamine at 303K. This modified fibrous membrane with a DEA group density of 2.9 mmol/g modified fiber adsorbed 490 mg/g modified fiber of bovine serum albumin (BSA).

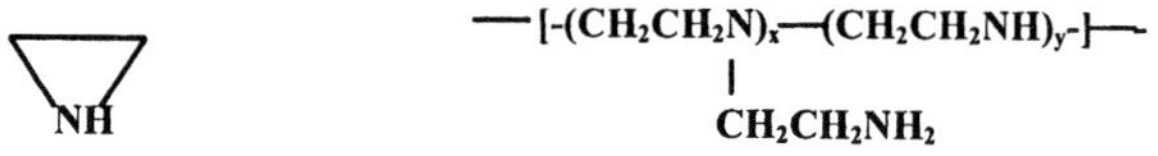

$$-[-(CH_2CH_2N)_x-(CH_2CH_2NH)_y-]-$$
$$|$$
$$CH_2CH_2NH_2$$

EI          PEI

*Scheme 3. Chemical structure of ethyleneimine (EI) and polyethyleneimine (PEI)*

Polyethyleneimine (PEI) *(Scheme 3)* was immobilized onto cellulose fibers by a crosslinking agent glutaraldehyde *(21)*. 10-60 wt% of PEI was attached by varying the mass ratio of PEI to cellulose. The cellulose-PEI fibers with 20-60 wt% of PEI adsorbed up to 10 mg endotoxin per g of modified fibers and could remove 90-100% endotoxin from 500 ng/ml endotoxin solution at pH 7.0 and ionic strengths lower than 0.4. In addition, the cellulose-PEI fibers with 20 wt% PEI showed selective removal activity toward endotoxins from various protein solutions at pH 7 and 0.05 ionic strength. This selective adsorption was due to simultaneous effects of cationic properties of amino groups and hydrophobic properties of the matrix alkyl chains as well as exclusive effects on protein molecules.

Kato and Ikada *(22)* studied the selective adsorption of proteins to their ligand bound PET fibers. PAA was grafted onto the ozone oxidized PET fibers to facilitate the covalent bonding of ligand proteins, *i.e.* human r-glubulin, *Staphylococcus aureus* protein A (SpA), rabbit immunoglobulin G (IgG), and human IgG obtained from yolk (IgY). The density of the ligand proteins was of the order of 1 mg/g fiber. The ligand bound PET fibers exhibited highly selective binding to the respective specific proteins without interference from other coexisting proteins.

The ionic nature of surface grafts has been found to affect protein adsorption on ozone- and photo-induced grafting of various non-ionic and ionic hydrophilic polymers on PET fibers *(23)*. Acrylamide (AAm) and methoxydiethylene glycol methacrylate (MDEGM) were used as non-ionic monomers, while acrylic acid (AA) and methacryloiloxyethyl phosphate (MOEP) were utilized as anionic monomers, and dimethylaminoethyl methacrylte (DMAEMA) as a cationic monomer. Adsorption of proteins with different isoelectric points, such as serum albumin, acidic immunoglobulin G, basic immunoglobulin G, and basic fibroblast growth factor, showed that surfaces grafted with positively charged polymer chains attracted negatively charged proteins and vice versa. The surface grafted with the same charge sign as the proteins exhibited repulsion. Non-ionic polymer chains on the fiber surfaces rejected protein adsorption, regardless of the isoelectric point of the proteins.

Our group has investigated the incorporation of proteins in fibers in two general directions. One involves physical encapsulation of proteins by incorporating them in the bulk of ultrafine porous fibers *(24,25)*. We have demonstrated that proteins, *i.e.*, casein and lipase enzyme, can be encapsulated into polymer matrixes and processed into sub-micrometer size fibers via electrospinning. Protein loading could be as high as 80% in these fibers. Entrapped lipase exhibits higher catalytic activity towards hydrolyzing olive oil than that in the cast films under the same condition. The other approach bound proteins to fibers by adding chemical functionalities to fiber surfaces.

Two examples from the second approach are presented here. Both add covalently bonded polymeric chains to ultra-fine cellulose fiber surfaces. One tethers enzyme from the surface by amphiphilic linear PEG spacers that carry reactive end groups to form covalent bonds with enzyme proteins. The other adds polyelectrolyte PAA grafts that are sufficiently polar to attract enzymes via secondary forces. Both surface polymer chains are compatible with aqueous and organic media.

# Experimental

## Ultra-fine Cellulose Fibers

15 wt% cellulose acetate (CA) was prepared by dissolving CA in a 1:2 mass ratio of $N,N$-dimethylacetamide (DMAc):acetone mixture *(26)*. For electrospinning, this CA solution was placed into a glass capillary with a stainless steel electrode immersed and connected to a power supply (Gamma High Supply, ES 30-0.1 P). A 10 kV was applied to generate a charged jet, which splayed into finer jets and collected onto a grounded Aluminum collector at a distance of 7 inch. Deacetylation was carried out in 0.05M NaOH aqueous solution at ambient temperature for 24 h. Rinsing CA fibers by deionized water stopped the reaction. The cellulose fibers were dried at 80 C under vacuum for 10 hrs.

## Enzyme and Catalytic Activity

Lipase, a triacylglycerol ester hydrolyses (EC 3.1.1.3, Candida rugosa), with an average diameter of 500 nm was used in this study. Active site residues of this lipase include SER, GLU and HIS, which allows amine groups to participate in the coupling reaction without affecting the active sites. The activity assay was performed using olive oil as the substrate as illustrated in *Scheme 4 (25)*. The substrate, lipase sample and buffer were mixed and incubated at preset temperature under constant shaking at 60 rpm. At designated assay time, the incubation solution was heated to 100°C to denature the lipase. Copper(II) nitrate aqueous solution and chloroform were added to extract the fatty acids liberated from hydrolysis of olive oil to chloroform layer in the form of copper complex. The amount of fatty acid in the complex form was determined spectrophotometrically at 436nm (HITACHI U-2000 Spectrophotometer) with sodium dithyldithiocarbamate as color indicator. The activity was expressed as μmol of fatty acid released per hour.

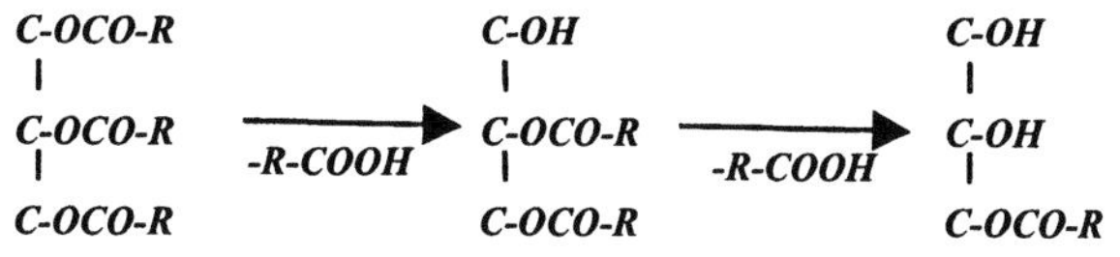

```
C-OCO-R              C-OH                  C-OH
|                    |                     |
C-OCO-R    ──────▶   C-OCO-R    ──────▶    C-OH
|          -R-COOH   |          -R-COOH    |
C-OCO-R              C-OCO-R               C-OCO-R
```

*Scheme 4. Lipase assay using olive oil as the substrate*

## Immobilization Mechanisms

### I: Adsorption

Poly(acrylic acid) (PAA) was grafted onto the ultra-fine cellulose fiber surfaces via ceric ion initiation *(Scheme 5)*. The carboxylic acid concentration could be controlled by varying monomer AA and/or ceric ion concentrations, and was determined by NaOH-HCl titration using phenophthlein as the indicator. Unless otherwise indicated, fibers with 3.6 mmol of COOH per g cellulose ($[AA]/[Ce(IV)]$=30) were used for enzyme adsorption and assay. The PAA activated cellulose fibers were immersed in 1.0 mg/ml lipase solution (pH=7.0) at room temperature for 24 hrs. The lipase adsorbed cellulose fibers were rinsed by pH 7 buffer and deionized water to remove loose lipase, then dried under vacuum at room temperature for 12 hrs.

### II: Covalent Binding

PEG-diacylchloride (COCl-PEG-COCl) was attached to ultra-fine cellulose fiber surfaces via ester bond formation between the COCl of the PEG-diacylchloride and the OH on the cellulose *(Scheme 6)*. The quantities of COCl or COOH could be optimized by varying the COCl/OH molar ratio and PEG chain length. Based on the quantification of COOH by the silver o-nitrophenolate method *(27)*, the PEG attached cellulose fibers contained up to 1.0 mmol of COOH per g of cellulose unless specified otherwise. Lipase was covalently bound with the PEG-attached cellulose fibers in pH 4 buffer in the presence of a carbodiimide (EDC) coupling agent and incubated for 7 hrs. Lipase bound fibers were then rinsed in buffers of increasing pH of 4, 7, and 10, then dried under vacuum at ambient temperature for 12 hrs.

## Results and Discussion

### Carboxylic Quantity and Catalytic Activity of Bound Enzyme

The quantity of COOH introduced by PAA grafting on the ultra-fine cellulose fiber was from 0.76 to 40.9 mmol/g cellulose under the experimental conditions. In the PEG grafting approach, the amount of COOH on the fibers was less, ranging from 0.14 to 1.00 mmol/g cellulose, with increasing COCl/OH molar ratios from 1 to 20. In spite of the significant differences in the quantities of COOH and the activity levels, the activity of bound lipase increased with increasing quantity of carboxylic acid on the fiber surfaces with either approach *(Figure 1)*. The redox initiated grafting of PAA produced significantly greater COOH on the fiber surfaces. Lipase bound to such surfaces also exhibited significant activities. The activities of covalently bonded lipases increased more closely proportional to the quantities of the PEG chains whereas the catalytic efficiency of the adsorbed lipases appeared to lessen. In fact, the catalytic efficiency of the lipase, calculated by activity per COOH, showed opposite

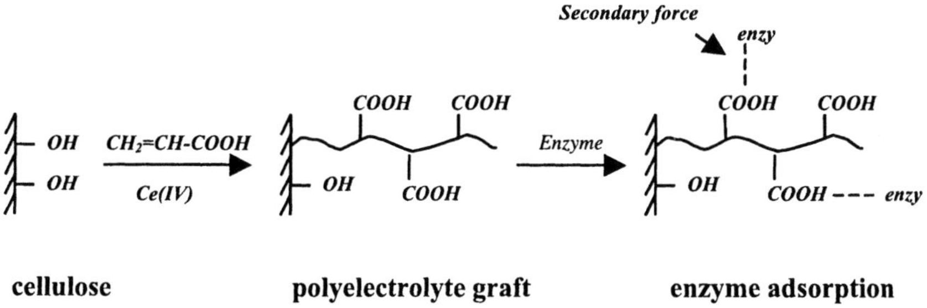

*Scheme 5. Surface grafting of polyelectrolyte and subsequent enzyme adsorption*

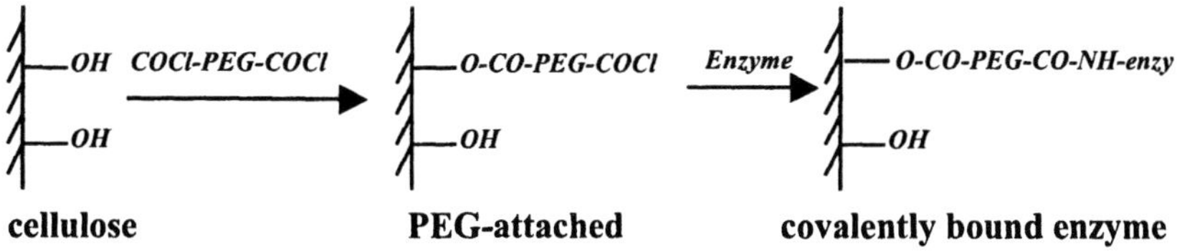

*Scheme 6. Attachment of PEG and subsequent enzyme binding*

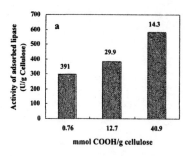 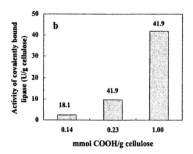

*Figure 1. Effect of carboxylic acid quantity on the activity of bound lipase: (a) adsorbed on PAA-grafted ([AA]/[Ce(IV)]=120 ); (b) covalently bonded on PEG-grafted cellulose fibers. Numbers above the columns represent the efficiency of bound lipase (unit per mmol of COOH).*

trends between these two approaches. The covalently bound lipase exhibited increasing catalytic efficiency with increasing COOH whereas the opposite held with the adsorbed counterparts. It should be noted that the efficiency of the lipase adsorbed at a low 0.76 mmol COOH per g of cellulose (PAA-grafted) was 391 U per mmol of COOH, one magnitude higher than that of the highest PEG-grafted.

**Stability Upon Exposure to Organic Solvents**

Both free and bound lipases were exposed to various organic solvents, then their activities were assayed. The enzyme activities following the exposure to organic solvents were normalized by the activity of that never exposed to the solvents. The lipase bound by both immobilization approaches had greatly improved stability against chloroform, hexane, and toluene *(Figure 2)*. For the free lipase, exposures to these organic solvents reduced the activity by 60-70% after 12 h and additional 5-10% after 24 h. On the contrary, lipase adsorbed on the PAA-grafted cellulose only lost 15-30% of activity following a 12 h exposure. The covalently bound lipase retained 62% and 34% of original activity after exposure to toluene and hexane for 24 h. The presence of organic solvents can change the conformation of enzyme molecules by altering the hydrophobic and hydrophilic domains, leading eventually to their denaturation. Covalent bonding or physical adsorption of enzyme proteins to solid supports can help to retain the favorable protein conformation and/or shield the protein from or decrease the accessibility of the denaturing agents.

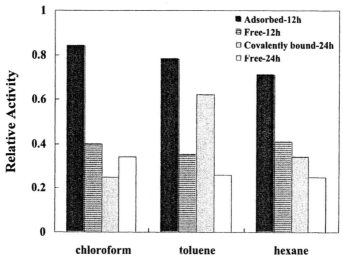

*Figure 2. Relative activity of free and cellulose fiber bound lipase ($pH=8.5, 30°C$) following exposure to organic solvents.*

**Thermal Stability**

Enzyme activities at elevated temperatures were assayed at temperatures between 20 C and 80 C. The activity of both free and bound lipases was normalized by the individual highest activity. The activity of the free lipase was strongly dependent on temperature, the optimum being observed between 25 C and 40 C *(Figure 3)*. The adsorbed lipase showed similar behavior between 20 C and 40 C, but suffered a higher loss in activity than the free lipase at 50 C. The only slight advantage of the adsorbed lipase was its higher activity at 20 C. The covalently bonded lipase exhibited the same activities as in the free form in the optimal 25 C to 40 C range. At 60 C, the bound lipase on the PEG-attached cellulose had a most impressively imporved activity, *i.e.*, 3 times higher than that at 30 C. Furthermore, nearly 80% of the activity was retained at 80 C. This demonstrates that the lipase covalently bonded to cellulose via a spacer possessed much better thermal stability than its free form.

**pH Stability**

The activities of free and bound lipases were assayed in buffer solutions of varying pH values (pH 2-12). The optimal pH for the free lipase to function was between 7 and 8.5 *(Figure 4)*. In general, the bound enzymes did not show much different catalytic behavior except for a slightly improvement at pH 6. Both free and bound lipases were completely deactivated under basic media (pH>10).

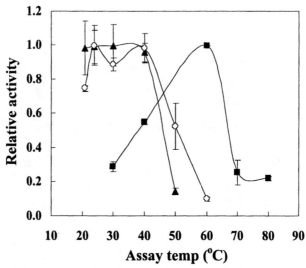

*Figure 3. Relative activity ( pH 8.5) of free lipase (○ ) and bound lipase on PAA-grafted (▲) and PEG-grafted (■) cellulose fibers at various temperatures.*

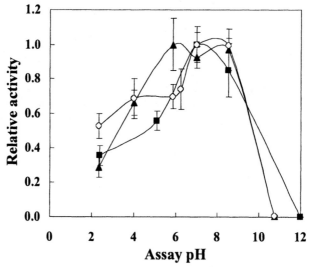

*Figure 4. Relative activity (30°C) of free lipase (○) and adsorbed lipase on PAA grafted (▲) and covalently bound lipase on PEG grafted (■) cellulose fibers under various assay pHs*

**Cyclic Activity**

The activities of the bound lipases were assayed in consecutive cycles *(Figure 5)*. The adsorbed lipase exhibited 50% and 30% of the initial activity, whereas the covalently bound lipase exhibited 100% and 50% of the original activity in second and third cycles, respectively. The decreased activities of both bound enzymes are due to the loss of enzymes. The loss of the adsorbed lipase is thought to be from diffusion driven by the concentrational difference. The enzyme loss from the PEG-bonded fibers is likely due to the hydrolysis of the ester linkages between the PEG and cellulose support than amide bond between lipase molecule and PEG. The activity of the adsorbed lipase at cycle 4 was higher than that of the covalently bound form.

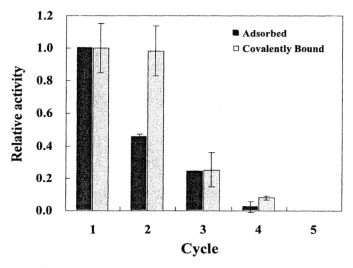

*Figure 5. Cyclic activity (pH 8.5, 30°C) of bound lipase on cellulose fibers.*

## Conclusions

Two approaches have been reported to immobilize enzyme on ultra-fine cellulose fibers, *i.e.*, adsorption and covalent binding. Both approaches improved the protein binding ability of fiber surfaces by adding chemical functionalities via covalently bonded polymeric chains to ultra-fine cellulose fiber surfaces. Adsorption of enzyme proteins on fiber surfaces was accomplished by ceric ion-initiated graft polymerization of electrolyte acrylic acid monomer and the subsequent enzyme adsorption via secondary forces.

Covalent bonding of enzyme was achieved by tethering enzyme molecules from the surface by amphiphilic linear PEG spacers that carry reactive end groups to form covalent bonds with enzyme proteins. Both surface polymer chains are compatible with aqueous and organic media.

The adsorbed lipase on the PAA grafted cellulose fibers exhibited significantly higher immobilization efficiency, *i.e.*, up to 391 U/mmol COOH, than the covalently bound lipase via PEG spacers. The stability of bound enzymes to chloroform, hexane, and toluene was improved upon the immobilization via both methods. The covalently bound enzymes exhibited much better thermal stability than the adsorbed and free form. The relative activity of all free, adsorbed, and covalently bound lipase showed similar pattern in the range of assay pH from 2 to 12. In addition, both adsorbed and covalently bound lipases can be repeatedly used for 4 cycles.

## Acknowledgements

The financial support from the National Textile Center (Project. M02-CD05) and the University of California at Davis is greatly appreciated.

## References

1. Bahar, T.; Celebi, S.S.T. *Enzyme Microb. Technol.* **1998**, *23*, 301-304.
2. Eldin, M.S.M.; Schroen, C.G,P.H.; Jassen, A.E.M.; Mita, D.G.; Tramper, J. *J. Mol. Catal. B: Enzymatic* **2000**, *10*, 445-451.
3. Chen, J.P.; Su, D.R. *Biotechnol. Prog.* **2001**, *17*, 369-375.
4. Wang, P.; Sergeeva, M.V.; Lim, L.; Dordick, J.S. *Nature Biotechnol.* **1997**, *15*, 789-793.
5. Albayrak, N.; Yang, S.T. *Enzyme and Microbial Technology* **2002**, *31*, 371-383.
6. Albayrak, N.; Yang, S.T. *Biotechnol. Prog.* **2002**, *18*, 240-251.
7. Moeschel,K.; Nouaimi, M.; Steinbrenner, C.; Bisswanger, H. *Biotechnol. Bioeng.* **2003**, *82*, 190-199.
8. Nouaimi, M.; Moschel, K.; Bisswanger, H. *Enzyme and Microbial Technology* **2001**, *29*, 567-574.
9. Magne, V.; Amounas, M.; Innocent, C.; Dejean, E.; Seta, P. *Desalination* **2002**, *144*, 163-166.
10. Amounas, M.; Magne, V.; Innocent, C.; Dejean, E.; Seta, P. *Enzyme and microbial technology* **2002**, *31*, 171-178.
11. Jia, H.F.; Zhu, G.Y.; Vugrinovich, B.; Kataphinan, W.; Reneker, D.H.; Wang, P. *Biotechnol. Prog.* **2002**, *18*, 1027-1032.
12. Harris, J.M. *In Poly(ethylene glycol) chemistry, biotechnical and biomedical applications*; Harris, J., Ed.; Plenum Press: New York, 1992.
13. Ross, E.A.; Branham, M.L.; Tebbett, I.R. *J. Biomed. Mater. Res.* **2000**, *51*, 29-36.
14. Shriver-Lake, L.S.; Donner, B.; Edelstein, R.; Breslin, K.; Bhatia, S.K.; Ligler, F.S. *Biosensors & Bioelectronics* **1997**, *12*, 1101-1106.

15. Nakayama, Y.; Matsuda, T. *Langmuir* **1999**, *15*, 5560-5566.
16. Ulbricht, M.; Riedel, Ml. *Biomaterials* **1998**, *19*, 1229-1237.
17. Steffens, G.C.M.; Nothdurft, L.; Buse, G.; Thissen, H.; Hocker, H.; Klee, D. *Biomaterials* **2002**, *23*, 3523-3531.
18. Kulik, E.A.; Kato, K.; Ivanchenko, M.I. Ikada, Y. *Biomaterials* **1993**, *14*, 763-769.
19. Fontaine, L.; Lemele, T.; Brosse, J.C.; Sennyey, G.; Senet, J.P.; Wattiez, D. *Macromol. Chem. Phys.* **2002**, *203*, 1377-1384.
20. Tsuneda, S.; Saito, K.; Furusaki, S.; Sugo, T. *J. Chromatography A* **1995**, *689*, 211-218.
21. Morimoto, S.; Sakata, M.; Iwata, T.; Esaki, A.; Hirayama, C. *Polym. J.* **1995**, *27(8)*, 831-839.
22. Kato, K.; Ikada, Y. *Biotechnol. Bioengng.* **1995**, *47*, 557-566.
23. Kato, K.; Sano, S.; Ikada, Y. *Colloids and Surfaces B: Biointerfaces* **1995**, *4*, 221-230.
24. Wang, Y.H; Hsieh, Y.L. Enzyme Immobilization to Ultrafine Cellulose Fibers via Amphiphilic Polyethylene Glycol (PEG) Spacer; to be submitted.
25. Xie, J.B.; Hsieh, Y.L. *J. Mat. Sci.* **2003**, *38*, 2125-2133.
26. Liu, H.Q.; Hsieh, Y.L. *J. Polym. Sci.: Part B: Polymer Physics* **2002**, *40*, 2119-2129.
27. Campbell, H. J.; Francis, T. *Textile Research Journal*, **1965**, 260-270

## Chapter 6

# Nondestructive Regioselective Modification of Laccase by Linear-Dendritic Copolymers: Enhanced Oxidation of Benzo-α-Pyrene in Water

Ivan Gitsov[1,2,*], Kevin Lambrych[2], Peng Lu[3], James Nakas[3,*], Joseph Ryan[2], and Stuart Tanenbaum[2,*]

[1]The Michael M. Szwarc Polymer Research Institute, [2]Department of Chemistry, and [3]Department of Enviromental and Forest Biology, College of Enviromental Science and Forestry, State University of New York, Syracuse, NY 13210

This chapter describes a novel approach towards enzyme modification which avoids generation or disruption of covalent chemical linkages or genetic intervention. A unique feature of this strategy is that a glycoenzyme, substrates and applicable cofactors are confined within the well-defined nanoporous regions of a regular micelle constructed by an amphiphilic linear-dendritic block copolymer containing water-soluble linear fragments and hydrophobic dendritic blocks. This new construction principle is illustrated with the enzyme laccase from *Trametes versicolor*. The complexes were shown to be capable of converting the essentially non-substrate veratryl alcohol to veratryl aldehyde. Further, the activity of this laccase complex in aqueous media, enhanced with selected hydrophobic N-hydroxy mediators, provided the expected diquinone products from the essentially insoluble model compound benzo-α-pyrene. Control probes with laccase, mediators and conventional linear-linear copolymers with comparable molecular weight characteristics failed to evince such substrate oxidation. These results presage application of glycoenzyme/linear-dendritic complexes for the biotransformation of variety of precursors, not normally enzyme substrates by virtue of their insolubility under physiological conditions, to useful intermediates.

The application of water-soluble, unmodified enzymes for practical purposes is often hampered by their protease susceptibility, thermal instability and inactivation by the intermediates, products formed or pH of the medium. One of the main thrusts in our research is the construction of micellar complexes of enzymes and amphiphilic block copolymers that contain linear and perfectly branched (dendritic) segments and their catalytic evaluation in aqueous media. Unique feature of these "supermolecules" is that the *water-soluble* biocatalyst, *hydrophobic* substrates and eventual mediator compounds are confined within the core of a *normal* micelle. The hydrophobic dendritic fragments have the primary function to serve as enzyme surface anchoring devices for the copolymer molecules while the hydrophilic linear portion of the copolymer will ensure the aqueous solubility and the stability of the entire construction. The complex is schematically presented in Fig. 1.

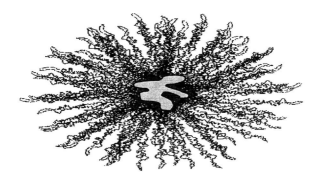

*Figure 1. Schematic cross-section of a linear-dendritic enzyme complex. The linear portion of the copolymer (black, wavy lines) extends into the aqueous phase while the monodendritic blocks of the copolymer (black, near the center) are anchored at the surface of the enzyme (grey feature at the center).*

The goal of this study is to illustrate the outlined approach with the enzyme laccase and amphiphilic block copolymers containing linear poly(ethylene glycol), PEG and dendritic poly(benzyl ether)s. Laccases are water soluble Cu-containing oxidative enzymes of relaxed specificities which can be further broadened by the addition of low molecular weight redox mediators (*1*). The catalytic activity of laccase in water-miscible solvents (*2*) or in reverse micelles (*3*) for the oxidation of various phenols in organic media is well known, the process being facilitated by the partial solubility of the phenol derivatives in both organic and aqueous solvents. However, the previously suggested application of this versatile enzyme in the oxidative radical bioremediation of xenobiotic polyaromatic hydrocarbons (PAHs) represents a substantially more challenging task because of their intrinsic insolubility in water and the necessity to use mediator compounds (*4*). Several prior strategies were aimed to improve the contact of the essentially water-insoluble PAHs with the enzyme: the substrates were solubilized by initial derivatization with PEG (*5*), metabolized in the presence of surfactants (*6*), or high concentrations of water-miscible organic solvents were used (*7*). To our knowledge, the use of laccase-normal micelle complexes for the oxidation of PAHs in aqueous medium has not been reported. In this study we explore the catalytic activity of laccase from *Trametes versicolor* embedded into normal micelles formed by an ABA block copolymer with two second-generation poly(benzyl ether) monodendrons (blocks A) and a PEG fragment of molecular mass 5000 Da (block B), Figure 2, in the presence of a highly hydrophobic mediator N-hydroxynaphthalimide (NHN), Figure 3A. The selection of the amphiphilic copolymer and mediator is based on our previous experience with the efficient encapsulation of substantial amounts of polyaromatic compounds by linear-dendritic supermolecules in water (*8*), the enhanced physisorption of the same copolymers onto cellulose surfaces (*9*) and the established mediator efficiency of N-hydroxy imide derivatives(*10*).

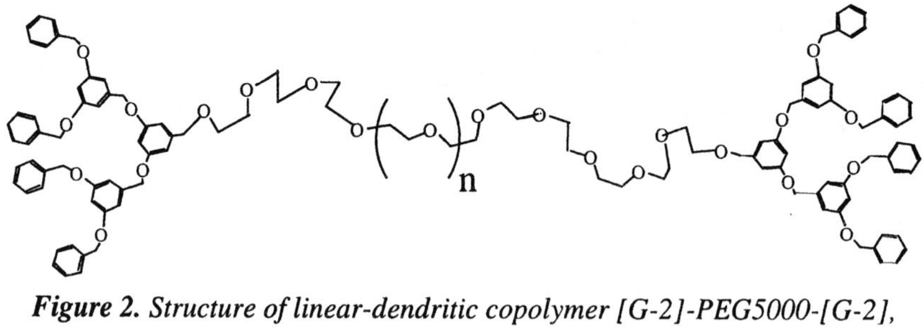

***Figure 2.*** *Structure of linear-dendritic copolymer [G-2]-PEG5000-[G-2], n = 113*

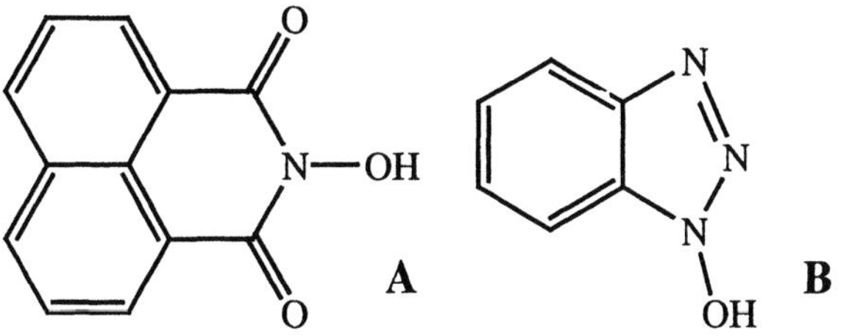

*Figure 3. Structure of the mediators A: N-Hydroxynaphthalimide (NHN); B: 1-hydroxybenzotriazole (HBT)*

## Experimental

**Materials.** Poly(ethylene glycol), PEG with molecular weight 5,000 Da and $M_w/M_n = 1.03$ was obtained from Polysciences, Inc. and was stored at -10°C under nitrogen atmosphere prior to use. The hybrid linear-dendritic block copolymer, [G-2]-PEG5000-[G-2] (Figure 2) was synthesized from PEG with nominal molecular weight 5,000 and two second-generation poly(benzyl ether) monodendrons with a benzyl bromide moiety at their 'focal' points (*11*). Experimental details are given elsewhere (*12*). Veratryl alcohol (3,4-dimethoxybenzyl alcohol), VA, the PAH – benzo-α-pyrene (BP), the model mediators 1-hydroxybenzotriazole (HBT), N-hydroxynaphthalimide (IMD-4) and the nonionic polymeric surfactants Brij-35, and Igepal CO-990 were purchased from Aldrich, and were used without further purification. Laccase was produced using a novel field isolate of *Trametes versicolor* from forest soil, after an induction with 600 μM 2,5-xylidine. A purification scheme was developed which included filtering followed by precipitation with acetone to remove any extracellular polysaccharide (*13*). The enzyme-containing solution was then subjected to ultrafiltration (twice, 30,000 molecular mass cutoff) and subsequently purified by hydrophobic interaction chromatography (HIC), Table 1. Laccase-active fractions were subsequently applied to a gel permeation column (BioGel P-60) as the final phase of purification. All experiments were performed in deionized (DI) water (18.3 MΩ) using stage-IV enzyme (Table 1).

**Instrumentation.** Size-exclusion chromatography (SEC) analyses were performed on a SEC line consisting of Waters 2690 Alliance system, 996 Waters photodiode-array (PDA) detector, DAWN-DSP multi-angle laser light scattering (MALLS) detector equipped with a 30mW, 488 nm argon-ion laser (Wyatt Technology Corporation), T50 differential viscometer (Viscotek Corporation) and Optilab DSP interferometric refractometer (Wyatt Technology Corporation) connected in series. The separations were achieved at 40°C across a set of three

5 µm Aquagel-OH columns (Polymer Laboratories) in water containing 0.1 M $NaN_3$ as a bacteriostatic agent. The spectroscopic measurements were accomplished with a DU 640B UV-Vis spectrometer (Beckman) at room temperature with a scan speed of 1200 nm/min over a range of 190-700 nm.

**TABLE 1. Purification of laccase from *Trametes versicolor***

| Stage | Volume (mL) | Protein (mg/mL) | Activity (µkat/mL) | Specific Activity (µkat/mg) | Recovery (%) |
|---|---|---|---|---|---|
| I. Supernatant after removing mycelia | 2550 | 0.576 | 0.647 | 1.12 | 100 |
| II. After acetone precipitation | 2800 | 0.51 | 0.527 | 1.03 | 89 |
| III. After ultrafiltration | 16.5 | 21.3 | 58.0 | 2.72 | 58 |
| HIC resolution | 8.10 | 3.10 | 41.9 | 13.52 | 21 |

**General Procedure for the Laccase Modification.** Two mL of a laccase solution (approximately 2 mg/mL) in a mixture of citrate buffer and glycerol with pH = 6 were placed in 5mL vials and combined with precalculated amounts of dry copolymer (0.007 g) and the corresponding redox mediator. The mixture was then allowed to stir in a refrigerator at 10°C for 12 h and then equilibrate for additional 8 h at room temperature. The integrity of the micellar enzyme complex and its composition were analyzed by SEC with multiple detection according to a known procedure (8). Mixtures of laccase without linear-dendritic copolymer or mediator were prepared and analyzed under identical conditions. The biological activity of the resulting complexes (Table 2) was determined by measuring the absorbance change at 420 nm using 5 mM of 2,2'-azino-bis-(3-ethylbenzothiazoline-6-sulfonic acid), ABTS, as a substrate in 50 mM sodium acetate buffer (pH = 4.0) at 30°C (14).

**General Procedure for the Biotransformation of BP with Laccase/Linear-dendritic complexes.** Fine powder of BP (0.010 g) was added to preformed laccase/[G-2]-PEG5000-[G-2] complexes and the mixture was stirred for additional 12 h. The mediator was added last and the reaction was followed by UV-Vis spectroscopy over extended time period. Control mixtures of laccase and BP with linear copolymers (Brij-35 and Igepal CO-990) and the same mediator were prepared and analyzed spectroscopically for biocatalytic activity under identical conditions.

Table 2. Specific Activity of [G-2]-PEG5000-[G-2]/Laccase Complexes

| [Copolymer] (mg/mL) | [Enzyme] (mg/mL) | Specific Activity ($\mu$kat/mg) |
|---|---|---|
| 0.065 | 1.75 | 2.07 |
| 6.454 | 2.77 | 1.59 |
| 0 | 1.72 | 2.72 |

## Results and Discussion

Initially we studied the interactions of laccase with [G2]-PEG5k-[G2] in aqueous media. In all cases investigated the copolymer is added as dry powder to the buffered solution of the enzyme in order to direct it towards enzyme surface sorption instead of forming "empty" linear-dendritic micelles.

The integrity of the micellar enzyme complex and its composition are analyzed by size-exclusion chromatography (SEC) with multiple detection, Figure 4. The results obtained confirm our previous findings(8) that in water [G2]-PEG5k-[G2] (Figure 2) forms micelles with molecular mass of $3.5 \times 10^6$ Da as measured by on-line by multi-angle laser light scattering (MALLS) at 488 nm. These species elute at 16.8 mL (Figure 4a). The laccase isolated from *Trametes versicolor* shows multimodal elution profile indicating the presence of three, possibly four, fractions with slightly different hydrodynamic volumes, the main fraction eluting at 18 mL, Figure 4c. The average molecular mass ($M_w$) obtained by MALLS for the mixture of isoforms is $(7.9 \pm 0.2) \times 10^4$ Da, a result in fair agreement with previously published data for the same enzyme (15). The small-molecular-mass mediator N-hydroxybenzotriazole (HBT, Figure 3B) elutes much later – at 29.6 mL, Figure 4d. Surprisingly, the mixture of laccase, linear-dendritic copolymer and HBT prepared at the same concentration of components shows no individual peaks for the [G2]-PEG5k-[G2] and HBT, but presented an elution pattern seemingly similar to that of the isolated laccase, Figure 4b. The MALLS analysis shows, however, that the $M_w$ of the total mixture is $(2.9 \pm 1.0) \times 10^6$ Da – a value much closer to the molecular mass of a supramolecular complex than a single enzyme macromolecule. The UV analysis of the main fraction eluting at 18.02 mL performed with an on-line photodiode array detector confirms the presence of both the copolymer and HBT within the enzyme peak, Figure 5. It should be mentioned that the intrinsic UV spectrum of the enzyme itself is subtracted in order to reveal the much weaker characteristic UV absorbency of [G2]-PEG5k-[G2] and HBT.

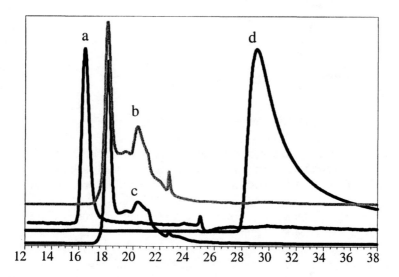

*Figure 4.* Size exclusion chromatography traces of (a) – [G2]-PEG5k-[G2], (b) – laccase/copolymer/mediator complex, (c) pure laccase in buffer solution, (d) pure HBT.

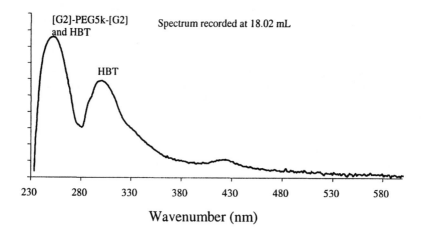

*Figure 5.* UV-spectra of the main fraction eluting at 18.02 mL recorded on-line using Waters 996 photodiode array detector.

The results from the SEC investigation combined with MALLS and UV analyses clearly indicate the formation of a supramolecular complex between the laccase, the linear-dendritic copolymer and the mediator. The similarity in the elution profiles of the resolved enzyme and its complex is explained by selective flat-form sorption of [G-2]-PEG5k-[G-2] only to the carbohydrate regions surrounding the three main protein domains (16) in the laccase macromolecule. Thus the overall hydrodynamic volume of the complex will be minimally affected and the elution volumes of the supermolecule and the individual enzyme will be rather similar, Figure 6.

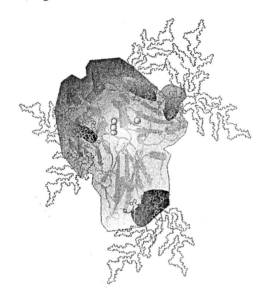

*Figure 6. Schematic diagram of the complex between laccase and a linear-dendritic ABA copolymer (shown as wavy lines around the enzyme). The protein portion of the enzyme is presented as the large area with a border(in grey), the carbohydrate domains are depicted as four smaller areas (in darker grey), and the four copper ions are shown as four small circles (in the middle)*

The evaluation of the specific activity of laccase shows that it is not markedly diminished by the addition of the linear-dendritic copolymer (Table 2). It should be noted, however, that ABTS is a highly polar compound and its distribution balance between the dendritic domains on the laccase surface and the surrounding water will be strongly shifted towards the aqueous phase ultimately affecting the specific activity of the complexes. Therefore the more hydrophobic HBT (Figure 3B) is chosen as the mediator for the next model experiment to compare the biocatalytic activity of the complex laccase/[G2]-PEG5k-[G2] to that of the native enzyme. Table 3 outlines the reaction conditions for the oxidation of veratryl alcohol (a substrate with relatively good solubility in water) to veratryl aldehyde.

**Table 3. Reaction Conditions for the Biocatalyzed Oxidation of VA**

| Vial | Reagents | [Reagent] (mg/mL) | Temp. (°C) | Amount Aldehyde (mg/mL) | Conversion (%) |
|---|---|---|---|---|---|
| 1 | Laccase | 1.03 | 25 | 17.97 | 31.97 |
|   | [G2]-PEG5k[G2] | 6.97 |   |   |   |
|   | HBT | 6.05 |   |   |   |
|   | VA | 56.21 |   |   |   |
| 2 | Laccase | 1.03 | 10 | 4.77 | 8.31 |
|   | [G2]-PEG5k[G2] | 6.97 |   |   |   |
|   | HBT | 5.90 |   |   |   |
|   | VA | 57.39 |   |   |   |
| 3 | Laccase | 1.03 | 10 | 2.60 | 4.68 |
|   | [G2]-PEG5k[G2] | 0 |   |   |   |
|   | HBT | 5.90 |   |   |   |
|   | VA | 55.60 |   |   |   |
| 4 | Laccase | 1.03 | 10 | 0 | 0 |
|   | [G2]-PEG5k[G2] | 0 |   |   |   |
|   | HBT | 0 |   |   |   |
|   | VA | 52.16 |   |   |   |
| 5 | Laccase | 0 | 10 | 0 | 0 |
|   | [G2]-PEG5k[G2] | 0 |   |   |   |
|   | HBT | 0 |   |   |   |
|   | VA | 51.77 |   |   |   |

Five combinations of reagents are utilized. The copolymer and the mediator are added to 2mL of the laccase/buffer solution (1mg/mL) placed in 5mL vials and the resulting mixture was vigorously stirred for 5 min. The veratryl alcohol is added and vials 2 through 4 are stirred at 10°C for 24 h. Vial 1 is allowed to stir at 25°C for 24 h. Purpald (Aldrich) in 1 N NaOH is used as an indicator to assess the concentration of veratryl aldehyde produced during the reaction (*17*).

The titration endpoint is determined by the color change of the analyte solution from a light yellow to a deep purple. The catalytic activity of the laccase is substantially increased at higher temperature and in the presence of the copolymer (Table 3, vials 1 and 2 vs. vial 3). The enzyme is not active without a mediator, Table 3, vial 4. It should be also noted that the conversion yields achieved with the enzyme/linear-dendritic copolymer complex are comparable to those reported previously for the same non-phenolic compound and *Trametes villosa* laccase also in combination with HBT (*18*).

In extension to the above experiments the usefulness of the laccase/[G2]-PEG5k-[G2] complex for biotransformations of highly hydrophobic substrates is evaluated with BP and HBT. The reaction is followed by UV-Vis spectroscopy, Figure 7. It is seen that, despite the notable difference in the intensities, the spectral characteristics of reaction mixtures with BP (Figure 7A) are rather similar to those of mixtures without BP (Figure 7B). Obviously HBT is not spectroscopically suitable as mediator for the analysis of laccase mediated transformations of BP, since previous studies have shown that it can also be oxidized by laccase, and the reaction product(s) possess absorption maxima in the same spectral region as the BP quinones (*19, 20*).

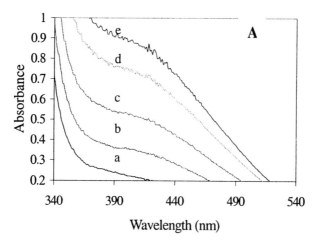

*Figure 7 A: UV-Vis spectra of aqueous reaction mixture containing laccase (1.03 mg/mL), [G2]-PEG5k-[G2] ($10^{-4}$M), BP($2 \times 10^{-4}$M) and HBT ($10^{-4}$M). a: 10 min; b: 20 min; c: 30 min; d: 40 min; e: 50 min*

NHN (Figure 3A) seems to be more suitable mediator for this investigation. It belongs to the class of π-electron rich N-hydroxy compounds that might have favorable redox potential (*21*) and substantial hydrophobicity (*i.e.* it would be favorably sequestered from the water into the dendritic fragments of the copolymer).

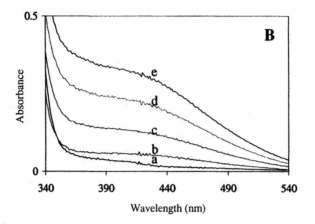

*Figure 7 B:* UV-Vis spectra of aqueous reaction mixture containing laccase (1.03 mg/mL), [G2]-PEG5k-[G2] ($10^{-4}M$), and only HBT ($10^{-4}M$). a: 10 min; b: 20 min; c: 30 min; d: 40 min; e: 50 min.

Indeed, a separate experiment confirms that at the concentrations used in this study micelles formed by [G2]-PEG5k-[G2] can bind IMD-4 at a ratio 0.45 mol per 1 mol copolymer. Analogously BP can be encapsulated at ratio 0.1 mol per 1 mol copolymer. The biotransformation of BP is followed by UV-Vis spectroscopy. Parallel measurements without the linear-dendritic copolymer are also made under identical reaction conditions. The spectroscopic analyses reveal that the mixtures containing laccase modified with [G2]-PEG5k-[G2] show a broad absorbance in the region 430 – 480 nm increasing with time, Figure 8A. It is known that the three most probable BP oxidation products: 6,12-, 1,6- and 3,6-benzo-α-pyrene diquinone have characteristic absorption maxima at 425 nm, 440 nm and 470 nm, respectively (20). Since the starting materials absorb at shorter wavelengths ($\lambda_{max}$ < 350 nm, see Figure 5) the appearance of this absorption could be unambiguously attributed to the formation of the BP oxidation isomers mentioned above. In a sharp contrast the mixtures without linear-dendritic copolymer do not show absorbance between 430 nm and 480 nm, Figure 8B.

The second key issue that needs to be addressed by this study is the necessity to use a linear-dendritic copolymer instead of the more simple and inexpensive linear-linear copolymers. The importance of the macromolecular architecture of the copolymer absorbed on the laccase carbohydrate domains is investigated using two common nonionic polymeric surfactants – Brij-35 and Igepal CO-990. They have similar hydrophilic/hydrophobic ratio as [G2]-PEG5k-[G2] and contain PEG chains of comparable length, but the hydrophobic fragments are linear.

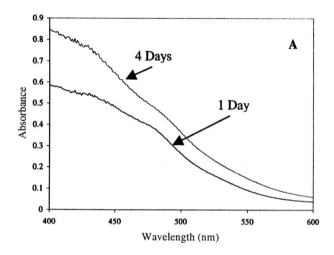

*Figure 8 A:* UV-Vis spectra of aqueous reaction mixture containing laccase (1.03 mg/mL), [G2]-PEG5k-[G2] ($10^{-4}$M), BP($2\times10^{-4}$M) and NHN ($10^{-4}$M).

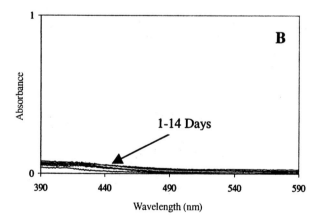

*Figure 8 B:* UV-Vis spectra of aqueous reaction mixture containing laccase (1.03 mg/mL), BP($2\times10^{-4}$M) and NHN ($10^{-4}$M).

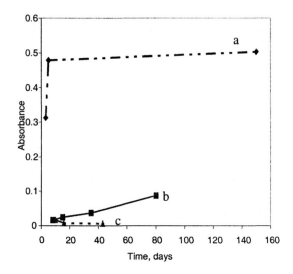

*Figure 9:* Time dependence of the absorption at 430 nm measured in aqueous reaction mixtures containing laccase (1.03 mg/mL), BP($2\times10^{-4}$M) and NHN ($10^{-4}$M) with a: [G2]-PEG5k-[G2] ($10^{-4}$M); b: Brij-35 ($10^{-4}$M); c: Igepal CO-990 ($10^{-4}$M)

Figure 9 demonstrates that [G2]-PEG5k[G2] notably facilitates the oxidation of BP in comparison to the linear analogues. The result is not surprising in view of the reported superior PAH binding capacity of the nanoporous hydrophobic domains formed by this linear-dendritic copolymer in water (8). It is assumed in analogy with the micelles that the dendritic fragments anchored at the carbohydrate surfaces are able to passively bind and transport both NHN and BP in close proximity to the active center in the laccase structure. Thus the linear-dendritic laccase complexes detailed here may be used to facilitate the transformation of those substrates that have unfavorable size and solubility, issues which were raised by D'Acunzo et al. (22) for laccase-catalyzed oxidations.

## Conclusions

In line with our the previous findings that linear-dendritic block copolymers demonstrated an affinity for polysaccharides and sequestered bulky hydrophobic organic compounds, the data of the current investigation show that laccase surface-modified with the amphiphilic copolymer [G2]-PEG5k-[G2] is capable of oxidative transformation of highly hydrophobic PAHs (benzo-α-pyrene) in the presence of hydrophobic N-hydroxy mediators. The operating parameters of such mediated substrate conversions – oxidations or polymerizations, effected by this particular enzyme have yet to be worked out. While the process in this instance is still relatively kinetically slow in comparison to totally aqueous enzymology, it offers promising possibilities for the constructive biotransformation of precursors, not normally considered substrates because of their bulk and solubility, into more valuable intermediates under "green chemistry" conditions.

## Acknowledgements

The financial support for this project provided by The Dreyfus Foundation (Senior Mentor Award to S.T.) and The Research Corporation (Cottrell Scholar Award to I.G.) is gratefully acknowledged.

## References

1. (a) Messerschmidt, A. *Multi-Copper Oxidases*. World Scientific: Singapore, 1997; (b) for a recent review on redox mediators see Call, H.P.; Mücke, I. *J. Biotechnol.* **1997**, *53*, 163
2. Rodakiewicz-Nowak, J.; Kasture, S.M.; Dudek, B.; Haber, J. *J. Mol. Catal. B: Enzymatic* **2000**, *11*, 1
3. Rodakiewicz-Nowak, J.; Kasture, S.M.; Dudek, B.; Haber, J. *J. Mol. Catal. B: Enzymatic* **2000**, *11*, 1
4. (a) Collins, P.J.; Kotterman, M.J.J.; Field, J.A.; Dobson, A.D.W. *Appl. Environ. Microbiol.* **1996**, *62*, 4563; (b) Johannes, C.; Majcherczyk, A.; Hüttermann, A. *Appl. Microbiol. Biotechnol.* **1996**, *46*, 313
5. Majcherczyk, A.; Johannnes, C. *Biochim. Biophys. Acta* **2000**, *1474*, 157
6. Alcalde, M.; Bulter, T.; Arnold, F.H. *J. Biomol. Scr.* **2002**, *7*, 547
7. Cho, S.J.; Park, S.J.; Lim, J.S.; Rhee, Y.H.; Shin, K.S. *Biotechnol. Lett.* **2002**, *24*, 1337
8. Gitsov, I.; Lambrych, K.R.; Remnant, V.E.; Pracitto, R. *J. Polym. Sci.: Part A: Polym. Chem.* **2000**, *38*, 2711
9. Fréchet, J.M.J.; Gitsov, I.; Monteil, Th.; Rochat, S.; Sassi, J.F.; Vergelati, C.; Yu, D. *Chem. Mater.* **1999**, *11*, 1267
10. Omori, S., Lai, Y.-Z., Nakas, J. P., Lu, P., Tanenbaum, S.W. unpublished results
11. Hawker, C.J.; Fréchet, J.M.J., *J. Am. Chem. Soc.* **1990**, *112*, 7638

12. Gitsov, I.; Wooley, K.L.; Fréchet, J.M.J., *Angew. Chem., Int. Ed. Eng.* **1992**, *31*, 1200
13. Lu, P.; Nakas, J. P.; Tanenbaum, S.W., unpublished results
14. (a) Wolfenden, B.S.; Wilson, R.L. *J. Chem. Soc., Perkin Trans. II* **1982**, 802; (b) Johannes, C.; Majcherczyk, A. *J. Biotechnol.* **2000**, *78*, 193
15. Brown, M.A.; Zhao, Zh.; Mauk, A.G. *Inorg. Chim. Acta* **2002**, *331*, 232
16. Piontek, K.; Antorini, M.; Choinowski, T. *J. Biol. Chem.* **2002**, *277*, 3763
17. Quesenberry, M.S.; Lee, Y.C. *Anal. Biochem.* **1996**, *234*, 50
18. Fabbrini, M.; Galli, C.; Gentili, *J. Mol. Cat. B: Enzymatic* **2002**, *18*, 169
19. Ander, P.; Messner, K. *Biotechnol. Techniques* **1998**, *12*, 191
20. (a) Haemmerli, S.D.; Lisola, M.S.A.; Sanglard, D.; Fiechter, A. *J. Biol. Chem.* **1986**, *261*, 6900; (b) Rama, R.; Mougin, C.; Boyer, F.-D.; Kollmann, A.; Malosse, C.; Sigoillot, J.-C. *Biotechnol. Lett.* **1998**, *20*, 1101
21. Xu, F.; Deussen, H.-J.W.; Lopez, B.; Lam, L.; Li, K. *Eur. J. Biochem.* **2001**, *268*, 4169
22. D'Acunzo, F.; Galli, C.; Masci, B. *Eur. J. Biochem.* **2002**, *269*, 5330

# Novel Biomaterials

Chapter 7

# Functional Hydrogel–Biomineral Composites Inspired by Natural Bone

Jie Song[1] and Carolyn R. Bertozzi[1–3]

[1]Materials Sciences Division, Lawrence Berkeley National Laboratory,
[2]Departments of Chemistry and Molecular and Cell Biology, and
[3]Howard Hughes Medical Institution, University of California, Berkeley, CA 94720

A urea-mediated mineralization technique was developed to enable the formation of pHEMA-based hydrogel-calcium phosphate composites with excellent polymer-mineral interfacial adhesion strength that is desirable for bone mimics. This mineralization method was also applied to generate more sophisticated composites containing functional hydrogels that possess anionic groups mimicking the extracellular matrix proteins in bone.

Bone is a hierarchical composite material that consists of carbonated calcium phosphates (CP) and protein matrices. The active organic matrix of mineralized tissues typically consists of a structural component (e.g. type I collagen) and acidic mineral nucleating proteins (e.g. bone sialoprotein rich in glutamate, or phosphophoryn rich in phosphoserine and aspartate).[1-3] They play essential roles in defining both the highly efficient nucleation (and subsequent high-affinity mineral integration) process and the toughness of bone. Matrix-mediated nucleation in nature is believed to occur by an epitaxial mechanism, in which a lattice match between the anionic organic matrix and the nascent crystal lowers the interfacial freee-energy barrier to critical nucleus formation.[2] In addtion, the presence of protein matrices also provides the necessary toughness complementary to the strong yet brittle mineral composition. Unlike the apatite crystals that cannot dissipate much energy, the 3-dimensional macromolecular network provides toughening mechanisms, such as divalent cation-based ionic bridges between two anionic sites on the protein scaffold, to deter crack propagation.[4]

The development of a new generation of bone-like composite materials with improved mechanical properties and enhanced biocompatibility calls for a biomimetic synthetic approach using natural bone as a guide. Our strategy involves the generation of functional polymer scaffolds displaying surface ligands that mimic critical extracellular matrix components of bone, and their high-affinity integration with biominerals. Specifically, these functional polymer scaffolds are derived from poly(2-hydroxyethyl methacrylate), or pHEMA, which is known to be biocompatible and closely mimic the hydrogel-like property of collagen. The polymerization of HEMA and its co-monomers is water compatible, allowing incorporation of anionic ligands that mimic the acidic matrix proteins regulating mineral growth,[2,3] and biological epitopes such as the tripeptide RGD[5] that promote cellular adhesion. The general assembling strategy of pHEMA-based functional hydrogel network and the biomimetic ligands copolymerized with HEMA are illustrated in Figure 1.

Here we introduce a novel mineralization method that leads to rapid, high-affinity integration of calcium phosphate with pHEMA-based hydrogels, a key step in the fabrication of functional bone-like composites.

## Experimental

**Hydrogel preparation.** pHEMA and its copolymers containing various anionic residues were prepared via radical polymerization as previously reported.[6] The formed gels were washed extensively with water to ensure the complete removal of unreacted monomers before they were used for mineralization and further physical characterizations.

**Mineralization of hydrogels with the urea-mediated process.** Hydroxyapatite, or HA, (2.95 g) was suspended into 200 mL of Milli-Q water

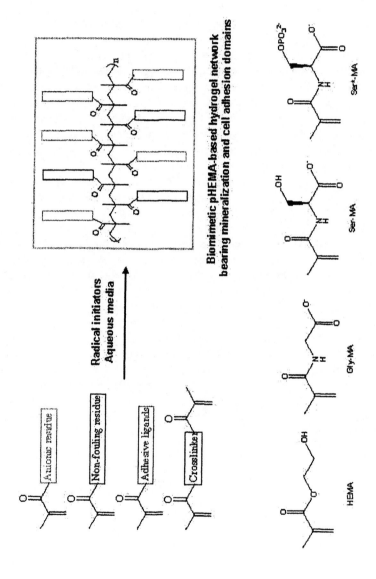

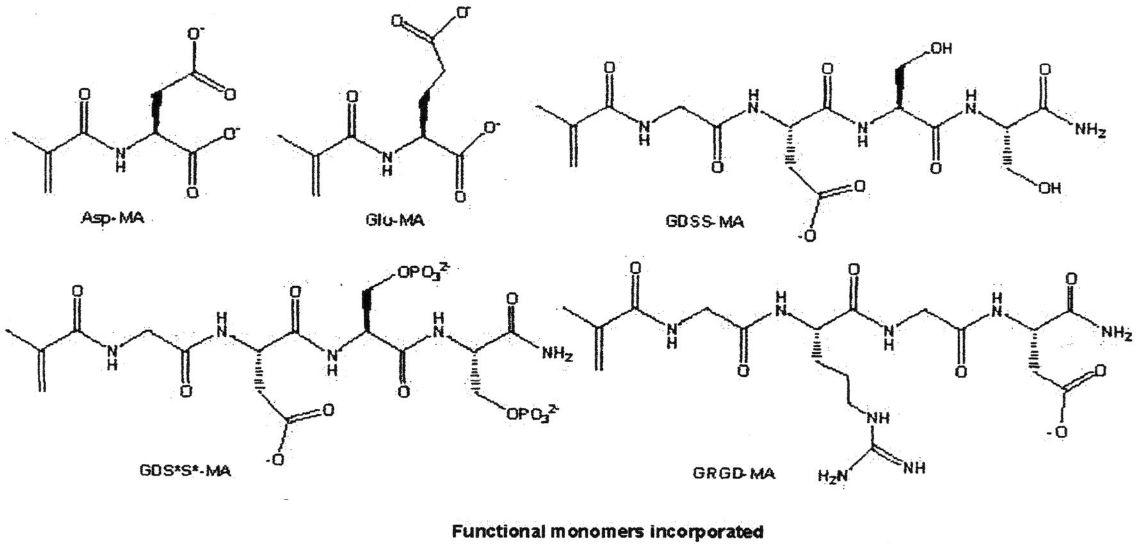

**Functional monomers incorporated**

*Figure 1. Assembling of a hydrogel scaffold containing multiple functional domains (top) and the anionic or adhesive monomers to be copolymerized with HEMA (bottom). Note that the anionic monomers differ in both the type of anionic residues and the number of anionic charges carried per side chain.*

with stirring, and 2 M HCl was added sequentially until all the HA suspension was dissolved at a final pH of 2.5-3. Urea (24 g) was then dissolved into the solution to reach a concentration of 2 M. Each hydrogel strip was then immersed into the acidic HA-urea stock solution. The solution was heated to 95 °C with varied linear heating rates (0.1 to 1.0 °C/min) without agitation of the mineralization solution and maintained at that temperature overnight when necessary (for preparation of composites with thicker CP layers).

**SEM-EDS.** All SEM micrographs of freeze-dried hydrogels and hydrogel-mineral composites were obtained with a ISI-DS 13OC dual stage SEM with associated EDS. Samples were either coated with Au or Pt on a BAL-TEC, SCD 050 sputter coater to achieve optimal imaging results, or coated with carbon for EDS analysis.

**XRD.** The crystallinity of the mineral phase of the composites was evaluated by XRD with a Siemens D500 instrument using Cu K$\alpha$ radiation.

**Evaluation of mineral-hydrogel interfacial adhesion.** In order to evaluate the adherence of the mineral layers attached to pHEMA hydrogels, the relative crack resistance was qualitatively evaluated by indentation. The indentation test was performed on the freeze-dried composite using a Vickers indentor (Micromet, Buehler, Ltd., USA). Loads from 5 to 15 Newtons were applied for 20 seconds for each measurement. After indentation, the samples where analyzed by SEM in order to check for delamination.

## Results and Discussion

Calcium apatites are known to promote bone apposition and differentiation of mesenchymal cells to osteoblasts.[7] In this work, synthetic HA was used in the fabrication of hydrogel-based bonelike composite materials. HA has limited solubility in water at neutral and basic pH but is highly soluble at acidic pH.[8] Based on this property, we devised a urea-mediated solution precipitation technique, in which a segment of pHEMA hydrogel was soaked in an acidic solution of HA containing a high concentration of urea. Upon gradual heating (without stirring) from room temperature to 95 °C, urea started to decompose and the pH slowly increased (around pH 8). Under these conditions, some hydrolysis of the 2-hydroxyethyl esters occurred, promoting heterogeneous nucleation and 2-dimensional growth of a thin CP layer at the hydrogel surface.[6]

The strong affinity between calcium and the *in situ* generated acidic surface of pHEMA led to the 2-dimensional outward growth and eventual merge of CP from individual nucleation sites (the bright centers indicated by arrows in Fig. 2A & 2B). The calibrated EDS area analysis performed on the mineral surface of the composite revealed a Ca/P ratio (1.6±0.1) similar to that of synthetic HA (Fig. 2C). X-ray diffraction (XRD) analysis performed on the mineralized pHEMA composite indicated that the CP layer was either nanocrystalline or amorphous (data not shown). The adhesion strength of the CP layer to the gel surface was qualitatively evaluated by microindentation analysis performed on the surface of the freeze-dried hydrogel-CP composite. No delamination of the mineral layer was observed by SEM (Fig. 2D) even after Vickers indentations

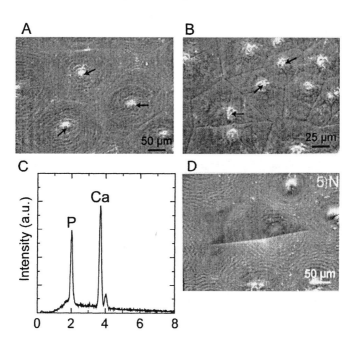

*Figure 2. Morphology, chemical composition and microindentation analysis of calcium phosphate layer grown on the surface of pHEMA via the urea-mediated process. The mineralization process lasted for 2 hours, with an average heating rate around 0.6 °/min. (A) & (B): SEM showing different patterns of 2-dimensional circular outward growth of mineral layers from multiple nucleation sites (indicated by arrows) on the acidic surface of pHEMA. Note the full coverage of the hydrogel surface with mineral layers and the sharp edges separating neighboring domains. (C): SEM-associated EDS area analysis of the mineral layer shown in micrograph A, confirming the chemical composition and Ca/P ratio that is typical for HA. (D): SEM showing an indent formed on the surface of mineralized pHEMA using a Vickers microindenter with a load of 5 N. The mineral layer did not delaminate.*

with loads up to 15 N, an indication of good adhesion at the mineral-gel interface. This represents a major improvement over the widely used simulated body fluid mineralization method, which results in flake-like crystal apatite coatings that tend to delaminate easily upon drying.[9]

Several notable features of this mineralization procedure include: 1) increasing pH and temperature during the process promotes the hydrolysis of the ethyl ester side chains of pHEMA and leads to the *in-situ* generation of an acidic surface and a partially acidic interior that has high affinity for calcium ions; 2) the high affinity between the calcium ions and the exposed carboxylate groups at the gel surface translates into a low interfacial energy between the hydrogel and calcium phosphate, and consequently, a low energy barrier for the heterogeneous nucleation of mineral on the hydrogel surface; 3) the thermo-decomposition of urea allows a homogeneous variation of pH across the solution, avoiding a sudden local pH change that is commonly observed with strong base-induced heterogeneous precipitation.

We also investigated how external factors, such as heating rate, the agitation of the mineral stock solution and the duration of the process, may affect the outcome of the gel-CP composite formation.[10]

We first demonstrated that avoiding direct stirring of the HA-urea mineral stock solution, which promotes homogeneous precipitation of HA across the solution, is essential in achieving the desired heterogeneous nucleation and growth of CP on the *in situ* generated acidic gel surface.

We then applied a range of linear heating rates (1.0 °C/min to 0.1 °C/min, from room temperature to 95 °C) to prepare the gel-mineral composites. We showed that a relatively fast heating rate such as 1.0 °C/min did not lead to a level of mineralization of the pHEMA gel that was detectable by either SEM or the associated EDS analysis (data not shown). When a 0.5 °C/min heating rate was applied, the formation of circular mineral layers on the hydrogel surface was observed (Fig. 3A), with similar 2-dimensional outward growth pattern formed around individual nucleation sites. When the linear heating rate was lowered to 0.2 °C/min, the surface of the hydrogel-mineral composite was fully covered with well-merged circular CP layers (Fig. 3B), suggesting that a slower heating rate and a more sufficient overall mineralization time promote the formation of better-merged CP layers on the gel surface. The most pronounced feature resulted from the slowest linear heating rate attempted (0.1 °C/min), however, was the dramatic increase of the number of nucleation sites formed on the pHEMA hydrogel surface (Fig. 3C). Longer exposure of the pHEMA gel to any given pH during the urea-mediated process is likely to lead to a more sufficient hydrolysis of the ethyl ester side chains, resulting in increased numbers of surface carboxylates that could serve as tight calcium ion binders and initial nucleation sites.

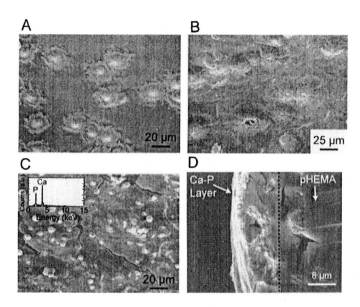

*Figure 3. Effects of varied heating rates and duration of the urea-mediated mineralization process on the formation of pHEMA-CP composites. (A): SEM showing 2-dimensional outward growth of mineral rings from multiple nucleation sites with a linear heating rate of 0.5 °C/min. (B): SEM showing 2-dimensional circular outward growth and merge of mineral layers from multiple nucleation sites with a linear heating rate of 0.2 °C/min. Note the full coverage of the hydrogel surface with the mineral layer. (C): SEM showing the full coverage of mineral layers on the surface with a linear heating rate of 0.1 °C/min. Note the overwhelming amount of nucleation sites scattered throughout the merged mineral domains. The inset shows the EDS area analysis performed over the same fully mineralized surface, confirming the chemical composition and Ca/P ratio that is typical for HA. (D): SEM showing the view of a cross-section of the pHEMA-mineral composite after extended mineralization (10 h after reaching 95 °C). The sample stage was tilted 45°. Note the micron scale thickness of the mineral layer and the fine integration at the mineral-gel interface.*

Finally, we examined the possibility of forming thicker CP layers over the pHEMA gel surface by extending the mineralization time for another 10-12 hours after reaching 95 °C. Mineral coatings with thicknesses up to several microns were obtained, with good integration at the mineral-gel interface as shown in a cross-section image of the composite (Fig. 2D).

This urea-mediated mineralization strategy can be extended to other calcium phosphates and pHEMA based functional hydrogel copolymers. The in vivo resorption rates of calcium phosphates vary greatly. For instance, crystalline hydroxyapatite (HA) is hardly soluble and its resorption could take years while tricalcium phosphate is more soluble and its resorption typically occurs in months.[11] The mineralization method we developed can be applied to a range of calcium phosphates (CPs) to produce hydrogel-CP composites with tunable in vivo bio-resorbability. Further, as shown in Figure 4, when the method was applied to the mineralizaton of copolymers of HEMA containing anionic residues that mimic the typical acidic sequences in bone sialoprotein (BSP) or phosphophoryn in dentin, similar mineralization patterns were observed (Fig. 4). Two-dimensional circular mineral growth and the saturation of the gel surface with amorphous CP layer preceded the surface-independent growth of crystalline calcium apatite. As shown in Figure 4B, during fracture of the composite, the CP layer did not delaminate, again suggesting excellent adhesion at the gel-apatite interface.

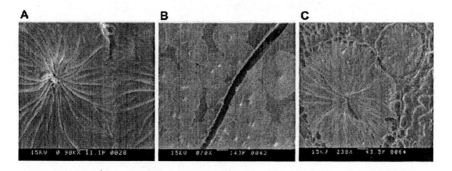

*Figure 4. A urea-mediated mineralization process leads to high-affinity growth of CP on pHEMA-based hydrogel copolymers displaying various anionic residues. Hydrogels mineralized here were pHEMA with 5% Glu-MA (A), 5% Gly-MA (B) and 5% Ser-MA (C). Note that the deliberate fracturing of the composite (B) did not lead to delamination of any circular CP domains, suggesting an excellent gel-mineral interfacial adhesion strength.*

It is worth noting, however, the distinction of subtle differences between the effects of various anionic residues (shown in the bottom panel of Figure 1) on urea-mediated mineral growth patterns is difficult using the pHEMA-based scaffold. The in situ hydrolysis of the 2-hydroxyethyl side chains of pHEMA exposes new surface carboxylates that are powerful competing mineral-nucleators. This issue is now being addressed by copolymerizing biomimetic anionic methacrylamide monomers with an analog of HEMA that is resistant to hydrolysis under the urea mediated mineralization condition. The new system allows us to investigate the role individual anionic residue plays in template driven mineralization without the interference of competing mineral nucleators generated in situ during the urea-mediated thermal treatment.

Finally, the pHEMA-based functional hydrogel–CP composites developed here are subjects of extensive in vitro and in vivo evaluations. Our preliminary in vitro cell culture evaluations have shown that osteoblasts can adhere and proliferate over the functional hydrogel-CP composites containing up to 10% anionic residues. We are also interested to learn how the underlying anionic mineral binding motifs of the composite would direct new mineral deposition by osteoblasts once the amorphous (or nanocrystalline) CP layer is resorbed by osteoclasts upon implantation.

## Conclusions

We have developed a novel template-driven nucleation and mineral growth process to enable high-affinity integration of calcium phosphate with pHEMA-type functional hydrogel scaffolds. This mineralization technique exposes carboxylate groups on the surface of crosslinked pHEMA or its copolymers displaying biomimetic mineral nucleating ligands, promoting high-affinity nucleation and growth of CP on the gel surface. It provides a foundation for integrating template-driven biomineralization with the versatile properties of 3-dimensional hydrogel scaffolds, and opens the door for application of pHEMA in the design of functional bone-like composites.

## Acknowledgements

This work was supported by the Laboratory Directed Research and Development Program of Lawrence Berkeley National Laboratory under the Department of Energy Contract No. DE-AC03-76SF00098.

## References

1. Lowenstam, H. A.; Weiner, S. *On Biomineralization*; Oxford University Press: Oxford, 1989.
2. Hunter, G. K.; Goldberg, H. A. *Biochem. J.* **1994**, *302*, 175-179.

3. George, A.; Bannon, L.; Sabsay, B.; Dillon, J. W.; Malone, J.; Veis, A.; Jenkins, N. A.; Gillbert, D. J.; Copeland, N. G. *J. Biol. Chem.* **1996**, *271*, 32869-32873.
4. Thompson, J. B.; Kindt, J. H.; Drake, B.; Hansma, H. G.; Morse, D. E.; Hansma, P. K. *Nature* **2001**, *414*, 773-776.
5. Massia, S. P.; Hubbell, J. A. *Ann. N.Y. Acad. Sci.* **1990**, *589*, 261-270.
6. Song, J.; Saiz, E.; Bertozzi, C. R. *J. Am. Chem. Soc.* **2003**, *125*, 1236-1243.
7. Darimont, G. L.; Cloots, R.; Heinen, E.; Seidel, L.; Legrand, R. *Biomaterials* **2002**, *23*, 2569-2575.
8. Nancollas, G. H.; Zhang, J. In *Hydroxyapatite and Related Materials*; Brown, P. W., Constantnz, B., Eds.; CRC: Boca raton, 1994; pp73.
9. Gomez-Vega, J. M.; Saiz, E.; Tomsia, A. P. *J. Biomed. Mater. Res.* **1999**, *46*, 549-559.
10. Song, J.; Saiz, E.; Bertozzi, C. R. *J. Eur. Ceram. Soc.* **2003**, *23*, 2905-2919.
11. Delloye, C.; Cnockaert, N.; Cornu, O. *Acta Orthop. Belg.* **2003**, *69*, 1-8.

Chapter 8

# Biomimetic Approach to Biomaterials: Amino Acid-Residue-Specific Enzymes for Protein Grafting and Cross-Linking

Fianhong Chen[1,2], David A. Small[1,3], Martin K. McDermott[4], William E. Bentley[1,3], and Gregory F. Payne[1,2,*]

[1]Center for Biosystems Research, University of Maryland Biotechnology Institute, 5115 Plant Sciences Building, College Park, MD 20742−4450
[2]Department of Chemical and Biochemical Engineering, University of Maryland Baltimore County, 1000 Hilltop Circle, Baltimore, MD 21250
[3]Department of Chemical Engineering, University of Maryland at College Park, College Park, MD 20742
[4]Division of Mechanics and Materials Science, Office of Science and Technology, Food and Drug Administration, 9200 Corporate Boulevard, HFZ−150, Rockville, MD 20850
*Corresponding author: payne@umbi.umd.edu

> Nature creates a range of functional materials using proteins and polysaccharides as starting materials, and enzymes as assembly catalysts. Inspired by nature, we are examining how proteins and polysaccharides can be enzymatically assembled into conjugates and crosslinked networks. Specifically, we used tyrosinase to conjugate proteins to the polysaccharide chitosan, and a microbial transglutaminase to catalyze protein crosslinking. We review results from our studies and suggest how the unique properties of the resulting biomaterials can be exploited in medical applications.

## 1. Introduction

Important mechanical functions are performed in nature by co-polymers of proteins and polysaccharides, and by crosslinked protein networks. For instance, mucins and proteoglycans are complex protein-polysaccharide conjugates that confer viscoelastic properties important for the lubrication and protection of organs and joints*(1-4)*. Protein crosslinking is important in functions that range from the setting of the mussel's water-resistant adhesive to the coagulation of blood*(5-10)*. These biopolymeric materials serve as inspirations for the generation of high performance, safe and biodegradable materials. However, methods for generating these materials in the laboratory are limited by the complexity of the systems*(11-15)*.

We use a biomimetic approach to create functional materials. Specifically, we; (i) begin with nature-derived proteins and polysaccharides, (ii) use enzymes to generate protein-polysaccharide conjugates or crosslinked protein networks, and (iii) obtain materials with useful functional properties. The protein and polysaccharide in our studies are gelatin and chitosan. Gelatin is obtained from collagen, the predominant structural protein in mammals. Chitosan is derived from the polysaccharide chitin, which is the major structural polymer in the integument of insects and crustaceans. The enzymes in our studies are tyrosinase and transglutaminase. We use these enzymes to conjugate gelatin to chitosan, and to crosslink gelatin. The resulting materials have distinct chemical or viscoelastic properties. These properties, along with the biocompatibility of the starting materials and mild processing conditions suggest the potential of this biomimetic approach for the construction of a variety of high performance and safe materials that should be especially well-suited for medical applications. Part 2 of this review describes our results with tyrosinase, while results with transglutaminase are described in Part 3.

## 2. Tyrosinase-catalyzed protein grafting to a polysaccharide

Tyrosinases are ubiquitous oxidative enzymes that use molecular oxygen to convert phenols into quinones. These quinones are reactive intermediates that undergo further reactions with various nucleophiles. In nature, tyrosinase-catalyzed reactions are responsible for the browning of food*(16)*, the setting of mussel glue*(5-10)* and the hardening of insect shells*(17-19)*.

Tyrosinase activity is not limited to low molecular weight substrates, these enzymes can also oxidize the phenolic moieties of peptides and proteins (i.e. tyrosine or dihydroxyphenylalanine residues). This amino-acid-residue-specific oxidation leads to the generation of quinone residues that are "activated" to undergo reactions with nucleophilies. In our studies, we generate these quinone residues in the presence of chitosan, a linear polysaccharide with nucleophilic amines at nearly every repeating unit (chitosan is a co-polymer of glucosamine and N-acetylglucosamine). As illustrated by the reaction below, we use tyrosinase to initiate the grafting of peptides and proteins onto chitosan's backbone.

To demonstrate that tyrosinase initiates covalent grafting, we preformed reactions with model, low molecular weight compounds. Specifically, we reacted tyrosinase with the dipeptide Tyr-Ala in the presence of the monosacchardie glucosamine, and analyzed the products using electospray mass spectrometry. Product spectra were consistent with the formation of an adduct between the dipeptide and the monosccharide. Evidence for peptide grating to the chitosan backbone was obtained from physical (i.e. rheololgical) studies *(20)*.

## 2.1. Grafting of an open chain structural protein (gelatin) to chitosan

To examine the potential of tyrosinase to react with higher molecular weight substrates, we studied the common industrial protein gelatin. Gelatin is derived from the mammalian structural protein collagen and consists of a large number of Gly-X-Y tripeptide repeats. These repeats allow individual random coil gelatin chains to associate to form triple helices. This coil-to-triple helix transition occurs at low temperature (typically less than 20° C) and leads to the physical network junctions that are responsible for gelatin's ability to undergo thermally-reversible gel formation. Interestingly, there are only a small number of tyrosyl residues in gelatin (0.3% of the total residues) and these residues appear to be exclusively located in the telopeptide regions*(21,22)*. We anticipated that these tyrosyl residues would be accessible for tyrosinase-catalyzed oxidation because gelatin has an open chain structure.

Initial studies demonstrated that gelatin's tyrosine residues can be oxidized by tyrosinase, and the "activated" quinone residues undergo further reaction with chitosan*(23)*. When tyrosinase is added to a blend of gelatin and chitosan, two changes are visually observed as illustrated in the left-hand image of Figure 1. First, the color of the solution changes from colorless to pink indicating that tyrosinase can oxidize gelatin's tyrosine residues. Second, the gelatin-chitosan blend undergoes a transition from a solution to a viscoelastic gel during the course of the tyrosinase-initiated reaction. The "control" on the right in Figure 1 shows that in the absence of tyrosinase, the gelatin-chitosan blend remains a colorless solution. The qualitative results in Figure 1 are supported by more rigorous chemical and rheological measurements that indicate tyrosinase can initiate the covalent grafting of gelatin onto the chitosan backbone*(23)*.

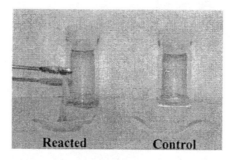

*Figure 1. Tyrosinase-catalyzed gel formation of gelatin-chitosan mixture. The "Reacted" sample contained gelatin (5.0 wt%), chitosan (0.5 wt%) and tyrosinase (60 U/ml) and was reacted at 35°C for 30 min. The "Control" contained gelatin and chitosan but no tyrosinase.*

In nature, branched co-polymers of proteins and polysaccharides organize to form networks with distinct viscoelastic properties. The complexity of nature's protein-polysaccharide conjugates has impeded their structural characterization and limits our understanding of their network structures. Similarly, the structure and network organization of the gelatin-chitosan conjugates are unknown. In Scheme 1, we propose a network structure for the tyrosinase-catalyzed gelatin-chitosan gels. Specifically, we believe gelatin chains are grafted to the chitosan backbone at a single point. Single-point attachment is consistent with the fact that an average gelatin chain has less than 1 tyrosine residue. We propose that the grafted gelatin chains can interact with a second chitosan chain to form network junctions through either entanglements or physical associations (e.g. hydrophobic or electrostatic interactions).

One aspect of the network structure proposed in the left of Scheme 1 is that chitosan is integral to the network. The importance of chitosan is demonstrated by separate experiments. In one experiment, reactions were performed with gelatin and tyrosinase (but not chitosan) and we observed a change in color consistent with tyrosinase-catalyzed oxidation of gelatin - but no gel network was formed. In the second experiment, we incubated gelatin and chitosan with tyrosinase to form a gel, then further reacted this gel with the chitosan-hydrolyzing enzyme, chitosanase, and observed that the network was immediately broken.

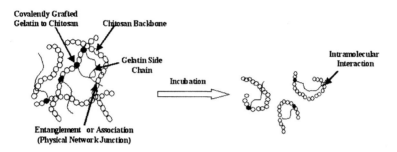

*Scheme 1. Proposed network structure of tyrosinase-catalyzed gelatin-chitosan gels and possible explanation for their transient nature.*

A second aspect of the proposed network structure is that the grafted gelatin chains appear to retain substantial mobility. In fact, the tyrosinase-catalyzed gelatin-chitosan gels are observed to break spontaneously over the course of several hours or days depending on temperature and composition. Dynamic light scattering studies indicate that both gel formation and gel breakage follow a power law behavior consistent with percolation theories(24). We propose in Scheme 1 that gel breakage occurs because the intermolecular physical associations responsible for network crosslinking are replaced by intramolecular associations between the grafted gelatin chain and its chitosan backbone.

A potential medical application for tyrosinase-catalyzed gelatin-chitosan gels is as an emergency burn dressing. The ability to generate these gels *in situ* over the course of 20-30 minutes suggests that such dressings could be formed directly on the damaged tissue. The hemostatic and antimicrobial activities of chitosan, as well as the ability of gelatin to promote cell attachment may facilitate healing. Finally, the transient nature of the network allows the gel to "dissolve" and may eliminate the need to "change" the dressing – thus avoiding further tissue damage.

## 2.2. Grafting of a compact protein (GFP) to chitosan

Unlike gelatin, many functional proteins (enzymes or antibodies) have more compact (e.g. globular) conformations. Potentially, these compact proteins may have few surface tyrosine residues that are accessible for tyrosinase reaction. Here, we examined Green Fluorescent Protein (GFP) as a model protein with compact structure. Figure 2 shows that GFP has nine tyrosine residues but only three are present on the surface and potentially accessible for the tyrosinase-catalyzed oxidation. We specifically studied two GFP fusion proteins. One GFP fusion contains a hexahistidine tail at the N terminus (His)$_6$-GFP. The (His)$_6$ tail is routinely used in biochemistry to facilitate protein purification. The second GFP fusion contains both (His)$_6$ tail on the N terminus and a pentatyrosine tail on the C terminus (His)$_6$-GFP-(Tyr)$_5$. This (Tyr)$_5$ tail was designed to increase the number of accessible tyrosine residues on GFP surface.

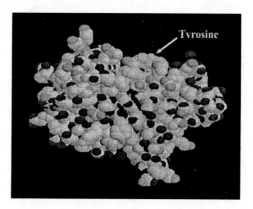

*Figure 2. Three-dimensional structure of Green Fluorescent Protein (GFP)*

To examine whether GFP was covalently conjugated onto chitosan, we added (His)$_6$-GFP fusion proteins into chitosan solution at pH 6. After tyrosinase was added and the solution was incubated at room temperature over night, a slight brownish color was observed indicating reaction of the Tyr residues. To demonstrate that the oxidized Tyr residues react with chitosan to generate a covalent conjugate, a series of experiments were designed as shown in Scheme 2. After incubating the "Sample" with tyrosinase, the pH of the solution was raised to 8 (shown in the lower part of Scheme 2). At this pH, chitosan becomes insoluble. Presumably, if GFP is covalently grafted onto the chitosan chain, this conjugate will also precipitate at the higher pH. We examined the fluorescence intensity and observed that only 27% of the original fluorescence remained in the supernatant. The precipitate (or pellet) was extensively washed and then re-dissolved in an acetic acid solution. The re-solubilized pellet was observed to have substantial fluorescence. A "Control" experiment (shown in the upper part of Scheme 2) was conducted with chitosan and GFP (but not tyrosinase). After raising the pH of the "Control", over 90% of the original fluorescence was observed to remain in the supernatant. Little fluorescence was observed in the re-solubilized pellet for this control. This result indicates that any physically-bound GFP can be removed by washing the chitosan precipitate. Comparison of the results from the "Sample" and "Control" indicates that tyrosinase-initiates the covalent grafting of the GFP fusion protein to chitosan. Further experiments showed that when we introduced a $(Tyr)_5$ tail to the $(His)_6$-GFP fusion (i.e. $(His)_6$-GFP-$(Tyr)_5$), the grafting efficiency was nearly doubled presumably because more Tyr residues are accessible for tyrosinase-catalyzed oxidation(25).

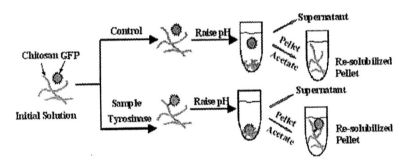

*Scheme 2. Experiment to demonstrate that tyrosinase-catalyzes the grafting of GFP to chitosan*

The GFP-chitosan conjugate has interesting pH-responsive properties characteristic of the polysaccharide chitosan(25). To demonstrate this pH sensitive property, we dissolved GFP-chitosan conjugate in a pH 5 solution. Figure 3 shows that as the pH of the solution was increased from 5 to 6, the GFP-chitosan conjugate began to precipitate from solution, and the supernatant began to loose fluorescence. When the pH was raised to 7.2, most of the conjugate was precipitated and the supernatant had little fluorescence.

One potential application for a pH-responsive protein-chitosan conjugate is as an injectable gel for controlled release of a protein drug. Conceivably, a

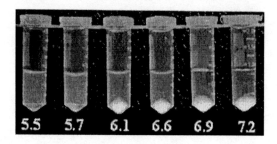

*Figure 3. Tyrosinase-catalyzed grafting of GFP to chitosan confers pH-dependent solubility to GFP*

conjugate of a protein drug and chitosan could be injected as a moderately acidic solution. Upon contact with body fluids of higher pH, the conjugate could form a gel matrix that immobilizes the protein drug at the localized site of injection(26). Over time as chitosan is degraded, the protein drug should be gradually released.

## 3. Transglutaminase catalyzed protein crosslinking

Transglutaminases are acyl-transfer enzymes that catalyze transamidation reactions between the glutamine and lysine residues of proteins to yield N-ε-(γ-glutamyl)lysine crosslinks(27,28). Mammalian transglutaminases typically require calcium for activity, are present in various tissues, and perform a range of functions. For instance, the blood factor XIIIa is a transglutaminase that catalyzes the fibrin crosslinking reactions in the last stages of the blood coagulation cascade(29,30).

A calcium-independent microbial transglutaminase (mTG) recently became available and has been investigated in various food applications (31-33). mTG may be appropriate for a broader range of applications because its calcium-independence makes it more convenient than tissue transglutaminases. mTG is known to react with gelatin chains to yield covalently crosslinked networks(34). In previous studies, we observed that the addition of chitosan to gelatin solutions accelerated transglutaminase-catalyzed gel formation and resulted in stronger gels, although we have no direct evidence that the gelatin and chitosan are covalently linked(35).

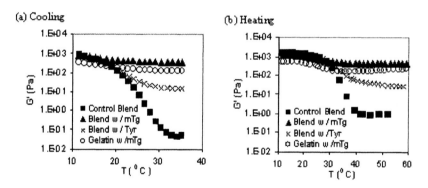

The networks generated from transglutaminase-catalyzed crosslinking are different from those obtained by either cooling gelatin solutions or reacting gelatin and chitosan with tyrosinase. To gain insights into the various networks generated in our studies, we blended gelatin (5%) with chitosan (0.32 %) and formed gels by cooling, or by reacting the blends with either tyrosinase or transglutaminase. The solid squares in Figure 4a show the storage modulus (G') of gelatin-chitosan blend as it is cooled from 37 °C. At the higher temperatures, G' is small but as the temperature is lowered near gelatin's coil-to-triple helix transition temperature, G' is observed to increase markedly. By cooling, these gelatin-chitosan blends undergo a sol-gel conversion with the triple helices serving as the physical network junctions. The solid squares in Figure 4b show how G' varies as the blend is heated above gelatin's helix-to-coil transition temperature. The drop in G' upon heating illustrates that the triple helix network junctions are "melted" and the blend returns to a solution state.

*Figure 4. Thermal behavior of gelatin (5%) and chitosan (0.32%) blends. Some blends were reacted with tyrosinase (Tyr; 60 U/ml) or microbial transglutaminase (mTG; 10 U/g-gelatin). A gelatin (5%) solution reacted with mTG was also analyzed. Oscillatory tests were performed with a controlled stress of 0.5 Pa and a frequency of 0.1 Hz. (a) Cooling test. -0.3°C/min. (b) Heating test. 1°C/min.*

As discussed previously, when tyrosinase is reacted with gelatin and chitosan, a hydrogel is formed even when temperatures exceed gelatin's coil-to-helix transition temperature. This is illustrated in Figure 4a by the fact that this sample (designated Blend w/ Tyr) has a significant G' even before cooling. The proposed network structure for this gel is illustrated in Scheme 1 where a grafted gelatin chain forms a physical interaction that serves as a network junction. Figure 4a shows that the tyrosinase-catalyzed gelatin-chitosan gel is strengthened upon cooling below gelatin's coil-to-helix transition suggesting that this strengthening is due to the formation of additional triple helix network junctions. Figure 4b shows that heating above gelatin's transition temperature weakens, but does not break, the tyrosinase-catalyzed gelatin-chitosan gel. Presumably, heating "melts" the triple helix network junctions without destroying the physical junctions formed with the grafted gelatin chains. The above behavior indicates that tyrosinase-catalyzed reactions do not disrupt the ability of gelatin chains to undergo thermally-reversible coil-to-helix transitions.

When the gelatin-chitosan blend is reacted with transglutaminase, a strong gel is formed, and the triangles in Figure 4 show that the gel's properties are largely independent of cooling or heating near gelatin's transition temperatures. Similar behavior is observed for gels formed by reacting gelatin with transglutaminase in the absence of chitosan (circles in Figure 4). This behavior is consistent with mTG catalyzing a covalent crosslinking reaction that destroys the ability of gelatin chains to undergo further coil-to-helix transitions. The results in Figure 4 also show that chitosan contributes to strengthening the network generated from the mTG-catalyzed crosslinking of gelatin.

One potential application of mTG-catalyzed crosslinking is as a biomimetic soft tissue adhesive analogous to fibrin sealants. Current fibrin sealants employ protein components derived from blood and utilize the transglutaminase activity of factor XIIIa to catalyze protein (i.e. fibrin) crosslinking. It may be possible to develop an alternative soft tissue adhesive based on gelatin and mTG that avoids the need for blood proteins.

A second potential application is for tissue engineering where mTG could be used for entrapping cells within a crosslinked gelatin network. In initial studies, we inoculated *Escherichia coli (E. coli)* bacterial cells into standard growth medium with gelatin (10%), and then added mTG. Gel formation occurred over the course of a couple hours and Figure 5 shows cells entrapped within this crosslinked gelatin matrix. The entrapped cells are viable and can grow and respond to their environment. Additionally, the crosslinked protein matrix can be degraded by proteases and the cells can be released *(36)*.

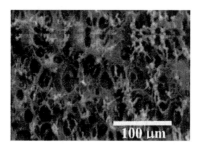

*Figure 5. SEM photograph of E. coli cells that had been entrapped in mTG-catalyzed gelatin gel matrix.*

In summary, we have shown that nature-derived polymers can be enzymatically modified to generate materials with interesting chemical and mechanical properties. The benign nature of both the raw materials and the enzyme-catalyzed reaction steps illustrates the potential for creating safe and environmentally-friendly materials that offer unique performance attributes (e.g. stimuli-responsive solubility or mechanical strength). We have attempted to highlight this potential by suggesting a range of possible medical applications.

## Acknowledgements

Financial support was provided by the United States Department of Agriculture (2001-35504-10667) and the National Science Foundation (grant BES-0114790).

## References

1. Sellers, L. A., Allen, A., Morris, E. R. & Ross-Murphy, S. B. *Carbohydrate Res.* **1988**, 178 93-110.
2. Iozzo, R. V. *Annual Review Biochemistry* **1998**, 67 609-652.
3. Tanihara, H., Inatani, M., Koga, T., Yano, T. & Kimura, A. *Cornea* **2002**, 21 (Suppl. 2), S62-S69.
4. Carlson, E. C. et al. *Biochemistry Journal* **2003**, 369 461-468.
5. Holl, S. M., Hansen, D., H., W. J. & Schaefer, J. *Arch. Biochem. Biophys.* **1993**, 302 255-258.

6. Yu, M. & Deming, T. J. *Macromolecules* **1998**, 31 4739-4745.
7. Hansen, C., Corcoran, S. G. & Waite, J. H. *Langmuir* **1998**, 14 1139-1147.
8. McDowell, L. M., Burzio, L. A., Waite, J. H. & Schaefer, J. *J. Biol. Chem.* **1999**, 274 20293-20295.
9. Yu, M., Hwang, J. & Deming, T. J. *JACS* **1999**, 121 5825-5826.
10. Burzio, L. A. & Waite, J. H. *Biochem.* **2000**, 39 11147-11153.
11. Williams, S. J. & Davies, G. J. *Trends Biotechnol* **2001**, 19 356-362.
12. Helenius, A. & Aebi, M. *Science* **2001**, 291 2364-2369.
13. Sears, P. & Wong, C.-H. *Science* **2001**, 291 2344-2350.
14. Dell, A. & Morris. *Science* **2001**, 291 2351-2356.
15. Alper, J. *Science* **2001**, 291 2338-2343.
16. Aberg, C. M., Chen, T. & Payne, G. F. *J. of Polymers and the Environment* **2002**, 10 (3), 77-84.
17. Peter, M. G. *Angew. Chem. Int. Ed. Engl.* **1989**, 28 555-570.
18. Sugumaran, M. *Adv. Insect Physiol.* **1988**, 21 179-231.
19. Andersen, S. O., Peter, M. G. & Roepstorff, P. *Comp. Biochem. Physiol.* **1996**, 113B 689-705.
20. Aberg, C. M., Chen, T., Olimnide, A., Raghavan, S. R. & Payne, G. F. *J. of Agricultural & Food Chem.* **2004**, (in press).
21. Mayo, K. H. *Biopolymers (Peptide Sci.)* **1996**, 40 (4), 359-370.
22. King, G., Brown, E. M. & Chen, J. M. *Prot. Eng.* **1996**, 9 (1), 43-49.
23. Chen, T., Embree, H. D., Wu, L.-Q. & Payne, G. F. *Biopolymers* **2002**, 64 292-302.
24. Kostko, A. F., Chen, T., Payne, G. F. & Anisimov, M. A. *Physica A* **2003**, 323 124-138.
25. Chen, T. et al. *Langmuir* **2003**, 19 (22), 9382-9386.
26. Gutowska, A., Jeong, B. & Jasionowski, M. *The Anatomical Record* **2001**, 263 342-349.
27. Greenberg, C. S., Birckbichler, P. J. & Rice, R. H. *FASEB J* **1991**, 5 (15), 3071-3077.
28. Motoki, M. & Seguro, K. *Trends in Food Science & Technology* **1998**, 9 204-210.
29. Aeschlimann, D. & Paulsson, M. *Thromb. Haemost* **1994**, 71 402-415.
30. Pisano, J. J., Finlayson, J. S. & Peyton, M. P. *Sci.* **1968**, 160 892-893.
31. Dickinson, E. & Yamamoto, Y. *J. Agric. Food Chem.* **1996**, 44 1371-1377.
32. Faergemand, M., Murray, B. S. & Dickinson, E. *J. Agric. Food Chem.* **1997**, 45 (7), 2514-2519.
33. Sakamoto, H., Kumazawa, Y. & Motoki, M. *J. Food Sci.* **1994**, 59 (4), 866-871.
34. Babin, H. & Dickinson, E. *Food Hydrocolloids* **2001**, 15 271-276.
35. Chen, T., Embree, H. D., Brown, E. M., Taylor, M. M. & Payne, G. F. *Biomaterials* **2003**, 24 2831-2841.
36. Chen, T., Small, D., McDermott, M. K., Bentley, W. E. & Payne, G. F. *Biomacromolecules* **2003**, 4 (6), 1558-1563.

Chapter 9

# Biodegradable Films from Pectin–Starch and Pectin–Poly(vinyl alcohol)

Marshall L. Fishman, and David R. Coffin

Eastern Regional Research Center, Agricultural Research Service, U.S. Department of Agriculture, 600 East Mermaid Lane, Wyndmoor, PA 19038

Blends of pectin and high amylose starch, and pectin and poly(vinyl alcohol) (PVOH), both with and without added plasticizer, were used to make solution cast films and characterized to quantify their properties. They were investigated using dynamic mechanical analysis, scanning electron microscopy, tensile testing, FT-IR, oxygen permeability, and solubility studies. The films were found to have high modulus and strength values, with the properties depending strongly on composition. Glycerol was found to be effective as a plasticizer, and PVOH was shown to impart ductility to the pectin. Significant changes in morphology were seen with changes in composition by SEM. Hydrogen bonding of pectin in the film is affected by temperature and glycerol. Oxygen permeability of these films is much lower than found in commercial plastic films.

The study of biodegradable films is a rapidly growing area of research due to their ability to degrade rapidly when disposed of in a landfill once the useful life of the product is over. In addition, they come from renewable resources whose use reduces stress on the environment. Some polysaccharides which have been investigated as biodegradable materials are starch, alginate, carrageenan, cellulose ethers, and pectin. We have been actively involved in studying pectin and its properties for over ten years, both in solution and as solid films. This article summarizes the literature on pectin films, and describes the work we have done on both pectin/starch and pectin/poly(vinyl alcohol) films.

Pectins are water soluble polysaccharides found in the cell walls of higher plants. They have found extensive use in processed foods as a result of their gelling properties. Because pectin is a film forming material it has desirable physical and mechanical properties.

Studies on pectin films were done as early as 1936 (*1*). Generally these studies involved derivatized pectins and the use of polyvalent cations such as calcium. Much of the work was on coatings for foods. Schultz et al. (*2, 3*) prepared films from low methoxyl (<11%) pectin and found that tensile strength for these films decreased with increasing methoxyl level. The films all had similar tensile strengths on the order of 90 MPa with and without added calcium. More recently, Hind et al. (*4*) studied blends of pectin and (carboxymethyl) cellulose for use as cigarette papers.

Our work has focused on high methoxyl pectin. In our studies we have investigated the preparation of plasticized and unplasticized pectin films, both with starch and with poly(vinyl alcohol), and characterized them using dynamic mechanical analysis, tensile measurements, scanning electron microscopy, solubility, and Fourier transfer infrared spectroscopy (*4, 5, 6, 7, 8, 9, 10*). We have also investigated the extrusion and characterization of pectin/starch blends (*11*).

Starch too is inherently biodegradable and has been investigated as a component in biodegradable films for applications such as agricultural mulch and pharmaceutical caplets (*12, 13, 14*). Amylose films, modified with amylopectin, were made and characterized over forty years ago (*15*). These films had physical properties comparable to those of commercial plastic films (tensile strength approximately 60 MPa), although they were water sensitive and embrittled easily. More recent work (*14*) added thermoplastics at fairly high levels to obtain materials that are highly flexible and do not embrittle. In addition to its biodegradability, incorporation of high amylose starch significantly lowers the cost of pectin films in that it is 90-95% cheaper than food grade pectin.

Poly(vinyl alcohol) is somewhat biodegradable. Thus it is an excellent candidate for blending with pectin to form materials with improved properties and with relatively little sacrifice in biodegradability over pectin alone. Poly(vinyl alcohol) polymers (PVOH) are strong tough materials that can be easily obtained in a variety of molecular weights. They are made by hydrolysis of poly(vinyl acetate) and can be made with any convenient level of unhydrolyzed acetate side chains. Differences in the amount of residual acetate groups present can change the physical properties of PVOH significantly.

## Experimental

### Materials

MexPec 1200, MexPec 1400, and MexPec 1500 which had which had degrees of methyl esterification of 65%, 71%, and 74% respectively were obtained from Grindsted Products, Kansas City, KS, and were used as received. The MexPec 1500 was a lime pectin, while the other two materials were identified as citrus pectins.

Amylomaize VII (70% amylose, 30% amylopectin) was provided by American Maize Co. (now Cerestar) of Hammond, IN, and was used as received. It is now sold as Amylogel 03003.

Poly(vinyl alcohol) samples, glycerol, urea, and poly(ethylene) glycol were obtained from Aldrich Corp., Milwaukee, WI. Non-polymeric chemicals were of ACS reagent grade. The water used was HPLC grade prepared using a Modulab Polisher I or a Barnstead Thermolyne water system.

### Film Preparation

Pectin/starch films were prepared by mixing solutions of pectin (and glycerol or other plasticizer) with gelatinized starch solutions and casting them on a polycarbonate plate using a Microm film applicator (Paul N. Gardner Co., Pompano Beach, FL), and allowing the films to air dry overnight. Then the films were dried under vacuum for at least 30 min at room temperature. Films were removed from the coating plates with a razor blade.

Gelatinized starch solutions were prepared by mixing the appropriate amount of starch (0.05-0.67 g) with 10 mL of HPLC grade water in a Parr microwave bomb (Parr Instruments Co., Moline, IL) and heating in a 700 watt Amana Model R321T Radarange microwave oven for 3 min at 50% of full power. The gelatinized starch solutions were cooled in a water bath at room temperature for 25 min and then added to the pectin solutions with stirring. More details are given in references 5 and 6.

Solutions of pectin/poly(vinyl alcohol) and the cast films thereof were prepared in a fashion similar to that of pectin films (8, 10). The solutions were typically made at either 6% or 7% by weight depending on which PVOH was used.

## Mechanical Testing

Thermal dynamic mechanical analysis (TDMA) was done on a Rheometric Scientific RSA II Solids Analyzer (Piscataway, NJ) using a film testing fixture (5, 6). A nominal strain of 0.1% was used in all cases, with an applied frequency of 10 rad/sec (1.59 Hz). A temperature ramp of 10°C/min was used in all cases. Nominal dimensions of the samples were 6.4 mm × 38.1 mm. The gap between the jaws at the beginning of each test was 23.0 mm. Data analyses were carried out using Rheometrics RHIOS and Orchestrator software.

Tensile testing was done on an Instron model 1122 using the SMS TestWorks v. 3.1 software. Gauge lengths of 25.0 and 50.0 mm and sample widths of 3 to 6 mm were used. Sample thicknesses ranged from 0.06 to 0.2 mm. Crosshead speeds of 10 and 20 mm/min were used. Samples were stored at 50-52% RH overnight prior to testing. Some samples were stored at 15% or 0% RH.

## Microscopy

Rectangular strips of films (ca. 0.5 × 2 mm) were prepared for microscopy as described previously (7). Films were imaged with a JSM840A scanning electron microscope (JEOL, USA, Peabody, MA) in the secondary electron imaging mode at instrument magnifications ranging from 1,000X to 10,000X.

## FT-IR Spectroscopy

Films were cast on $CaF_2$ windows and spectra obtained using a Nicolet 740 FR-IR spectrometer equipped with a MCT-B detector and the Nicolet 660 data system as previously described (9).

## Dissolution of Films

Test films were cut into 1 centimeter squares from a template and dissolution kinetics in water were measured as described earlier (8, 9).

**Oxygen Permeability Testing**

Oxygen permeability testing was carried out by Mr. Paul Dell of the U.S. Army Natick Research, Development, and Engineering Center, Natick, MA, using a MOCON OX-TRAN 2/20 oxygen permeability tester (Modern Controls, Inc., Minneapolis, MN). The samples were tested at 22°C using dry air and dry nitrogen, and the relative humidity seen by the samples was less than 2% in all cases (6).

## Results and Discussion

**Pectin/Starch Films**

In the case of pectin/starch/glycerol films that we have investigated, pectin is the most abundant component because it is the film forming material. Initial experiments revealed that MexPec 1500 and Amylomaize VII gave the strongest films (5). The probable cause was that MexPec1500 had a higher molecular weight and higher degree of esterification than the other pectins investigated. Blends of waxy maize starch (100% amylopectin) and pectin failed to produce free standing films. Normal starch gave weaker films than Amylomaize VII (data not reported). MexPec 1500 films loaded with 20% glycerol, urea, poly(ethylene glycol) 300, or poly(ethylene glycol) 1000 were evaluated by dynamic mechanical analysis run for their ability to act as plasticizers. All four plasticizers showed about the same ability to plasticize over the temperature range of 20 to 130°C. Glycerol was chosen for further study because it is generally acceptable as a food additive.

We undertook systematic studies to find the effect of glycerol level on the properties of pectin by thermal dynamic mechanical analysis (TDMA) (9) (see Figure 1). Pure pectin films gave a practically flat storage modulus over the range of -100 to 200°C indicating a stiff material with no thermal transitions. Addition of 30% glycerol induces a glass transition ($T_g$) commencing at about -50°C. As the temperature increases the storage modulus (E') decreases. With increasing level of glycerol, $T_g$ commences at increasingly lower temperatures and E' decreases more rapidly with increasing temperature. With increasing glycerol level, two other thermal transitions emerge at about 50 and 140°C respectively. Because these secondary transitions occur at temperatures higher than the $T_g$ and become more prominent with increasing levels of glycerol, they may be transitions related to movement of whole molecules, so called $T_{\parallel}$ transitions (16). The existence of two transitions may indicate that two classes of pectin structures exist with different molar masses. We have shown that during its extraction, orange and lime pectin can exist as mixtures of high molar mass branched structures, and lower molar mass linear structures (17, 18).

These could be remnants of hydrogen bonded networks of pectin which exist in plants. These same transitions can be observed in pectin/starch/glycerol films, although the second and third transitions tend to merge into one broad transition at higher amounts of starch incorporation (6) (see Figure 2). The incorporation of starch in pectin/glycerol films is advantageous in that it increases the storage modulus of the film while allowing the film to maintain its flexibility. Furthermore it reduces the material cost of pectin based films. Similar trends were also seen for the temperature dependence of the loss modulus (data not shown).

All of the citrus pectin based films were thermally stable up to about 180°C, as was evidenced by samples which exhibited little or no color change if not heated beyond this temperature. However, samples heated to 190-200°C turned dark brown and tended to embrittle.

The gelatinization (disruption) of the starch granules and the resulting degree of solubilization are highly dependent on the time and temperature conditions to which the starch is exposed. Gelatinization was achieved by heating with microwave energy a dispersion of starch granules in water contained in a closed cell. The pressure generated at elevated temperatures in this cell gelatinized the starch. We conducted experiments to determine the influence of gelatinization conditions on the morphological structure and mechanical properties of the cast films prepared (7, 9). Blends of pectin, starch, and glycerol were made using the same methods in our previous work, but with one exception. Instead of using a constant gelatinization time for the starch, it was varied from 10 to 105 seconds. The microwave oven was run at full power rather than at 50% power used in the previous experiments. The films were made using a 60:40 pectin/starch ratio, and 30% glycerol as plasticizer.

The dynamic mechanical properties of the films were relatively unaffected by the time of gelatinization or whether the glycerol was mixed with water or starch. There were, however, very significant changes seen in the morphologies of the films.

Fracture faces of the films, whether or not the starch was gelatinized in the presence of glycerol, displayed three general types of structural organization that were linked to gelatinization times. The first type included films made from mixtures gelatinized for 0 to 20 sec. These contained fractured and non fractured starch granules ranging from 5-10 μm in diameter, which were uniformly distributed and embedded within a smooth continuous pectin matrix (Fig. 3B). The second type included films made from mixtures gelatinized for 30 to 60 sec. In these, starch granules or recognizable remnants were rarely found. Instead, the films were composed of a homogeneous granular matrix that often split unevenly along the planes of fracture (Fig. 3C). The third type of organization, present in films made from mixtures gelatinized for 75 to 105 sec, was characterized by a uniform and close-packed distribution of 1 μm diameter particles embedded in a homogeneous smooth matrix [Fig 3D]. Fracture planes

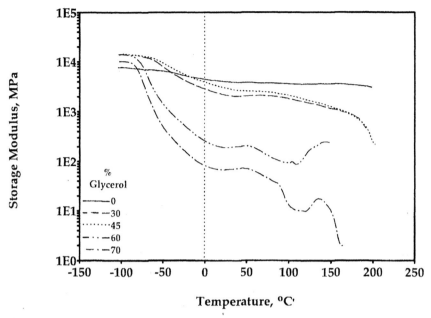

*Figure 1. Effect of glycerol level and temperature on the storage modulus of pectin films (9).* (Reproduced from reference 9. Copyright 1996.)

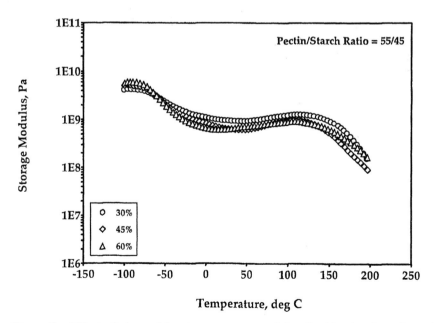

*Figure 2. Effect of glycerol level on storage modulus for blends of pectin and high amylose starch (6).* (Reproduced from reference 6. Copyright 1995.)

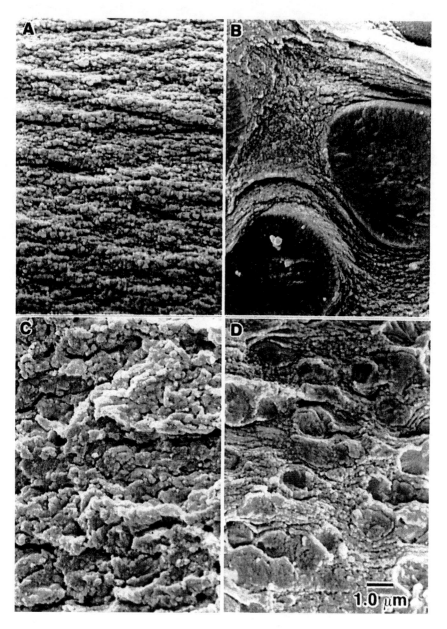

*Figure 3. Fracture surfaces of (A) pectin film; (B) P/S/G film 10 sec. gelatinization; (C) P/S/G film 45 sec. gelatinization; (D) P/S/G film, 105 sec. gelatinization (9).* (Reproduced from reference 9. Copyright 1996.)

appeared to go through the starch granule. These observations indicate a strong level of adhesion at the pectin-starch interface.

Recently, Figure 3D has been cited as evidence that starch was not completely gelatinized in our experiments (19). Nevertheless Figure 3C shows no evidence of intact starch granules. Furthermore, if one places starch granules on a heated optical microscope stage, one observes that that the starch granules gradually disappear, but eventually after cooling, the starch retrogrades into 1 μm particles comparable to those in Figure 3D. It was also suggested in that same reference that pectin and starch films are not real composites. The basis for this statement is the findings from atomic force microscopy which indicate that pectin/starch coatings are not completely miscible. This is the criteria for compatible blends (20). A composite is a polymeric material reinforced with a filler (21). In (9) we suggest that "these films (i.e. starch/pectin/glycerol films) are highly compatible starch in pectin composites." The basis for this statement is that there is great adhesion between starch granules and pectin (see Figure 3B) and the addition of starch improves the thermomechanical properties of pectin/glycerol films (see Figure 2). Furthermore we were unable to fabricate free standing films from pure high amylose starch.

Measurement of the tensile properties of the pectin/starch/glycerol films showed a significant effect of glycerol content (9). The Young's modulus of the films decreased dramatically, from about 2,800 MPa to about 100 MPa as the glycerol content was increased from 0 to 60%. Elongation to break increased from about 2% to around 25-30% over the same range, with the elongation increasing somewhat more rapidly above 30% glycerol. Tensile strength went through a maximum at 30% glycerol, with values around 35 MPa. At 30% glycerol, Young's modulus and tensile strength were increased by the addition of starch to pectin/glycerol films whereas elongation to break decreased. Tensile properties were found to be sensitive to humidity levels, with samples tested at 0% RH having modulus and tensile strength values more than twice those of samples tested at 52% RH.

FT-IR spectra (9) of pectin/starch/glycerol (P/S/G) films were taken to provide information concerning the microstructure around ester and carboxylic acid side chains in the backbone of pectin. Pectin, unlike starch and glycerol, exhibits infrared absorption bands in the regions near 1610 and 1740 cm$^{-1}$. For pectin dissolved in $D_2O$, absorption bands occur at about 1730, 1740, and 1607 cm$^{-1}$. These arise from the carboxylic acid C=O stretching band, the methyl ester C=O stretching band, and the antisymmetric COO$^-$ stretching band, respectively. The carboxylic acid band overlaps the ester band sufficiently that these appear as one broad band (22).

Figure 4 shows FT-IR spectra in the 1500-1800 cm$^{-1}$ spectral region for P/S/G films with components in the ratio 55:45:0 and 22:18:60. Two broad bands were observed, one with a peak in the 1609-1616 cm$^{-1}$ range, and one in the 1746-1750 cm$^{-1}$ range. Because the broad peak in the lower wavenumber

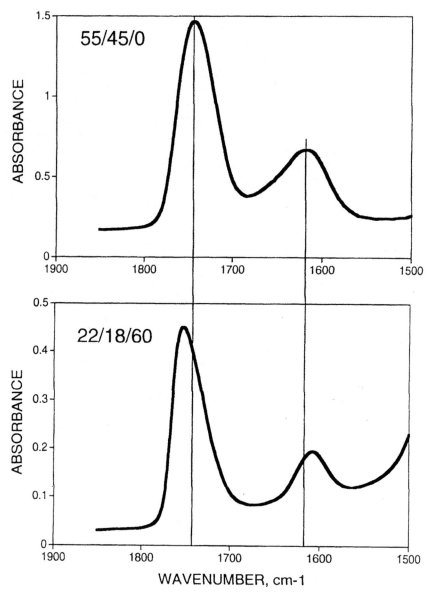

*Figure 4. Typical FT-IR spectra for P/S/G films at two compositions* (Reproduced from reference 9. Copyright 1996.)

range is attributable to carboxylate ions, the films were analyzed for metal ions by x-ray analysis. None were found, indicating their absence above 1000 ppm. Possibly a significant portion of the carboxyl protons are hydrated, thereby forming ion pairs with carboxylate ions.

Plots of wavenumber against temperature for films gelatinized for three different times, 10, 60, and 90 seconds revealed that the wavenumber for the carboxylate band increased about 5 cm$^{-1}$ from 25 to 150°C, whereas the methylester band decreased by 1 cm$^{-1}$ over the same range. In the case of metal salts of glucuronic acid, a configurational isomer of galacturonic acid, it has been demonstrated that the carboxylate ion forms stronger hydrogen bonds than the neutral acid (23). Since it is well established that increasing temperature breaks hydrogen bonds, the breaking of carboxylate hydrogen bonds must be related to the increase in wavenumber.

Plots of wavenumber against % glycerol revealed that increasing glycerol from 0 to 80% decreased the carboxylate band by 8 cm$^{-1}$, and increased the methylester band by 10 cm$^{-1}$. These changes were essentially independent of starch content. We hypothesize that increasing the amount of glycerol increases the number of pectin-glycerol bonds, some of these at the expense of pectin-pectin hydrogen bonds. This interpretation is consistent with dissociation of pectin by glycerol that we observed previously (24, 25).

Because these films are possible packaging films for food, their oxygen permeability is of considerable importance. Thus, oxygen permeability was measured on several films. The results are given in Table I, and the differences among them are considered to be significant. The values of 1.2 to 3.7 mL/m$^2$/day are considerably better than the values of 40-50 mL/m$^2$/day for poly(ethylene terephthalate) and 1,500-10,000 mL/m$^2$/day for polyethylene found in the literature (26). The addition of starch to the films resulted in a small further reduction of the oxygen permeability, while an increase in the glycerol level increased the permeability somewhat.

**Table I. Oxygen Permeability Data of MexPec1400/Amylomaize VII Blends**

| Pectin/Starch Ratio | O$_2$ Permeability | |
|---|---|---|
| | % Glycerol | (mL/m$^2$/day) |
| 100:0 | 30 | 1.7 |
| 80:20 | 30 | 1.2 |
| 100:0 | 50 | 3.7 |
| 80:20 | 50 | 2.2 |

Solubility studies were carried out on pectin/starch/glycerol (9). The time required for complete dissolution of a 1 cm$^2$ film was measured at various temperatures for several pectin/starch ratios. Glycerol content was constant at

30%. Dissolution time decreased with increasing temperature. At 30°C, the sample with the 55:45 pectin/starch ratio dissolved somewhat more slowly than the other three materials, possibly as a result of the greater insolubility of the starch component.

Recently, mixtures of pectin, starch and glycerol (PSG) were extruded rather than solution cast into films in an effort to reduce the cost of film fabrication (*11*). SEM images revealed that the temperature profile in the extruder (TP) and the amount of water present during extrusion could control the degree to which starch was gelatinized. TDMA revealed that moisture and TP during extrusion, and by inference that the amount of starch gelatinization had little affect on the mechanical properties of PSG films. TDMA also revealed that extruded films underwent the same thermal transitions as cast PSG films.

## Pectin/Poly(vinyl alcohol) Films

Six different poly(vinyl alcohol) samples were mixed with high methoxyl pectin to evaluate the effectiveness of pairing these two polymers. Two different molecular weights, nominally 50,000 and 124,000, were used. These were obtained with nominal degrees of ester hydrolysis of 87%, 98%, and 99+%. Blends with the pectin were made containing 10, 30, and 50% PVOH by weight with each sample. Each sample was tested using TDMA over the temperature range of –100 to +200°C at 10°C/min using the procedures described previously (*5, 6*) to determine the size and temperature of the thermal transitions present.

Pure PVOH films also were analyzed by DSC to further characterize the transitions (*8*). Each film exhibited a broad transition by DSC. The transition commenced at 19-23°C and ended at 63-66°C for 50,000 molar mass samples, and commenced at 20-27°C and ended at 57-60°C for the 124-186,000 molar mass samples. Thus the transition for the lower molar mass materials was somewhat broader than the thermal transition observed for the 124-186,000 molar mass materials. No thermal transitions were noted for the pure pectin up to 200°C.

TDMA of pure PVOH (*8*) revealed the presence of a glass transition which commenced in the vicinity of 5°C. $T_g$ was marked by a large and relatively sharp drop in E' and in the loss modulus (E") with temperature (*8*). All of the blends showed a distinct glass transition temperature attributable to the PVOH. The size of the transition was proportional to the amount of PVOH in the blend. The general features of the thermograms were similar for both molecular weights and for all three degrees of ester hydrolysis. Figure 5 is a plot of E" against temperature for PVOH with a 50,000 molecular weight and 99+%.

From an academic and practical point of view, an interesting feature of pectin/PVOH films is the ability to fabricate free standing films with all possible combination of pectin and PVOH (*10*). If we consider the blended film to be a pectin film modified by PVOH, the temperature at which the glass transition of the blend occurred was found to decrease as the amount of PVOH in the blend increased. Alternatively, if we consider the blend to be a PVOH film modified by pectin, then the addition of pectin to the PVOH resulted in an increase in the glass transition temperature of PVOH. In general, the glass transition temperature of PVOH in the blended films was found to be near 45°C when it was present at the 10% level by weight in the pectin. In pure PVOH films the $T_g$ was about 5°C as measured by the peak in the E" curve. The change in $T_g$ with composition for pectin/PVOH mixtures was essentially linear. Degree of ester hydrolysis appeared to have little effect on the $T_g$.

Shown in Figure 6 is the effect of glycerol on E" against temperature of a 70/30 blend of pectin and 99+% hydrolyzed PVOH with a molecular weight of 124-186,000. Little effect was observed when 5% glycerol by weight was added. Above this level, both the storage (not shown) and loss modulus were reduced with increasing glycerol concentration. The effect of glycerol on pectin/ PVOH films was comparable to glycerol on pectin/starch films (*10*).

In Figure 7 is summarized the effects of glycerol on the storage modulus of various pectin/PVOH ratio films (*10*). The PVOH is 99+% hydrolyzed and has a molecular weight range of 124 to 186,000. A blend of 70% pectin and 30% PVOH has a slightly lower modulus than pectin alone, indicating a possible plasticizing effect of PVOH on pectin. Such would be the case if pectin-PVOH interactions were lower than pectin-pectin interactions. Addition of 30 wt % glycerol to the pectin lowers its storage modulus somewhat more than does PVOH, and clearly introduces a glass transition, as indicated by the change in the slope of the modulus with temperature. A film with 49% pectin, 21% PVOH, and 30% glycerol gave a thermogram which more closely resembles PVOH than pectin. It appears from this curve that the plasticizing effects of PVOH and glycerol on pectin are synergistic

To better understand the distribution of pectin and PVOH in film blends, scanning electron micrographs of fracture surfaces of the films were obtained at 10,000X (*10*). Figure 8A reveals that the fracture plane of the pure pectin film was uneven and consisted of small ridges and crevices which were oriented parallel to the film plane. By way of comparison, in Figure 8E the fracture plane of pure PVOH (98% hydrolyzed, 124-186,000 mol. wt.) appears smooth except for a few small linear ridges which may represent irregularities in the fracture planes. Figure 8B-D reveal that the matrix of the fracture plane becomes increasingly smoother with increasingly lower ratios (wt: wt) of pectin to PVOH (i.e. 7:3, 1:1, 3:7 respectively). In Figure 8D asymmetric objects are visible, many of which are completely separated and appear to have left holes in the plastic matrix in which they were imbedded prior to fracture. These objects have a broad distribution of sizes, and appear to be up to 2-3 µm in length and 0.3 µm in width. Crevices or holes appear in all fracture plane images except

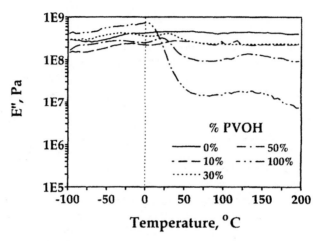

*Figure 5. Effect of PVOH content on loss modulus of films using 50,000 mol. wt PVOH with 99+% degree of hydrolysis*
(Reproduced from reference 8. Copyright 1996.)

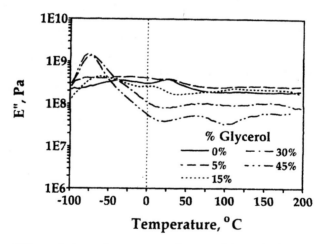

*Figure 6. Effect of glycerol content on loss modulus of 70:30 pectin/PVOH (124-186,000 mol. wt., 99+% degree of hydrolysis) films*
(Reproduced from reference 8. Copyright 1996.)

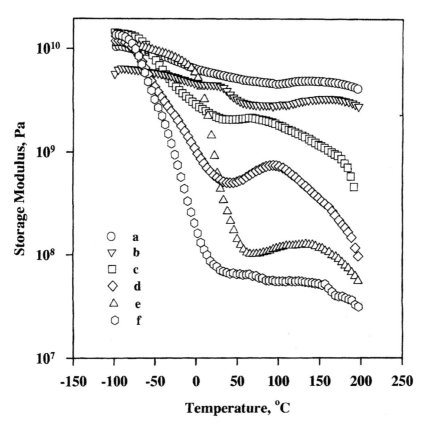

*Figure 7. Effect of glycerol and PVOH on storage modulus of films. (a) Pectin; (b) 70% pectin/30% PVOH; (c) 70% pectin/30% glycerol; (d) 49% pectin/21% PVOH/30% glycerol; (e) PVOH; (f) 70% PVOH/30% glycerol*
(Reproduced from reference 10. Copyright 1998.)

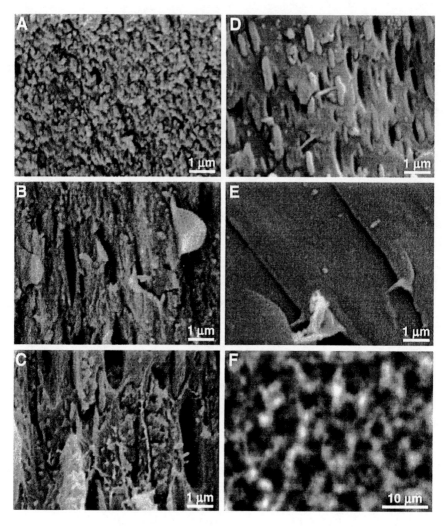

*Figure 8. SEM of freeze-fracture surfaces. (A) pectin; (B) 70:30 pectin/PVOH; (C) 50:50 pectin/ PVOH; (D) 30:70 pectin/PVOH; (E) PVOH; (F) phase contrast image, 50:50 pectin/PVOH*

(Reproduced from reference 10. Copyright 1998.)

for PVOH (Fig. 8E). Comparison of Figure 8A with Figure 8D reveals that these asymmetric structures have a smoother surface than the pectin structures in Figure 8A. This may indicate that these structures are pectin coated with PVOH. A set of micrographs made from comparable blends containing 30% glycerol were very similar.

A plausible interpretation of these fracture surface images is that pectin/PVOH mixtures form compatible composites at all ratios. Thus at high pectin:PVOH ratios they should behave as a pectin matrix with islands of PVOH, while the reverse would be expected at low pectin:PVOH ratios. These films for the most part appear transparent to the naked eye which is a necessary, but not sufficient condition for compatibility in polymer blends (27). The SEM images appear to dictate against the possibility of these films being true blends.

The pure PVOH films are flexible at room temperature and above, while the pure pectin films are brittle over a wide temperature range. Figure 9 shows the elongation to break of pectin/PVOH films made with the high molar mass PVOH over a range of compositions. Data were obtained at 52% RH (Figure 9A) and 15% RH (Figure 9B). At 52% RH, elongation increased gradually up to 50% PVOH, and then rapidly increased with further addition of PVOH. The behavior was similar at 15% RH, although the large increase did not begin until above 70% PVOH. Essentially the same pattern was observed with materials containing lower molar mass PVOH. This rapid change in flexibility above a certain composition probably signals the composition range at which the film changed from a pectin-filled composite to a PVOH-filled composite and is marked by a brittle to ductile transition in the mechanical properties of the films. The difference in the composition at which this occurred is most likely a result of the plasticizing effect of absorbed moisture on the composite.

In Figure 10, we have plotted Initial modulus (IM) against % PVOH in composite films containing low (Figure 10a) and high (Figure 10b) molar mass PVOH with different degrees of hydrolysis (10). These complex curve shapes were obtained by fitting the data to $4^{th}$ order polynomials. The $r^2$ values for all curves were greater than 0.96 indicating excellent "goodness of fit". Overall, IM decreased with % PVOH in the film. Furthermore, changes in curve slopes provide information concerning self and cross polymeric interactions. For example, local maxima at the high or low end of the curve could be interpreted as resulting from strong self interactions. In cases where two relative maxima are visible, the one with the higher relative maxima would be the stronger self interaction. Often minimums occur in the middle range of compositions. In those cases it would appear that pectin-pectin or PVOH-PVOH interactions would be weaker than pectin-PVOH interactions.

The dissolution rate of pectin/PVOH films with 30% or less PVOH was approximated by zero order kinetics (10). Activation energies were about 3-5 kcal/mole. At temperatures above 70°C, all combinations of pectin/PVOH films

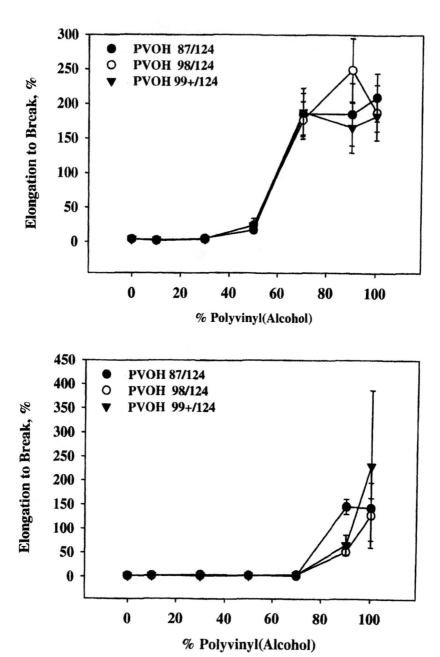

*Figure 9. Elongation to break of pectin/PVOH films. (A) 52% RH; (B) 15% RH (Reproduced from reference 10. Copyright 1998.)*

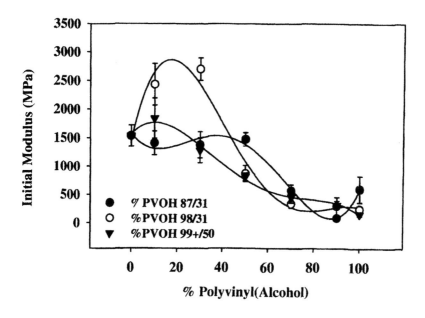

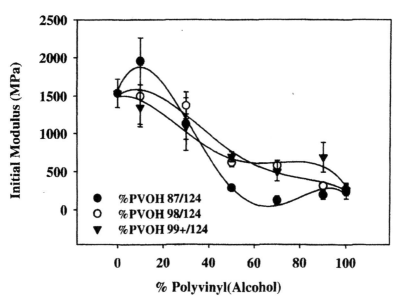

*Figure 10. Initial modulus of pectin/PVOH films. (A) lower molar mass films; (B) higher molar mass films* (Reproduced from reference 10. Copyright 1998.)

dissolved. Below 50°C only composite films with 70% or more pectin dissolved.

## Conclusions

These studies have shown that plasticized and unplasticized blends of pectin and starch, and pectin and poly(vinyl alcohol) can be made into strong, often fairly flexible films with tensile strengths on the order of 30-50 MPa, approximating those of commercial plastic films. Elongation to break can range from 1-2% to 25% in pectin/starch films to as much as 100-150% in blends with PVOH. Tensile strengths increased significantly for films tested at very low humidity levels. The dynamic mechanical properties of these films are strongly composition dependent. Addition of glycerol as a plasticizer imparts a sub-ambient glass transition temperature to the films. At higher glycerol concentrations it may impart a second order transition due to the motion of whole molecules. Both the storage modulus and loss modulus of the films are lowered by the addition of glycerol but the films remain flexible. However, as the level of high amylose starch in the films is increased, this plasticizing effect is lessened. The morphology of the pectin/starch films is strongly affected by the gelatinization conditions for the starch, although the effect on the mechanical properties is substantially less. Hydrogen bonding in the pectin/starch films is strongly affected by both temperature and glycerol content as shown by wavenumber shifts for the ester and carboxylate peak of the pectin. Blending of pectin with various poly(vinyl alcohol) polymers enabled the introduction of a brittle to ductile transition in the pectin with increasing PVOH levels. Both tensile properties and dynamic mechanical properties are strongly affected by the amount of PVOH in the blend, with the PVOH imparting a glass transition or melting behavior to the pectin in the range of 5-45°C. The pectin/starch films were rapidly soluble in water, with the rate of solubility increasing with increasing temperature. The pectin/PVOH showed similar solubility behavior, however below 70°C they were only soluble with up to 30% PVOH in the blend. Pectin/starch and pectin/starch/glycerol films exhibited a very low level of oxygen permeability compared to synthetic polymer films. The properties of these films we have studied compare very favorably to synthetic polymers in many cases and have a good potential to find significant uses in commercial applications.

## References

1. Henglein, F. A.; Schneider, G. *Chem. Ber.* **1936**, *69B*, 309-324.
2. Schultz, T. H.; Owens, H.S.; Maclay, W.D. *Colloid. Sci.* **1948**, *3*, 53-62.

3. Schultz, T. H.; Miers, J. C.; Owens, H. S.; Maclay, W. D. *J. Phys. Colloid Chem.* **1949**, *53*, 1320-1330.
4. Hind, J.; Hopkins, W. U.S. Patent 4,129,134, 1978.
5. Coffin, D. R.; Fishman, M. L. *J. Agric. Food Chem.* **1993**, *41*, 1192-1197.
6. Coffin, D. R.; Fishman, M. L. *J. Appl. Polym. Sci.* **1994**, *54*, 1311-1320.
7. Coffin, D. R.; Fishman, M. L.; Cooke, P. H. *J. Appl. Polym. Sci.* **1995**, *57*, 663-670.
8. Coffin, D. R.; Fishman, M. L.; Ly, T. V. *J. Appl. Polym. Sci.*, **1996**, *61*, 71-79.
9. Fishman, M. L.; Coffin, D. R.; Unruh, J. J.; Ly, T. *J. Macromol. Sci.-Pure Appl. Chem.* **1996**, *A33*, 639-654.
10. Fishman, M. L.; Coffin, D. R. *Carbohyd. Polym.* **1998**, *35*, 195-203.
11. Fishman, M. L.; Coffin, D. R.; Konstance, R. P.; Onwulata, C. I. *Carbohyd. Polym.* **2000**, *41*, 317-325.
12. Otey, F. H.; Westhoff, R. P.; Russell, C. R. *Ind. Eng. Chem. Prod. Res. Dev.* **1977**, *16*, 305-308.
13. Otey, F. H.; Westhoff, R. P.; Doane, W. M. *Ind. Eng. Chem. Prod. Res. Dev.* **1980**, *19*, 592-595.
14. Otey, F. H.; Westhoff, R. P.; Doane, W. M. *Ind. Eng. Chem. Res.* **1987**, *26*, 1659-1663.
15. Wolff, I. A.; David, H. A.; Cluskey, J. E.; Gundrum, L. J.; Rist, C. E. *Ind. Eng. Chem.* **1951**, *43*, 915-919.
16. Sperling, L. H. *Introduction to Physical Polymer Science*, Wiley: New York, 1992, p. 329.
17. Fishman, M. L.; Chau, H. K.; Hoagland, P. D.; Ayyad, K. *Carbohyd. Res.* **2000**, *323*, 126-138.
18. Fishman, M. L.; Chau, H. K.; Coffin, D. R.; Hotchkiss, A. T. In *Advances in Pectin and Pectinase Research*; Voragen, F; Schols, H.; Visser, R., Eds.; Kluwer Academic Publishers: Boston, MA, 2003; pp 107-122.
19. Dimantov, A; Kesselman, E.; Shimoni, E. *Food Hydrocolloids* **2004**, *18*, 29-37.
20. Krause, S. In *Polymer Blends*; Paul, D.R.; Newman, S., Eds.; Academic Press: New York, 1978; Vol. 1, pp. 16-113.
21. Chang, W. V.; WijayArathna, B.; Salovey, R. In *Polymer Blends and Composites in Multiphase Systems*; Han, C.D., Ed.; Advances in Chemistry Series 206, American Chemical Society: Washington, DC, 1984; pp. 233-259.
22. Bociek, S. M.; Welti, D. The quantitative analysis of uronic acid polymers by infrared spectroscopy. *Carbohyd. Res.* **1975**, *42*, 217-226.
23. Tajmir-Riahi, H.-A. *Carbohyd. Res.* **1984**, *125*, 13-20.
24. Fishman, M. L.; Cooke, P.; Levaj, B.; Gillespie, D. T.; Sondey, S. M.; Scorza, R. *Arch. Biochem. Biophys.* **1992**, *294*, 253-260.
25. Fishman, M. L; Cooke, P.; Hotchkiss, A.; Damert, W. *Carbohyd. Res.*, **1993**, *248*, 303-316.

26. Kaplan, D. L., et al., In *Biodegradable Materials and Packaging*, Ching, C.; Kaplan, D. L.; Thomas, E. L., Eds.; Technomic Publishing Co.: Lancaster, PA, 1993; pp 1-42.
27. MacKnight, W. J.; Karasz, F. E.; Fried, J. R. In *Polymer Blends*, Paul, D.R.; Newman, S., Eds.; Academic Press: New York, 1978; Vol. 1, pp 185-242.

## Chapter 10

# Synthesis of Zein Derivatives and Their Mechanical Properties

Atanu Biswas, David J. Sessa, Sherald H. Gordon, John W. Lawton, and J. L. Willett

Plant Polymer Research Unit, National Center for Agricultural Utilization Research, Agricultural Research Service, U.S. Department of Agriculture, 1815 North University Street, Peoria, IL 61604

Zein is a naturally occurring protein polymer, obtained as a product of industrial corn processing. It could be possibly used as a coating, ink, fiber, adhesive, textile, chewing gums, cosmetic and biodegradable plastics. Thus, we sought to develop a methodology to chemically modify the zein structure so that zein mechanical properties can be manipulated. A method to prepare acyl derivatives of zein was developed. Zein was dissolved in dimethyl formamide (DMF) and acylated with anhydrides and acid chlorides. The reactions were done by reacting zein DMF solution with anhydrides at 70C or at room temperature with acid chloride. The amine/ hydroxyl group of zein reacted to form ester/amide link. The structure was confirmed with proton NMR and IR spectra. These acetyl, benzoyl and butaryl amide/ester derivatives of zein were compression molded and their mechanical properties were measured. This study provided structure/ mechanical property relationships for these derivatives. It was found that chemical modifications by acetylation did have little impact on the mechanical properties. We also used dianhydrides to crosslink zein. At low level, i.e. 2-5% of dianhydrides zein showed significant improvement in tensile strength. When dianhydrides were used at higher level the products were crosslinked and insoluble.

© 2005 American Chemical Society

Zein *(1,2)* the alcohol soluble protein from corn may well be a commercially valuable product because it can be spun into fibers and it possesses ability to form tough adherent films. In the past, zein fibers *(3)* were used for garments, hats and other commercial applications, such as coatings. However, zein's chemical inertness and globular structure make molding articles difficult. Yet, it has not been possible to manipulate the structure of zein to make it more moldable or obtain favorable properties. One reason is the alcohol water that is commonly used as solvent is reactive towards most reagents. There have been reports for crosslinking zein in alcohol/water solutions with cyanuric chloride, formaldehyde, carbodiimide—and others *(4)*. However, DMF is one the very few non-reactive solvent for zein and reactions of zein in DMF solutions has not been studied. While our work was in progress Wu et. al. reported the synthesis of zein/nylon copolymer in DMF *(5,6)*. A method to prepare zein acetate *(7)* was invented in order to increase the water resistance, strength, and flexibility of zein films, coating or other bodies. However, this method is useful for making acetyl derivative only i.e. zein acetate. It would be desirable to devise a general method to modify zein structure.

Our objective was to develop a general method to prepare any amide or ester derivatives of zein. Zein has free alcohol and amine groups that are capable of reacting with anhydrides or acid chlorides. The common solvents for zein, such as alcohol/water mixtures prohibit such reactions because these solvents will react with anhydride and acid chloride. To circumvent this problem, we dissolved zein in dimethylformamide *(8)* and this clear solution was subjected to various reactions. Thus, our primary objective is to synthesize various esters/amides of zein by reacting with acid anhydrides and acid chlorides. Our second objective is to mold these derivatives into bars and the mechanical properties were determined.

## Discussion of Results

Reactions of zein: The reported chemical reactions of zein have been limited to zein dissolved in water or water/ alcohol mixture *(9)*. As these solvents themselves are reactive towards electrophilic reagents, not many reactions have been reported. We found that at 50°C a solution of greater than 50% could be obtained in DMF. This opened up the possibilities of wide variety of known reactions of hydroxy and mostly secondary amine functional groups to be applied to zein (Figure 1).

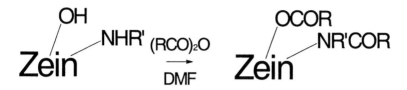

Figure 1 Examples of potential chemical derivatives of zein

We have made acyl, benzoyl and butyryl esters/amides of zein. These products were washed with ethyl acetate to remove unreacted acid chlorides or anhydrides. The NMR of these confirmed the acylation. When dianhydride, 1,2,4,5-benzenetetracarboxylic dianhydride (BTCD), was reacted in various levels, slightly crosslinked to highly crosslinked products were obtained. The highly crosslinked product gelled out of the DMF solution. However, the slightly crosslinked *(10,11)* zein was still soluble in DMF and we could make bars out of them by compression molding.

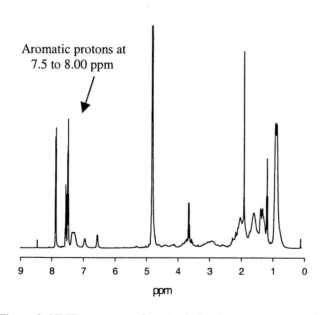

Figure 2 NMR spectrum of a zein derivative. The arrow points out the aromatic protons of benzoate group.

The NMR spectrum for zein in Figure 2 showed aromatic peaks in the region of 7.5-8.0 ppm., representing aromatic protons of benzoate group. The IR spectrum of the zein derivative in Figure 3 showed evidence of the ester linkage as a pronounced shoulder at 1731 cm$^{-1}$ on the amide I peak of the zein. Evidence of the reaction of benzoic anhydride with the zein also appeared at 714 cm$^{-1}$ from the mono-substituted benzene group in the derivative.

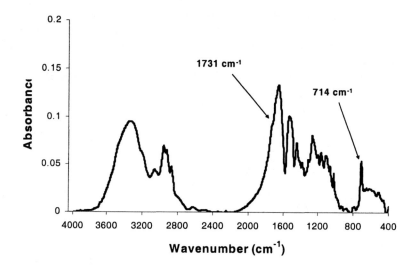

Figure 3 IR spectrum of zein derivative. Shoulder at 1731 cm$^{-1}$ of the amide peak of zein is evidence of an ester linkage.

Mechanical Properties of Native and Chemically-Modified Zein: Mechanical strength measurement was performed to evaluate the bulk properties of the tensile bars from both native and chemically-modified zeins. Triethylene glycol (TEG) is an excellent plasticizer *(12)* for zein. Native and chemically modified zeins were each blended with either 10 or 15% TEG in a Haake Rheocord 90 torque rheometer equipped with high shear roller blades to yield a taffy like matrix. The rheometer was stopped when the torque dropped and became constant.

Whether a longer residence time in the torque rheometer would impact on the mechanical properties of the compression-molded tensile bars has yet to be determined. The taffy-like blend was snipped into small pieces which were frozen with liquid nitrogen and ground in a Wiley mill. This ground mass was

Table 1: Tensile Properties of Native and Chemically Modified Zein[a]

| Treatment | % Plasticizer | % Moisture | % Elongation ± SD | Tensile Strength (MPa) ± SD |
|---|---|---|---|---|
| None | 10 | 5.3 | 18.4 ± 2.7 | 27.8 ± 1.5 |
| Benzoate | 10 | 4.3 | 25.5 ± 12.5 | 20.8 ± 6.8 |
| Benzoate | 15 | 3.8 | 25.7 ± 4.0 | 28.3 ± 2.1 |
| Benzoate | 15 | 7.4 | 18.9 ± 2.0 | 21.6 ± 3.3 |
| Acetate | 15 | 8.2 | 15.2 ± 2.5 | 17.3 ± 0.1 |
| Butyrylate | 15 | 5.6 | 46.6 ± 33.8 | 21.1± 3.3 |
| None | 15 | 6.2 | 24.3 ± 2.8 | 23.9 ± 0.2 |
| BTCD (0.25%)[b] | 15 | 5.0 | 27.2 ± 6.1 | 37.4 ± 0.9 |
| BTCD (0.50%)[b] | 15 | 5.0 | 23.1± 2.8 | 37.6 ± 1.0 |
| BTCD (1.00%)[b] | 15 | 5.2 | 18.0 ± 4.8 | 26.5 ± 1.7 |
| BTCD (2.00%)[b] | 15 | 5.0 | 19.7 ± 3.3 | 31.5 ± 0.8 |
| BTCD (3.00%)[b] | 15 | 4.9 | 19.7 ± 2.3 | 32.1 ± 1.0 |

[a]Samples stored at 50% RH for one week
[b]BTCD= 1,2,4,5-benzene tetracarboxylic dianhydride

sieved through 30 mesh screen to remove fines; samples retained on the sieve were compression molded into tensile bars. The grinding step may break down any network structure formed during the processing with the torque rheometer. Therefore, the mechanical properties of the tensile bars generated should reflect the properties related to the packing structure of the native derivatized zeins.

Mechanical properties of the native and chemically-modified zeins are given in Table 1. The chemical modification used will either add a side chain or cross-link the zein, either of which should increase the free volume (i.e. space between free molecules) of the protein. In general, an increase in the free volume of the macromolecule should result in a plasticized mass with increased elongation and decreased tensile strength. Zein that was acetylated had lower elongation and tensile strength despite an increased moisture content of 8.2% when compared with unmodified zein similarly processed. The benzoate derivative possessed similar elongation and slightly higher tensile strength than did unmodified zein. When concentration of derivatizing agent was increased to 20% both mechanical properties dropped below the unmodified zein. When zein is cross-linked with 1,2,4,5-benzene tetracarboxylic dianhydride, labeled BTCD, the sample with 0.25% BTCD gave no change in the % elongation but did show a significant increase in tensile strength of 37.4MPa compared with 23.9MPa for unmodified zein. With BTCD, this reactant becomes part of the zein macromolecule. Amounts above 0.25% gave diminished % elongations as well as tensile strengths. Despite the diminished tensile strengths observed they were all significantly higher than those obtained with unmodified zeins. Cross-linking zein with BTCD gave significant changes in tensile strength that merit further investigation.

## Experimental

Zein was obtained from Freeman Industries. The acid chlorides, anhydrides and the solvents were obtained from Aldrich Chemical. NMR solutions were prepared by warming a mixture of 10 mg zein esters and 10 mg of 40% sodium deuteroxide in one ml of deuterium oxide (NaOD and $D_2O$ were obtained from Cambridge Isotope Laboratories, Inc., Andover, MA).

NMR spectra were obtained by using Bruker Instruments DRX 400 spectrometer. FTIR spectra of samples pressed in KBr disks were measured on an FTS 6000 FTIR spectrometer (Digilab, Cambridge, CT) equipped with a DTGS detector. The absorbance spectra were measured at 4 $cm^{-1}$ resolution, signal-averaged over 32 scans and baseline corrected.

Mechanical Properties were measured with Haake Rheocord 90 equipped with high sheer roller blades; Carver Press, Model C with ASTM D 638 type V bar; Instron Model 4201.

Reaction procedure: Six gms. of zein was dissolved in 18 gms. of DMF while stirring with a magnetic stirrer at room temperature. To this solution, 6.5 gms. of benzoic anhydride (28 mmole) was added. After 15 minutes, 2.2 gms. (28 mmole) of pyridine was added. The reaction mixture was heated to 50°C,

stirred for 4 hours, subsequently cooled to room temperature and poured into 250 ml of water. The product, along with the excess reactants precipitated as a solid. It was filtered, washed with hot water and ground in a blender. As the DMF dissolved in water the pasty solid turned into a yellow powder, which was filtered and washed with dilute hydrochloric acid to remove pyridine. Finally the solid was washed three times with hot ethyl acetate to remove any organics such as benzoic acid or pyridine. We obtained 6.5 gms. of product as yellow solid. The other derivatives such as acetate, butyrate were prepared similarly. When acid chlorides were used they were added at 0°C and the reactions were done at room temperature.

Preparation of Tensile Bars by Compression Molding: As stated in Discussion of Result section blends of TEG with either native or chemically modified zein were prepared with a Haake Rheocord 90 torque rheometer. The taffy like matrix were frozen with liquid nitrogen and ground in a Wiley mill. The powder was molded into tensile bars with a Carver Press, Model C at temperatures 175°C for native zein and 240°C for chemically modified zein and pressures up to 10,000 lbs. for 20 minutes. The resulting tensile bars were conditioned for 1 week at 23°C and 50% relative humidity prior to testing mechanical properties with an Instron Universal tester of a crosshead speed of 1cm/minute. Spent tensile bars were evaluated for moisture by heating at 105°C for 4 hours in a forced draft oven.

## Conclusions

We have developed a simple method to prepare amide/ester of zein. Zein was dissolved in dimethylformamide in 25% concentration and in this solution the free hydroxyl and amine groups of zein readily reacted with electrophilic reagents such as anhydrides. Thus, we were able to prepare acyl, benzoyl, butyroyl amides/esters of zein. We used di-anhydrides to crosslink zein, thus increasing the strength and stability of articles made from it. Crosslinked zein could be useful for coating, ink, adhesives *(13)* and cross-linking zein with other di functional organic molecules merit further investigation. In contrast, the mechanical properties of acylated zein derivatives did not show much improvement over zein itself.

## Acknowledgements

The authors would like to thank Janet Berfield, Luke Neal and Benjamin Rocke for their excellent technical assistance and David Weisleder for NMR analyses.

## References

1. Lawton, J.W. *Cereal Chem.* **2002**, 79, 1-18.
2. Cheryan. M.; Shukla R. *Ind. Crops Prod.*, **2001**, 13, 171-192.
3. Sturken, O.U.S. Patent 2,361,713, **1944**.
4. Veatch, C. U.S. Patent 2,236,768, **1941**.
5. Wu, Q.X.; Sakabe. H.; Isobe, S. *Polymer*, **2003**, 44, 3901-3908.
6. Wu, Q.X.; Yoshino,T.; Sakabe,H.; Zhang,H.K.; Isobe,S. *Polymer*, **2003**, 44, 3909-3919.
7. Veatch, C. U.S. Patent 2,236,768, **1941**.

8. Danzer, L. A.; Rees, E. D. *Can. J. Biochem.,* **1976**, 54, 196-199.
9. Hagemeyer, Jr., H. J. U.S. Patent 2,401,685, **1946**.
10. Howland, D.W.; Reiner, R.A. *Paint and Varnish Prod.,* **1962** 52, 31-35, 78.
11. Pelosi, L F. U.S. Patent 5,596,080, **1997**.
12. Lawton, J.W. *Cereal Chem.,* **2004**, 81(1),1-5.
13. Satow, S. U. S. Patents 1,245,976; 1,245,981; 1,245,977; **1917**.

# Silicon-Containing Materials

## Chapter 11

# Protein-Mediated Bioinspired Mineralization

Siddharth V. Patwardhan[1,2,6], Kiyotaka Shiba[3,4], Christina Raab[5], Nicola Huesing[5], and Stephen J. Clarson[1,*]

[1]Department of Chemical and Materials Engineering, University of Cincinnati, Cincinnati, OH 45221
[2]Department of Materials Science and Engineering, University of Delaware, Newark, DE 19711
[3]Department of Protein Engineering, Cancer Institute, Japanese Foundation for Cancer Research and [4]CREST/JST, Toshima, Tokyo 170–8455, Japan
[5]Institute of Materials Chemistry, Vienna University of Technology, Getreidemarkt 9, A-1060 Vienna, Austria
[6]Current Address: Division of Chemistry, School of Science, Nottingham Trent University, Nottingham NG11 8NS, United Kingdom
*Corresponding author: Email: Stephen.Clarson@UC.Edu; Fax: 1–513–556–3473

Biomineralisation, due to its remarkable sophistication and hierarchical control in mineral 'shaping', is profoundly inspiring for the design of novel materials and new processes. Recent investigations have begun to elucidate the specific interactions of various biomolecules with their respective minerals *in vivo*. In order to better understand the roles of such (bio)molecules in (bio)mineralisation, various *in vitro* model systems have been examined recently. Here we report the bioinspired mineralisation of silica in the presence of a lysine and arginine rich α-helical synthetic protein YT320 that was tailor-made using genetic engineering. Furthermore, in order to examine the possible specificity (or perhaps the lack thereof) of this protein with silica, we also report studies of bioinspired mineralisation of germania. It is proposed that this protein facilitates and 'guides' the mineralisation through residue specific and conformationally directed interactions at the molecular level.

## Introduction

Biomineralisation[#] involves a complex combination of genetically controlled processes carried out *in vivo* that lead to the formation of highly ornate hierarchical biomineral structures that are highly species specific [1]. As an example, Figure 1 shows the fascinating beauty of patterned biosilica frustules of the diatom *Aulacoseira granulata*. The remarkable elegance of Nature is further illustrated by the fact that these processes typically occur under ambient conditions, unlike the corresponding laboratory methods where harsh conditions are often employed. Recent studies of the precise control of structure formation as seen in biomineralisation are beginning to reveal the possible role(s) of biomacromolecules at the molecular level [2-6]. For example, proteins have been isolated from the biogenic silica of grasses [3], sponges [4a], and diatoms [5] and were characterised. These proteins have been shown to facilitate silicification *in vitro* and thus it has been proposed that they are responsible for facilitating biosilicification in their respective species. These and many other

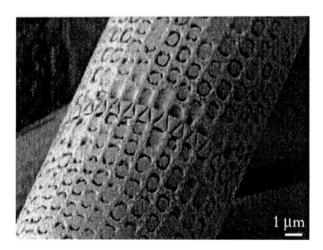

*Figure 1. SEM image of the ornate nanostructured biosilica of the diatom Aulacoseira granulata. The diatom samples were collected by Professor Miriam Steinitz-Kannan (Northern Kentucky University) from the Ohio River (USA) on August 7, 2002. Bar = 1 µm.*

findings have led to the development of bioinspired *in vitro* model systems where the role(s) of several (bio)macromolecules in mineralisation have been investigated. Such research provides insight into biomineralisation, allows for the development of novel materials, and leads to the discovery of new material processing technologies. Indeed, the interactions of various polypeptides, polyamino acids, block copolypeptides, block copolymers and polymers with various minerals in the process of mineralisation have been documented in the cases of systems such as barium-, calcium-, lead- and silicon-based minerals [4b,7-25]. As to the role of (bio)macromolecules in bioinspired and biomimetic mineralisation, various mechanisms have been proposed [5,8,19-21,24,25]. However, the exact nature of the interactions, such as any correlation between the properties of (bio)molecules and (bio)mineral morphogenesis; and any specificity between (bio)molecules and corresponding (bio)minerals, are still unclear and thus constitute a major scientific challenge.

In the case of biosilicification, proteins that are thought to direct biosilica deposition were found to contain high compositions of amino acids in the primary sequence carrying basic substituents [3a,5]. These residues are able to interact with the growing biosilica. *In vitro* studies with basic polyamino acids (e.g. polylysine and polyarginine) were found to affect the silicification kinetics and the structure of the resulting silica [8,9,14,19-21,23]. To date very little is known about the exact structure and behaviour of proteins isolated from living organisms that deposit biominerals and thus a clear and complete understanding of the mechanisms behind mineralisation remain to be unveiled.

pYT320
MRGSHHHHHHSSGWVD
PENLQAE
RKVLQGRMENLQAE
RKVLQGRMENLQAE
RKVLQGRMENLQAE
RKVLQGRMENLQAE
RKVLQGRMENLQAEP
QSIAGSYGKPASGG
MW    12.3
pI    9.86
alpha-rich in CD

*Figure 2. Amino acid primary sequence of the YT320 protein.*

In this research, we have carefully and intentionally chosen to investigate the use of a lysine and arginine rich protein YT320 of known and well defined secondary structure (Figure 2). In particular, we report the bioinspired mineralisation of silica in the presence the protein YT320. As it is known that germanium can be incorporated into biogenic silica; that the chemical characteristics of germanium and its resemblance to silicon have been described previously; and that it competes with silicon in biosilicification[26], it is our hypothesis that the various (bio)macromolecules that facilitate (bio)silicification may also influence related mineral formation *in vitro*, as shown in Scheme 1. Furthermore, it should be noted that although naturally occurring proteins [27], model peptides (e.g. the R5 peptide derived from the diatom *Cylindrotheca fusiformis* [5,7,15]), polyamino acids [8,9,14], and block copolypeptides [4b] have been studied in their respective *in vitro* model systems, the *in vitro* mineralisation utilising genetically engineered synthetic proteins has not been studied to date.

## Experimental

The protein YT320 was prepared by tandem polymerisation of the 42-bp long microgene, MG-15, whose first coding frame had similarity to a segment of the coiled coil α-helix of the natural protein (seryl-tRNA synthetase). The

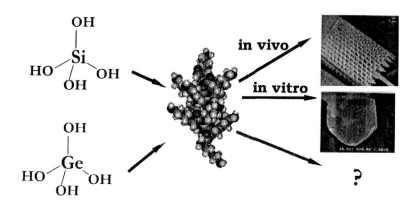

*Scheme 1. Silica formation and structure control has been observed in vivo (right top) and in vitro (right center); and the role(s) of (bio)macromolecules is being studied. Due to similarities in the sol-gel chemistry of silica and germanium dioxide, can the same (bio)macromolecules be used for structure formation of $GeO_2$?*

protein possessed a calculated pI of 9.86 and thus it is cationically charged at circumneutral pH (i.e. under the reaction conditions employed here). The protein YT320 has 5.5 repeats of the α-helix forming frame of the microgene, comprising about 80% of the protein and showed a typical α-helix-rich spectrum in Circular Dichroism (CD) analysis [28a]. Small Angle X-ray Scattering analysis has shown that protein YT320 does not fold as tightly as natural proteins but is rather more compact than if completely denatured. The protein has been crystallised, indicating repetitious artificial proteins can undergo a transition to a more ordered state by choosing appropriate conditions [28b].

Tetramethoxysilane (TMOS), used as a silica precursor, was obtained from Sigma-Aldrich. Water glass or sodium silicate (≥27% as $SiO_2$) was purchased from Riedel-de Haën. Ethylene glycol modified silane (EGMS) was synthesised as described below. For the preparation of EGMS, tetraethoxysilane (TEOS) was mixed with a four-fold molar excess of ethylene glycol and heated to $150^0$ C under argon. Continuous distillative removal of ethanol resulted in the formation of a clear oily liquid product [29]. This novel water soluble precursor does not require any catalyst for hydrolysis and condensation. In addition, ethylene glycol released upon hydrolysis of EGMS does not appear to be detrimental for biomolecules [30]. Precursors used for germania synthesis were germanium(IV) ethoxide and germanium(IV) isopropoxide and they were obtained from Sigma-Aldrich. Potassium phosphate buffer (pH 7) was used to maintain the reaction mixture at neutral pH and was obtained from Fisher.

Bioinspired silica formation assay was performed in the presence of the YT320 protein, starting with several silica precursors as listed in Table 1. Upon mixing a precursor solution with a protein solution of the desired concentration and buffering the reaction content to pH 7, silica precipitation was observed. The synthetic procedures were similar to those we have previously described [13,19,20]. Briefly, in potassium phosphate buffer (pH 7), 5 mg/ml of the YT320 protein solution (in the same buffer) was added and then 1 M pre-hydrolysed TMOS solution was added. The volume ratios of buffer: protein solution: silica precursor solution was 8:2:1 in the case of TMOS based synthesis. For the experiments under externally applied shear, the TMOS based reaction mixture was stirred for 5 minutes. The product was then centrifuged, washed and dried.

**Table 1. Silica precursors used for the silicification reactions.**

| Silica Precursors | Structure |
|---|---|
| Tetramethoxysilane (TMOS)[a] | $Si(OCH_3)_4$ |
| Water glass (sodium silicate)[a] | $Na_2Si_3O_7$ |
| Ethylene glycol modified silane (EGMS) | (structure) |

[a] TMOS and water glass were pre-hydrolysed.

When EGMS was used as the silica precursor, 3 µL EGMS was directly added to a mixture of 20 µL of a 5 mg/ml YT320 protein solution and 80 µL buffer. In the case of water glass as the silica precursor, 1 ml water glass was hydrolysed in 1 mM HCl before use. 10 µL of this water glass solution was mixed with 20 µL protein solution (5 mg/ml) and was then buffered.

After 5 minutes of reaction time in each case, the solution was centrifuged and washed with DI water three times. A few drops of washed solutions were placed on sample holders in each case and left to dry under ambient conditions. The dried samples were either coated by evaporation of gold-palladium alloy or sputter coated with gold and they were further characterised by Field Emission-Scanning Electron Microscope (FE-SEM) for product morphology, which was also equipped with an Energy Dispersive Spectroscopy (EDS) facility for elemental analysis. In the control experiments without using the protein, no silica precipitation was observed even over a 24 hour period, as was described previously [13].

For germania synthesis, 2 µL of either germanium (IV) ethoxide or germanium (IV) *i*-propoxide was directly added to a mixture of 20 µL of a 5 mg/ml YT320 protein solution and 80 µL buffer. The mixture was left for 5 minutes to react and was then washed and characterised as described above. No precipitation was observed in control experiments carried out without any protein.

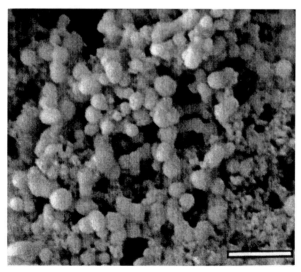

*Figure 3. Representative SEM images of YT320 mediated silica particles synthesised using pre-hydrolysed TMOS under static conditions. Bar = 1 µm.*

*Figure 4.* Elongated silica structures using pre-hydrolysed TMOS and YT320 formed upon subjecting the medium to shear for five minutes. Highlighted areas from (a) and (b) are presented at higher magnifications in (b) and (c) respectively. Bar = 1 μm.

## Results and Discussion

To our knowledge this is the first attempt demonstrating the use of a genetically engineered synthetic protein for mineralisation of silica and germania. The results obtained for each respective system are presented and discussed below.

The SEM analysis of the silica particles revealed that spherical or sphere-like particles were successfully synthesised (Figure 3). In the case of the tetramethoxysilane (TMOS) based system, well-defined spherical silica particles were observed with average sizes of approximately 300 nm. Upon subjecting the reaction mixture to the externally applied shear, ellipsoidal (Figure 4), fibre-like (Figure 5a, b) and fused (see arrowhead in Figure 5c) particles were seen in addition to spherical particles. The formation of non-spherical elongated morphologies as observed herein is in accordance with our previous investigations for related systems [7,22]. The (bio)macromolecule conformation in solution may be responsible for the formation of such elongated structures [10,11]. When ethylene glycol modified silane (EGMS) was used as the silica precursor, similar to the TMOS based system, formation of approximately 300 nm sized spherical silica particles were synthesised (see Figure 6). Formation of

*Figure 5. Fiber-like silica structures using pre-hydrolysed TMOS and YT320 formed upon subjecting the medium to shear for five minutes. Highlighted area in (a) is presented at higher magnification in (b). The arrowhead in (c) indicates the formation of 'fused' structures. Bar = 1 µm.*

interconnected networks were observed instead when water glass was used as the silica precursor. The formation of silica was further confirmed by silicon and oxygen signatures as observed in Energy Dispersive Spectroscopy (EDS) (Figure 6). The incorporation of YT320 into silica was also verified by the carbon and nitrogen peaks in the EDS. The aluminium and magnesium peaks correspond to the sample holder while the phosphorous peak may be due to the buffer used.

In order to extend our investigations on the role(s) of (bio)macromolecules in bioinspired mineralisation, we have studied the *in vitro* mineralisation of germania in presence of several (bio)macromolecules. The YT320 protein (reported herein) was found to facilitate germania particle formation (Figure 7). It is noted that no precipitation was observed in the absence of the protein. The elemental analysis confirmed the presence of germanium and oxygen indicating the formation of germania. The carbon and nitrogen signatures suggest the occlusion of the protein. In addition, a synthetic polymer (polyallylamine hydrochloride), that has been previously studied for silica precipitation, was also found to direct the formation of germania *in vitro*, thus further supporting our hypothesis [31]. In a recent study, Morse and co-workers have shown that the silicatein protein isolated from the silica spicules of sponge *Tethya aurantia* was able to catalyse the formation of titania [32]. It was further reported that silicatein may stabilise anatase – a polymorph of titania that is typically formed only at higher temperatures in the absence of silicatein protein. Their findings on silicatein mediated titania synthesis and the data presented in this paper on the

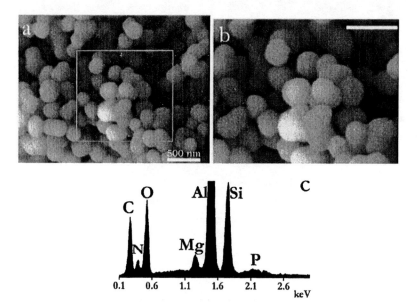

*Figure 6. (a), (b) SEM and (c) EDS images of YT320 mediated silicification using EGMS as the silica precursor. The highlighted area in (a) is presented at higher magnification in (b). Bar = 500 nm.*

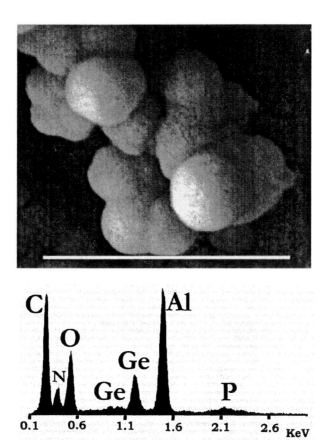

*Figure 7. Representative SEM micrograph and EDS spectrum for germania prepared by YT320. Bar = 1μm.*

role of a genetically engineered synthetic protein in the formation of silica and germania demonstrates the ability of such biomolecules to be tailored for mineralisation and thus proven to be technologically important, as hypothesized previously [20].

Metal alkoxides undergo polymerisation via hydrolysis and condensation to produce oligomers. These oligomers then grow in size and they act as nuclei. Particle formation occurs due to the growth in size of these nuclei either by precipitation of soluble species (monomers and dimers) or by coagulation of insoluble particles. At circumneutral pH, the particles possess a net negative surface charge. As mentioned above, the protein YT320 possessed a pI of 9.86

and thus it is cationically charged at pH 7 - the reaction conditions employed here. It is thus proposed that the protein under consideration here interacts electrostatically with the minerals at molecular level. It is interesting to note that only some biomolecules have an effect on mineralisation, while many others lack such ability as demonstrated by Livage and co-workers recently [33]. Hence the role to YT320 in successfully facilitating mineralisation as presented herein is of particular importance.

The activity of YT320 protein in the *in vitro* bioinspired mineralisation is proposed to be due to the organisation of the protein molecules in solution. This spatial arrangement is then followed by nucleation, catalysis and/or scaffolding of the inorganic structures around the protein molecules leading to phase separation as described in detail elsewhere [5,8,19-21,24-25]. This model further predicts that the three-dimensional organisation of such (bio)molecules in solution may govern the structures of the respective (bio)minerals which may be in turn controlled by chemical structure of the (bio)molecules. *In vivo*, as the primary sequence of proteins (i.e. chemical structure) and their structure and functions are genetically controlled, the formation of species-specific biomineral structures is presumably genetically governed. It should be kept in mind at this point that this simple model for the interactions of (bio)molecules and (bio)minerals is only a first approximation. Various parameters affecting biomineralisation *in vivo* such as the presence of other (macro)molecules, pH and concentration (chemical potential) gradients have yet to be fully considered. The results discussed herein on silica and germania synthesis in the presence of a carefully tailored protein supports our earlier hypothesis [20] that the effect of (bio)macromolecules on *in vitro* mineralisation can be somewhat generic rather than being specifically limited to a particular mineral system.

Various proteins of known sequence and structure are currently being investigated for their role in mineralisation. In addition, further studies pertaining to the solution behaviour of such (bio)molecules, their interactions with inorganic species at the molecular level and their exact roles (catalytic or scaffolding or both) in mineralisation are currently being undertaken [34].

## Conclusions

We report here for the first time, an investigation of a genetically engineered protein in mediating the bioinspired mineralisation of silicon-based and germanium-based systems. Ionic and hydrogen-bonding interactions between the organic and inorganic molecules in addition to chain conformations and self-assembly of the protein are thought to be key in such bioinspired mineralisation.

## Acknowledgements

We would like to thank Professor Miriam Steinitz-Kannan for kindly providing us with the diatom sample. S. V. P. thanks Professor Kristi L. Kiick and Professor Carole C. Perry for various helpful discussions. K. S. thanks Ms. Tamiko Minamizawa for her technical assistance.

## References

\# It should be noted that the word 'biomineralisation' is only used to indicate mineralisation *in vivo* and should not be confused with mineralisation *in vitro* using biomacromolecules. The later – *in vitro* mineralisation – can be appropriately categorised as 'bioinspired' or 'biomimetic' mineralisation.
1. Lowenstam, H. A. *Science* **1981**, *211*, 1126; Lowenstam, H. A.; Weiner, S. *On Biomineralization*; Oxford University Press: New York, 1989; Simkiss, K.; Wilbur, K. M. *Biomineralization*; Academic Press: San Diego, 1989; Baeuerlein, E., Ed. *Biomineralization*; Wiley-VCH: Chichester, 2000; Muller, W. E. G., Ed. *Silicon Biomineralization*; Springer Verlag, 2003.
2. Addadi, L.; Weiner, S. *Proc. Natl. Acad. Sci. USA* **1985**, *82*, 4110.
3. C Harrison, C. C. *Phytochemistry* **1996**, *41*, 37; (b) Perry, C. C.; Keeling-Tucker, T. *Colloid Polymer Science* **2003**, *281*, 652.
4. Shimizu, K.; Cha, J.; Stucky, G. D.; Morse, D. E. *Proc. Natl. Acad. Sci. USA* **1998**, *95*, 6234; (b) Cha J. N.; Stucky, G. D.; Morse, D. E.; Deming, T. J. *Nature* **2000**, *403*, 289.
5. Kroger, N.; Deutzmann, R.; Sumper, M. *Science* **1999**, *286*, 1129; Kroger, N.; Lorenz, S.; Brunner, E.; Sumper, M. *Science* **2002**, *298*, 584; Sumper, M. *Science* **2002**, *295*, 2430.
6. Arakaki, A.; Webb, J.; Matsunaga, T. *J. Biol. Chem.* **2003**, *278*, 8745.
7. Whitlock, P. W.; Naik, R. R.; Brott, L. L.; Clarson, S. J.; Tomlin, D. W.; Stone, M. O. *Polymer Preprints* **2001**, *42*, 252; Naik, R. R.; Whitlock, P. W.; Rodriguez, F.; Brott, L. L.; Glawe, D. D.; Clarson, S. J.; Stone, M. O. *Chem. Commun.* **2003**, *2*, 238; Brott, L. L.; Pikas, D. J.; Naik, R. R.; Kirkpatrick, S. M.; Tomlin, D. W.; Whitlock, P. W.; Clarson, S. J.; Stone, M. O. *Nature* **2001**, *413*, 291; Naik, R. R.; Brott, L. L.; Clarson, S. J.; Stone, M. O. *J. Nanosci. Nanotech.* **2002**, *2*, 95.
8. Coradin, T.; Livage, J. *Colloids Surf. B* **2001**, *21*, 329; Coradin, T.; Durupthy, O.; Livage, J. *Langmuir* **2002**, *18*, 2331; Coradin, T.; Lopez, P. J. *ChemBioChem* **2003**, *4*, 251.

9. Sudheendra, L.; Raju, A. R. *Mater. Res. Bull.* **2002**, *37*, 151; Mizutani, T.; Nagase, H.; Fujiwara, N.; Ogoshi, H. *Bull. Chem. Soc. Jpn.* **1998**, *71*, 2017.
10. Slocik, J. M.; Wright, D. W. *Biomacromol.* **2003**, *4*, 1135.
11. Sone, E. D.; Zubarev, E. R.; Stupp, S. I. *Angew. Chem. Int. Ed.* **2002**, *41*, 1706.
12. Patwardhan, S. V.; Clarson, S. J. *Polym. Bull.* **2002**, *48*, 367.
13. Patwardhan, S. V.; Mukherjee, N.; Clarson, S. J. *Silicon Chemistry* **2002**, *1*, 47.
14. Patwardhan, S. V.; Clarson, S. J. *Silicon Chemistry* **2002**, *1*, 207.
15. Patwardhan, S. V.; Clarson, S. J. *J. Inorg. Organomet. Polym.* **2002**, *12*, 109.
16. Patwardhan, S. V.; Clarson, S. J. *Mat. Sci. Eng. C.* **2003**, *23*, 495.
17. Patwardhan, S. V.; Clarson, S. J. *J. Inorg. Organomet. Polym.* **2003**, *13*, 49.
18. Patwardhan., S. V.; Durstock, M. F.; Clarson, S. J. In *Synthesis and Properties of Silicones and Silicone-Modified Materials*; Clarson, S. J.; Fitzgerald, J. J.; Owen, M. J.; Smith, S. D.; Van Dyke, M. E., Eds.; Oxford University Press, 2003; Vol. 838, pp 366-374.
19. Patwardhan, S. V.; Ph.D. Dissertation, Department of Materials Science and Engineering, University of Cincinnati, 2003.
20. Patwardhan, S. V.; Mukherjee, N.; Steinitz-Kannan, M.; Clarson, S. J. *Chem. Commun.* **2003**, *10*, 1122.
21. Patwardhan, S. V.; Clarson, S. J. *J. Inorg. Organomet. Polym.* **2003**, *13*, 193.
22. Patwardhan, S. V.; Mukherjee, N.; Clarson, S. J. *J. Inorg. Organomet. Polym.* **2001**, *11*, 117.
23. Patwardhan, S. V.; Mukherjee, N.; Clarson, S. J. *J. Inorg. Organomet. Polym.* **2001**, *11*, 193.
24. Vrieling, E. G.; Beelen, T. P. M.; van Santen, R. A.; Gieskes, W. W. C. *Angew. Chem. Int. Ed.* **2002**, *41*, 1543.
25. Yu, S.-H.; Colfen, H.; Hartmann, J.; Antonietti, M. *Adv. Functional Mater.* **2002**, *12*, 541; Yu, S.-H.; Colfen, H.; Antonietti, M. *J. Phys. Chem. B.* **2003**, *107*, 7396; Faul, C. F. J.; Antonietti, M. *Adv. Mater.* **2003**, *15*, 673.
26. Azam, F.; Volcani, B. E. In *Silicon and Siliceous Structures in Biological Systems*; Simpson, T. L.; Volcani, B. E., Eds.; Springer-Verlag: New York, 1981.
27. Coradin, T.; Coupe, A.; Livage, J. *Colloid Surf. B* **2003**, *29*, 189.
28. Shiba, K.; Takahashi, Y.; Noda, T. *J. Mol. Biol.* **2002**, *320*, 833; (b) Shiba, K.; Shirai, T.; Honma, T.; Noda, T. *Protein Engn.* **2003**, *16*, 57.

29. Mehrotra, R. C.; Narain, R. P. *Indian J. Chem.* **1966**, *9*, 431; Huesing, N.; Raab, C.; Torma, V.; Roig, A.; Peterlik, H. *Chem. Mater.* **2003**, *15*, 2690.
30. Gill, I.; Ballesteros, A. *J. Am. Chem. Soc.* **1998**, *120*, 8587.
31. Patwardhan, S. V.; Clarson, S. J. in preparation.
32. Sumerel, J. L.; Yang, W. J.; Kisailus, D.; Weaver, J. C.; Choi, J. H.; Morse, D. E. *Chem. Mater.* **2003**, *15*, 4804.
33. Coradin, T.; Coupe, A.; Livage, J. In *Biological Biomimetic Materials-Properties to Function*; Aizenberg, J.; McKittrick, J. M.; Orme, C. A., Eds.; MRS Symp Proc, 2002; Vol. 724, pp 147-152.
34. Patwardhan, S. V.; Shiba, K.; Belton, D.; Perry, C. C.; Clarson, S. J. Results presented at the Silicones and Silicone Modified Materials Symposium. American Chemical Society National Meeting, Anaheim, California, USA, March 31st, 2004 and manuscript in preparation.

## Chapter 12

# Biocatalysis of Siloxane Bonds

**Alan R. Bassindale[1], Kurt F. Brandstadt[2,*], Thomas H. Lane[2], and Peter G. Taylor[1]**

[1]Department of Chemistry, The Open University, Milton Keynes, United Kingdom
[2]Dow Corning Corporation, Midland, MI 48686

The intricate siliceous architectures of diatom species have inspired our exploration of biosilicification. In order to better understand the role of various proteins in the biosilicification process, a carefully chosen model study was performed to test the ability of homologous enzymes to catalyze the formation of molecules with a *single* siloxane bond during the *in vitro* hydrolysis and condensation of alkoxysilanes. This model study is believed to be the first rigorous study to demonstrate biocatalysis at silicon. Our data suggests that homologous lipase and protease enzymes catalyze the formation of siloxane bonds under mild conditions. In particular, the active site of trypsin, a proteolytic enzyme, was determined to selectively catalyze the *in vitro* condensation of silanols. Conversely, the reverse reaction was not favored. Furthermore, trypsin as well as several other proteins and polypeptides promoted the hydrolysis of alkoxysilanes in a non-specific manner. Given the selectivity and mild reaction conditions of enzymes, the opportunity to strategically use biocatalysts to synthesize novel hybrid materials with structural control and spatial order is promising.

Silicon is often regarded as an insignificant element in our carbon-based world. Few are aware that silicon is the second most abundant element in the crust of the earth *(1)* or that it is essential for growth and biological function in a variety of plant, animal, and microbial systems *(2,3)*. Silicate-based minerals account for more than 90% of all known terrestrial specimens *(4)*, which are actively involved in the global silica cycle. Biosilicification occurs on a globally vast scale *(5)* under mild conditions (e.g. neutral pH, low temperature) *(6)*. In fact, minute planktonic algae (diatoms) control the marine silica cycle. These single cell plants process an estimated 200-280 teramoles or gigatons of particulate silica every year *(7)*. Given the complex design and diversity of the siliceous skeletons of diatoms *(2)*, these organisms are considered paragons of nanostructural architecture *(8)*. The intricate siliceous architectures of diatom species have inspired the exploration of silica biosynthesis.

The study of the hydrolysis and condensation reactions during biosilicification is complicated due to the sensitivity of silica, silicates, and silicic acid to pH, concentration, and temperature *(2)*. As a silicic acid precursor, tetraalkoxysilanes are easily hydrolyzed and the resulting silanols are condensed during the formation of particulate silica. Silicatein (i.e. a protease isolated from the *Tethya aurantia* marine sponge) was documented *(9)* to catalyze the *in vitro* polycondensation of tetraethoxysilane as well as phenyl- and methyl-triethoxysilanes under mild conditions during the formation of siloxane precipitates. Based on site-directed mutagenesis results *(10)*, the enzymatic active site of silicatein (i.e. Ser-His-Asn) was determined to catalyze the hydrolysis and/or condensation of the alkoxysilanes during the biosilicification reactions. In contrast, polypeptides isolated from the *Cylindrotheca fusiformis* diatom (i.e. silaffin) and *Equisetum telmateia* plant (i.e. biopolymer) as well as biomimetic analogues (e.g. polyamines) were reported to catalyse the *in vitro* hydrolysis *(11,12)* or condensation *(11-17)* reactions during the formation of silica. Consequently, it has been shown an active site in a defined tertiary structure is not necessarily required. The data suggests that the reactions were catalyzed by non-specific peptide interactions with nucleophilic, basic, and cationic amino acid residues.

In order to better understand the role of various proteins in the biosilicification process, a carefully chosen model study was performed to test the ability of homologous enzymes to catalyze the formation of siloxane bonds during the *in vitro* hydrolysis and condensation of alkoxysilanes under mild conditions (Scheme 1). Given the complications of silicic acid analogues *(2)*, mono-functional silanes were chosen to focus on the formation of molecules with a *single* siloxane bond. It was understood that the *in vitro* biocatalyzed reactions might not be comparable to the natural *in vivo* reactions.

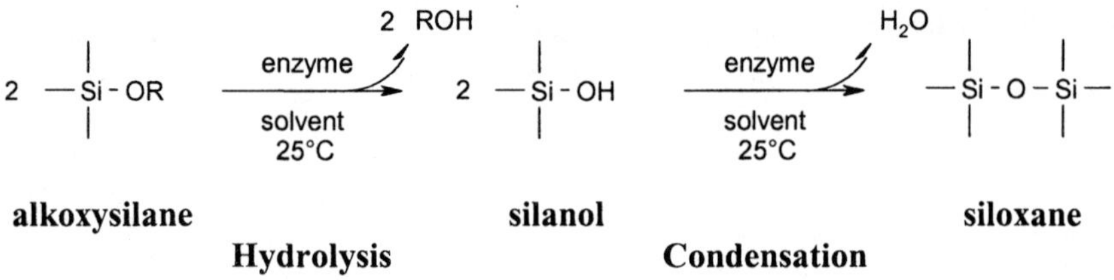

$$2 \overset{|}{\underset{|}{\text{Si}}}-\text{OR} \xrightarrow[\substack{\text{solvent}\\25°\text{C}}]{\text{enzyme},\ 2\ \text{ROH}} 2 \overset{|}{\underset{|}{\text{Si}}}-\text{OH} \xrightarrow[\substack{\text{solvent}\\25°\text{C}}]{\text{enzyme},\ \text{H}_2\text{O}} \overset{|}{\underset{|}{\text{Si}}}-\text{O}-\overset{|}{\underset{|}{\text{Si}}}$$

alkoxysilane                silanol                siloxane
            **Hydrolysis**            **Condensation**

***Scheme 1.*** *Enzyme-catalyzed siloxane bond formation.*

## Background

Previously, a protein isolated from the *Tethya aurantia* marine sponge (i.e. silicatein) *(9,10)* as well as polypeptides isolated from a *Cylindrotheca fusiformis* diatom (i.e. silaffin) *(13-16)*, an *Equisetum telmateia* plant (i.e. biopolymer) *(12)*, and a phage-display library *(17)* were reported to catalyze the polycondensation of silicic acid analogues during the formation of particulate silica. Despite the absence of long-range order following nucleation of silica on silicatein filaments *(18)*, silicatein was postulated *(19)* to control the architecture of silicon-functional materials. In addition, silicatein-mimetic diblock polypeptides were reported *(11)* to control the macroscopic structure of silica. Specifically, reduced and oxidized cysteine-functional lysine copolypeptides synthesized large aggregates (~600 nm) and packed silica columns, respectively. Comparatively, cysteine and lysine polypeptide blends catalyzed amorphous silica. Based on these observations, the diblock polypeptides were documented to catalyze the polycondensation of tetraethoxysilane, while "simultaneously directing the formation of ordered silica morphologies *(11)*". In review, the *in vitro* studies *(6,9,12-16)* of natural systems within the area of silica biosynthesis are complicated. These earlier mechanistic queries including biomimetic approaches *(11,17)* often failed to recognize the chemistry of silicic acid and its analogues *(2)*.

The study of the hydrolysis and condensation reactions during biosilicification is complicated due to the sensitivity of silica, silicates, and silicic acid to pH, concentration, and temperature *(2)*. As silicic acid precursors, tetraalkoxysilanes are easily hydrolyzed and the resulting silanols are condensed during the formation of particulate silica. During the formation of particulate silica, silicic acid condenses to form low molecular weight linear and cyclic molecules, which react to form a distribution of branched and crosslinked siloxanes prior to precipitating as aggregated- or sol-gels (Scheme 2). The formation of different three-dimensional gel networks is dependent on the solution and deposition of silica due to the pH of the medium as well as the presence or absence of salts. At low pH (< 7) or in the presence of salts, the

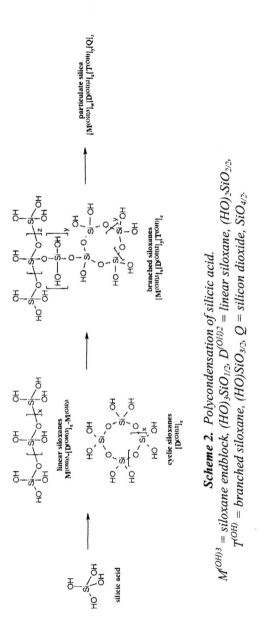

**Scheme 2.** *Polycondensation of silicic acid.*
$M^{(OH)3}$ = siloxane endblock, $(HO)_3SiO_{1/2}$, $D^{(OH)2}$ = linear siloxane, $(HO)_2SiO_{2/2}$, $T^{(OH)}$ = branched siloxane, $(HO)SiO_{3/2}$, $Q$ = silicon dioxide, $SiO_{4/2}$.

collision of spherical silica particles increases due to decreased ionic charges. This promotes the formation of aggregated silica chains and gel networks. Alternatively, the particle growth of spherical silica is enhanced due to increased dissolution and deposition under basic conditions. At pH values greater than 7 in the absence of salts, the repulsion of the negatively charged particles also limits aggregation during the formation of large spherical sol-gels. The reaction mechanisms for the acid- and base-catalyzed condensation of silanols are detailed in Scheme 3 *(20)*.

The acid-catalyzed polycondensation of tetraethoxysilane, a silicic acid precursor, results in the formation of an aggregated silica gel (Figure 1). Comparatively, the three-dimensional structure of the aggregated silica gel was similar to the natural morphologies of the silica particles catalyzed by proteins *(9)* and polypeptides *(12,13,17)*. Furthermore, the use of external force was demonstrated *(21)* to control the morphology (e.g. arched, fibrillar) of a silica material catalyzed by a biomimetic silaffin peptide (i.e. a non-modified Sil 1p R5 peptide). Different flow dynamics (e.g. gas bubbles, tubular shear) within the reaction mixture changed the reactive interfaces and, consequently, the directional growth and deposition of the material. Although silicatein was determined *(10)* to catalyze the formation of particulate silica and silsesquioxanes, it was also shown *(11-17)* that an active site in a defined tertiary structure was not necessarily required. This data suggests that the formation of silica is also catalyzed by interactions with non-specific peptides such as nucleophilic, basic, and cationic amino acid residues. Although research has progressed in the area of silica biosynthesis, the molecular mechanisms of these interactions are effectively unknown *(22)*.

## Experimental

Since mono-functional silanes were chosen as reactants to study the formation of molecules with a single siloxane bond, rigorous procedures were established to prepare glassware, as well as isolate and quantitatively analyze the reaction products by gas chromatography (GC) *(23)*. Prior to analysis, the aqueous reactions were extracted with tetrahydrofuran, while the organic reactions were directly analyzed. The GC analyses were performed on a Hewlett-Packard (Palo Alto, CA) 6890 plus gas chromatograph with a flame ionization detector. Dodecane was used as an internal standard to gravimetrically quantitate the chromatographic analyses. The samples were prepared at ~1% (w/w) product in a THF solution containing 1% (w/w) dodecane. Mass balances greater than 98% were routinely obtained. The

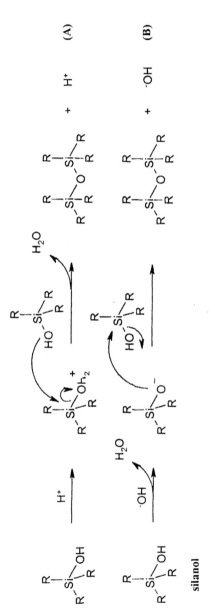

*Scheme 3.* Acid- (A) and base- (B) catalyzed silanol condensation.

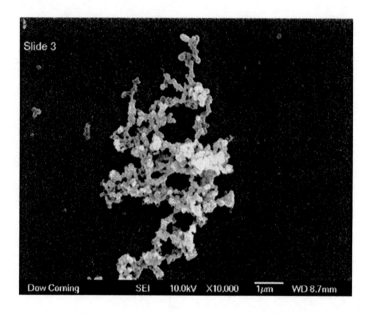

***Figure 1.*** *Acid-catalyzed formation of an aggregated silica gel.*

alkoxysilane, silanol, and disiloxane analytes were chromatographically resolved. The ability to resolve these analytes was necessary to differentiate between the role of an enzyme in the hydrolysis and condensation reactions during biosilicification. Comparatively, given the limitations of the product and resultant analyses, the *Tethya aurantia* marine sponge (i.e. silicatein) *(9)* and *Equisetum telmateia* plant (i.e. biopolymer) *(12)* studies including a silicatein-mimetic approach (i.e. diblock polypeptides) *(11)* were not able to differentiate between the role of the proteins or polypeptides in the hydrolysis and condensation reactions during biosilicification.

## Results and Discussion

Our model study *(23-25)* is believed to be the first rigorous study to demonstrate biocatalysis at silicon. This data suggests that homologous lipase and protease enzymes catalyze the formation of siloxane bonds under mild conditions.

**Enzyme-Catalyzed Condensation Study.** Based on the similarities of the proteins isolated from the *Tethya aurantia* marine sponge (silicatein), *Cylindrotheca fusiformis* diatom (silaffin), and *Equisetum telmateia* plant (biopolymer extract), a series of lipases and serine-proteases were selected as homologous proteolytic enzymes. In addition to catalyzing comparable reactions (i.e. the hydrolysis of amide and/or ester bonds *(26,27)*), the active sites in the hydrolase enzymes are composed of similar serine-histidine-aspartate catalytic triads *(28,29)*. In comparison to control reactions, mammalian, fungal, and bacterial lipases and proteases were screened with a model silanol, trimethylsilanol, in order to evaluate their ability to catalyze the formation of a siloxane bond. In this study, control reactions were defined as non-enzymatic reactions. Specifically, experiments conducted in the absence of a protein were defined as negative control reactions. Proteinaceous molecules such as bovine serum albumin and porcine γ-globulins were used to study non-specific protein catalysis. Given the broad selection of enzymes, the condensation study was performed in suitable organic and aqueous media. Toluene was chosen as a hydrophobic solvent in order to potentially promote the lipase-catalyzed interfacial reactions *(30)*. Water was selected as an alternate medium in order to evaluate the activity of solubilized enzymatic solutions. The reactions were formulated with an ~1000:1 silanol to enzyme mole ratio and conducted in inert glass vials at $25°C$ with magnetic stirring for six days. The reaction products were isolated and quantitatively analyzed by GC. The results are summarized in Table I. In comparison to negative control and non-specific protein (i.e. BSA) reactions, select lipases as well as trypsin and α-chymotrypsin were observed (√) to catalyze the condensation of trimethylsilanol during the formation of hexamethyldisiloxane (HMDS) under mild conditions. Conversely, the ability of the unchecked enzymes in Table I to catalyze the model condensation reaction was not substantially different than the control reactions. In review, the relative rate of condensation increased in water. As opposed to lipases, proteases will only interact with water soluble substrates *(31)*. The estimated water solubility of trimethylsilanol is 4.2% (i.e. 42.56 mg/mL) *(32)*.

**Table I.** Enzyme-catalyzed condensation study of trimethylsilanol after six days

| Enzyme | Dry Toluene | Wet Toluene | Water | Buffered pH 7 |
|---|---|---|---|---|
| negative control | | | | |
| bovine serum albumin | | | | |
| *Aspergillus niger* lipase | | | | |
| *Candida antarctica* lipase | | | √ | |
| *Candida antarctica* lipase B[1] | | | √ | |
| *Candida lipolytica* lipase | | | | |
| α-chymotrypsin | √ | √ | √ | √ |
| *Mucor javanicus* lipase | | | | |
| *Penicillium roqueforti* lipase | | | | |
| phospholipase A2 | | | | |
| porcine pancreatic lipase | | | | |
| *Pseudomonas cepacia* lipase | | | | |
| *Pseudomonas fluorescens* lipase | | | | |
| *Rhizomucor miehei* lipase | | | √ | |
| trypsin | √ | √ | √ | √ |
| wheat germ lipase | | √ | √ | |

[1] Candida antarctica lipase B was immobilized on acrylic resin beads (Novozyme® 435).
√ The checked enzymes catalyzed the condensation of > 10x more trimethylsilanol than the negative control reactions; as well as > 3 or 10x more trimethylsilanol than the BSA reactions in organic and aqueous media, respectively.

Reproduced with permission from reference 33. Copyright 2003 Elsevier.

**Protease-Catalyzed Condensation Study.** Based on the exceptional activity of trypsin and α-chymotrypsin from bovine pancreas (Table I), protease enzymes were identified as target catalysts. Consequently, a series of proteases (i.e. serine, cysteine, aspartic, and metallo) were selected in order to screen their ability to catalyze siloxane condensation with trimethylsilanol. The reactions were formulated with an ~1000:1 silanol to protein mole ratio and conducted in neutral media (pH 7.0) at 25°C for three hours. Comparatively, the original reactions conducted in the enzyme-catalyzed condensation study were performed for six days versus three hours in this study. Based on the estimated solubility of trimethylsilanol in water (42.56 mg/mL) *(32)*, the concentration of trimethylsilanol (~160 mg/mL) saturated the aqueous media and created two-phase reaction mixtures. Although the pH may not be optimal for the proteases *(27)*, a neutral pH was used to minimize acid- and base-catalyzed silanol condensation *(2)*. The reaction products were isolated and quantitatively analyzed by GC (Figure 2) *(25)*.

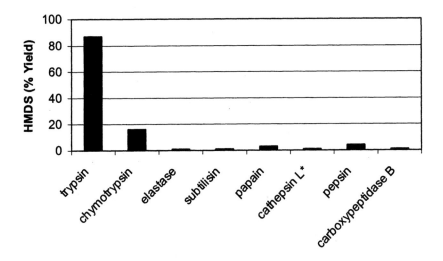

*Figure 2. Protease-catalyzed condensation study of trimethylsilanol after three hours.*

Trypsin preferentially catalyzed the condensation of trimethylsilanol under mild conditions (87% HMDS). Substantial condensation of trimethylsilanol was not observed in the negative control, non-specific protein (i.e. BSA, γ-globulins), small molecule (i.e. $CaCl_2$, imidazole, N-methylimidazole), and polypeptide (i.e. poly-L-lysine) reactions in comparison to the raw material (< 1% HMDS). Based on an impurity study with small molecule inhibitors *(23)*, the exceptional activity of trypsin and α-chymotrypsin observed in the original enzyme-catalyzed condensation study was soley due to a tryptic impurity. The tertiary and secondary structures of the digestive enzymes vary according to function. Given the specificity in the catalytic regions, the proteases are substrate selective. Notably, three sources of cathepsin L (Figure 2) did not catalyze the condensation reaction. In contrast, silicatein catalyzed the formation of particulate silica and silsesquioxanes. Although silicatein was documented to be highly homologous with cathepsin L (i.e. a cysteine-protease) (6), the marine protein lacked proteolytic activity as measured with synthetic chromogenic substrates (18). In addition, the nucleophilic residue within the enzymatic active site of silicatein (i.e. Ser-His-Asn) was determined to be serine (10). Correspondingly, the activities of the enzymes are dependent on the functionality of the non-natural organosilicon substrates.

**Trypsin-Catalyzed Hydrolysis and Condensation of Alkoxysilanes.** Based on the results of the condensation study, the role of *Bos taurus* or bovine pancreatic trypsin in the formation of molecules with a single siloxane bond was studied during the *in vitro* hydrolysis and condensation of a model alkoxysilane, trimethylethoxysilane. The reactions were formulated with an ~1000:1 alkoxysilane to trypsin mole ratio in a neutral medium (pH 7.0) and conducted at 25°C for 3 hours. Based on the estimated solubility of trimethylethoxysilane in water (1 mg/mL) (32), the concentration of trimethylethoxysilane (~160 mg/mL) saturated the aqueous media and created two-phase reaction mixtures. The reaction products were isolated and quantitatively analyzed by GC (Figure 3). Although various rates of hydrolysis were observed, substantial condensation of trimethylethoxysilane was not observed in the negative control, non-specific protein (i.e. BSA, γ-globulins), small molecule (i.e. $CaCl_2$, imidazole, N-methylimidazole), and polypeptide (i.e. poly-L-lysine) reactions in comparison to the raw material. Although BSA (9) and poly-L-lysine (11) were not observed to catalyze the condensation of tetraethoxysilane or trimethylethoxysilane (Figure 3), BSA and poly-L-lysine promoted the hydrolysis of trimethylethoxysilane and formation of trimethylsilanol at different rates in a neutral medium. In review, the trimethylsilanol and trimethylethoxysilane control reactions were comparable and highlighted the definitive scope of the model system.

In the presence of trypsin, trimethylethoxysilane was hydrolyzed (100%) and condensed (84%) during the formation of HMDS in a neutral medium at 25°C over three hours (Figure 3). Since the relative rate of condensation decreased at temperatures < 25°C (25), a time study of the trimethylethoxysilane reaction was conducted at 10°C for defined periods of time over three hours. The reaction products were isolated and quantitatively analyzed by GC (Figure 4).

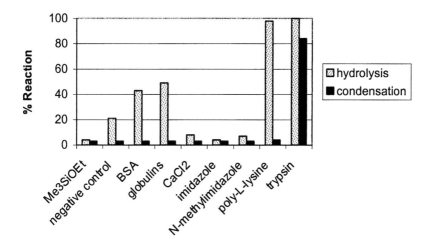

*Figure 3.* Hydrolysis and condensation control reactions of trimethylethoxysilane after three hours at 25 °C.

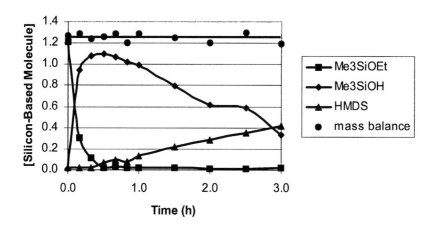

*Figure 4.* Trypsin-catalyzed hydrolysis and condensation of trimethylethoxysilane at 10 °C.

Based on the stoichiometry of the hydrolysis and condensation reactions (Scheme 1), two moles of trimethylethoxysilane were consumed in the formation of two moles of trimethylsilanol, which produced one mole of hexamethyldisiloxane (mass balance, Figure 4). The chromatographic data set acquired during the time study was analyzed in order to study the kinetics of the hydrolysis and condensation reactions catalyzed by trypsin at 10°C. Comparatively, trimethylethoxysilane was readily hydrolyzed within the initial 30 minutes and, subsequently, condensed during the formation of hexamethyldisiloxane. The turnover numbers ($k_{cat}$) in the hydrolysis (0.53 s$^{-1}$) and condensation (0.048 s$^{-1}$) reactions were calculated. Since trypsin may not be saturated due to the limited solubility of trimethylethoxysilane and trimethylsilanol in water, the turnover numbers were treated as relative values. Given relative turnover numbers equal to 0.53 and 0.048 s$^{-1}$, the time between each hydrolysis and condensation reaction catalyzed by trypsin was calculated to be 2 and 20 s at 10°C, respectively. Based on the relative turnover numbers, the rate of the trypsin-catalyzed hydrolysis of trimethylethoxysilane was one order of magnitude (ten times) faster than the condensation of trimethylsilanol at 10°C (Figure 5). Comparatively, the rate of the trypsin-catalyzed condensation of trimethylsilanol ($k_{cat}$ = 0.066 s$^{-1}$) at 25°C was approximately 38% faster than the reaction conducted at 10°C. In comparison to the maximum turnover numbers of other enzymes with their physiological substrates *(28)*, the turnover number of the trypsin-catalyzed condensation of trimethylsilanol was several orders of magnitude (i.e. 10-10,000,000) slower than the cited values.

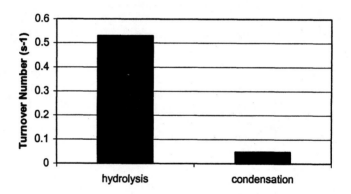

*Figure 5. Turnover numbers of the trypsin-catalyzed hydrolysis of trimethylethoxysilane and condensation of trimethylsilanol at 10°C.*

Since trypsin catalyzed the formation of siloxane bonds, alternate monofunctional alkoxysilanes were chosen as substrates to investigate the ability of trypsin to selectively catalyze the *in vitro* hydrolysis and condensation of organofunctional alkoxysilanes under mild conditions. 1,1-dimethyl-1-sila-2-oxacyclohexane and phenyldimethylethoxysilane were selected to study the activity of trypsin due to different interactions with the substrates. The two-phase reactions were formulated with a 4:1 monomer to enzyme weight ratio and conducted at 25°C for three hours. Based on the chromatographic results, trypsin catalyzed the ring-opening hydrolysis (81%) of 1,1-dimethyl-1-sila-2-oxacyclohexane and condensation (71%) of hydroxybutyldimethylsilanol during the formation of the carbinol-functional disiloxane (Scheme 4). Analogous to basic residues, the carbinol-functional silanol intermediate was hypothesized to be an acceptable substrate due to its ability to hydrogen bond with the aspartic acid residue within the binding domain *(28)* of the catalytic region of trypsin. Comparatively, phenyldimethylethoxysilane was hydrolyzed but not condensed in the presence of trypsin. The decreased enzymatic activity appeared to be due to the increased hydrophobicity and steric bulk of the phenyl-functional substrate.

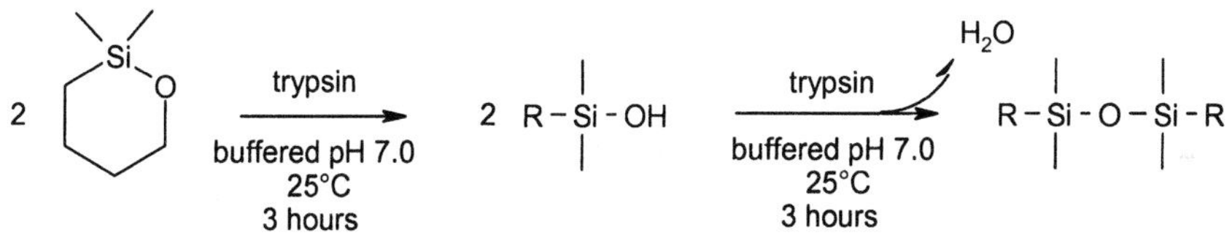

*Scheme 4. Trypsin-catalyzed hydrolysis and condensation of 1,1-dimethyl-1-sila-2-oxacyclohexane (R = hydroxybutyl).*

Previously, silicatein was determined *(9,10)* to catalyze the polycondensation of tetraethoxysilane during the formation of particulate silica. Despite fractional yields (< 0.005%) and insignificant conversions, the study focused on the analysis of the solid polycondensation products as opposed to a complete mass balance. Given the limitations of the product and the resultant analysis, the study was not able to differentiate between the role of silicatein in the hydrolysis and condensation reactions during biosilicification. Regardless, a general acid/base reaction mechanism was proposed to catalyze the hydrolysis reaction or "rate-limiting step *(9)*" during polycondensation. However, it should be noted that tetraalkoxysilanes are easily hydrolyzed and silanols are condensed during the formation of particulate silica due to their sensitivity to pH, concentration, and temperature *(2)*.

In comparison to a control reaction, trypsin reportedly did not catalyze the polycondensation of a silicic acid analogue, tetraethoxysilane, in an aqueous medium at pH 6.8 *(9)*. Similarly, trypsin did not hydrolyze or condense tetraethoxysilane in a replicate reaction formulated with a 4:1 monomer to enzyme weight ratio and conducted at 25°C for three hours.

In review, trypsin was observed to selectively catalyze the hydrolysis and condensation of some organo-functional alkoxysilanes under mild conditions.

**Inhibition Study.** A proteinaceous inhibition study was conducted to study the role of the enzymatic active site in the hydrolysis and condensation of trimethylethoxysilane. Prior to reaction, trypsin was independently inhibited with an excess amount of the Bowman-Birk inhibitor *(34)* (4:1 BBI to trypsin mole ratio) and the Popcorn inhibitor *(35)* (2:1 PCI to trypsin mole ratio) in stirred neutral media for two hours. Based on standard enzymatic activity assays *(36)*, trypsin was fully inhibited by the BBI (98%) and PCI (91%). The reactions were formulated with an ~1000:1 trimethylethoxysilane to trypsin mole ratio and conducted at 25°C for three hours. The reaction products were isolated and quantitatively analyzed by GC (Table II). Although the treated enzymes were observed to catalyze the hydrolysis of trimethylethoxysilane, the condensation of trimethylsilanol was completely inhibited in comparison to the control reactions. Notably, the rate of hydrolysis decreased in the presence of the BBI- and PCI-inhibited trypsin. Following thermal denaturation, the activity of trypsin was comparable to the proteinaceous inhibition experiments. Based on a standard enzymatic activity assay *(36)*, the relative decrease in the rate of silanol condensation correlated with the enhanced stability of trypsin at higher protein concentrations *(25)*. Consequently, it appears that non-specific interactions with trypsin including the active site promoted the hydrolysis of trimethylethoxysilane. Therefore, the active site of trypsin was determined to selectively catalyze the *in vitro* condensation of trimethylsilanol under mild conditions.

**Table II.** Proteinaceous inhibition of trypsin in the hydrolysis and condensation of trimethylethoxysilane at 25°C *(33)*.

| Reaction | % Yield | | |
|---|---|---|---|
| | $Me_3SiOEt$ | $Me_3SiOH$ | $Me_3SiOSiMe_3$ |
| $Me_3SiOEt$ | 96 | 1 | 3 |
| negative control | 79 | 18 | 3 |
| BBI | 67 | 27 | 3 |
| PCI | 64 | 30 | 3 |
| trypsin | 0 | 8 | 84 |
| trypsin + BBI | 24 | 74 | 3 |
| trypsin + PCI | 6 | 86 | 3 |
| boiled trypsin, 20 m | 1 | 64 | 35 |

Based on the estimated solubility of trimethylsilanol in water (42.56 mg/mL) *(32)*, the concentration of trimethylsilanol (~160 mg/mL) saturated the aqueous medium and created a two-phase reaction mixture. Since proteases will only interact with water-soluble substrates *(31)*, the trypsin-catalyzed condensation of trimethylsilanol was postulated to occur in the aqueous phase. Although the condensation reaction was conducted in water, the enzyme-catalyzed reaction was promoted by the phase separation of the product. The immiscibility of the product, hexamethyldisiloxane, changed the equilibrium *(37)* and promoted the condensation reaction in the presence of water. Since the aqueous medium was saturated with trimethylsilanol, the reactant would continue to enter the aqueous phase due to the dynamic equilibrium of the condensation reaction. In addition, the hydrolysis or reverse reaction would be severely hindered due to the immiscibility of the disiloxane product in the aqueous phase.

## Conclusions

Since previous studies *(6,9,11-17)* were complicated mechanistic queries due to the use of silicic acid analogues, this model study is believed to be the first rigorous study to demonstrate biocatalysis at silicon. Our data suggests that homologous lipase and protease enzymes catalyze the formation of siloxane bonds under mild conditions. In particular, the active site of *Bos Taurus* or bovine pancreatic trypsin, a proteolytic enzyme, was determined to selectively catalyze the *in vitro* condensation of silanols. Conversely, although trypsin would theoretically catalyze of the hydrolysis of a siloxane bond due to the law of microscopic reversibility, the reverse reaction was not favored. Furthermore, trypsin as well as several other proteins and polypeptides promoted the hydrolysis of alkoxysilanes in a non-specific manner. Based on the relative turnover numbers, the rate of the trypsin-catalyzed hydrolysis of trimethylethoxysilane was one order of magnitude (ten times) faster than the condensation of trimethylsilanol at $10°C$. Comparatively, the rate of the trypsin-catalyzed condensation of trimethylsilanol at $25°C$ was approximately 38% faster than the reaction conducted at $10°C$. In comparison to the maximum turnover numbers of other enzymes with their physiological substrates *(28)*, the turnover number of the trypsin-catalyzed condensation of trimethylsilanol was several orders of magnitude (i.e. 10-10,000,000) slower than the cited values. Given the selectivity and mild reaction conditions of enzymes, the opportunity to strategically use biocatalysts to synthesize novel hybrid materials with structural control and spatial order is promising.

## Acknowledgements

The authors acknowledge colleagues at the Dow Corning Corporation, The Open University, and Genencor International, Inc. (Palo Alto, CA) for their insightful consultations. Furthermore, the Dow Corning Corporation is acknowledged for their financial support.

## References

(1) Foth, H. D. *Fundamentals of soil science*; 6 ed.; John Wiley & Sons: New York, 1978.
(2) Iler, R. K. *The chemistry of silica: Solubility, polymerization, colloid and surface properties, and biochemistry*; John Wiley & Sons: New York, 1979.
(3) Voronkov, M. G. *Silicon in living systems*; Ellis Horwood Limited: Chichester, 1987.
(4) Perry, C. C.; Keeling-Tucker, T. *J Biol Inorg Chem* **2000**, *5*, 537-550.
(5) Lowenstam, H. A. *Science* **1981**, *211*, 1126-1131.
(6) Shimizu, K.; Cha, J.; Stucky, G. D.; Morse, D. E. *PNAS* **1998**, *95*, 6234-6238.
(7) Nelson, D. M.; Treguer, P.; Brzezinski, M. A.; Leynaert, A.; Queguiner, B. *Global Biogeochem. Cycles* **1995**, *9(3)*, 359-372.
(8) Mann, S. *Nature* **1993**, *365*, 499-505.
(9) Cha, J. N.; Shimizu, K.; Zhou, Y.; Christiansen, S. C.; Chmelka, B. F.; Stucky, G. D.; Morse, D. E. *PNAS* **1999**, *96*, 361-365.
(10) Zhou, Y.; Shimizu, K.; Cha, J. N.; Stucky, G. D.; Morse, D. E. *Angew. Chem. Int. Ed.* **1999**, *38(6)*, 780-782.
(11) Cha, J. N.; Stucky, G. D.; Morse, D. E.; Deming, T. J. *Nature* **2000**, *403*, 289-292.
(12) Perry, C. C.; Keeling-Tucker, T. *J. Chem. Soc. Chem. Commun.* **1998**, 2587-2588.
(13) Kroger, N.; Deutzmann, R.; Sumper, M. *Science* **1999**, *286*, 1129-1132.
(14) Kroger, N.; Deutzmann, R.; Bergsdorf, C.; Sumper, M. *PNAS* **2000**, *97(26)*, 14133-14138.
(15) Kroger, N.; Deutzmann, R.; Sumper, M. *J. Biol. Chem.* **2001**, *276(28)*, 26066-26070.
(16) Kroger, N.; Lorenz, S.; Brunner, E.; Sumper, M. *Science* **2002**, *298*, 584-586.
(17) Naik, R. R.; Brott, L. L.; Clarson, S. J.; Stone, M. O. *J. Nanosci. Nanotech.* **2002**, *2(1)*, 95-100.

(18) Cha, J. N.; Shimizu, K.; Zhou, Y.; Christiansen, S. C.; Chmelka, B. F.; Deming, T. J.; Stucky, G. D.; Morse, D. E. *Mat. Res. Soc. Symp. Proc.* **2000**, *599*, 239-248.
(19) Morse, D. E. *TIBTECH* **1999**, *17(6)*, 230-232.
(20) Eaborn, C. *Organosilicon compounds*; Butterworths Scientific Publications: London, 1960.
(21) Naik, R. R.; Whitlock, P. W.; Rodriguez, F.; Brott, L. L.; Glawe, D. D.; Clarson, S. J.; Stone, M. O. *J. Chem. Soc. Chem. Commun.* **2003**, 238-239.
(22) Morse, D. E. In *The chemistry of organic silicon compounds*; Rappoport, Z., Apeloig, Y., Eds.; John Wiley & Sons: New York, 2001; Vol. 3, pp 805-819.
(23) Bassindale, A. R.; Brandstadt, K. F.; Lane, T. H.; Taylor, P. G. *J. Inorganic Biochemistry* **2003**, *96*, 401-406.
(24) Bassindale, A. R.; Brandstadt, K. F.; Lane, T. H.; Taylor, P. G. *Polymer Preprints* **2003**, *44(2)*, 570-571.
(25) Brandstadt, K. F. Ph.D. Thesis, The Open University, Milton Keynes, England, Dec. 2003.
(26) Burtis, C. A.; Ashwood, E. R. *Clinical chemistry*; 2 ed.; W.B. Saunders: Philadelphia, 1994.
(27) Barrett, A. J.; Rawlings, N. D.; Woessner, J. F., Eds. *Handbook of proteolytic enzymes*; Academic Press: San Diego, 1998.
(28) Stryer, L. *Biochemistry*; 3 ed.; W. H. Freeman and Company: New York, 1988.
(29) Dodson, G.; Wlodawer, A. *TIBS* **1998**, 347-352.
(30) Lagocki, J. W.; Law, J. H.; Kezdy, F. J. *J. Biol. Chem.* **1973**, *248(2)*, 580-587.
(31) Bornscheuer, U. T.; Kazlauskas, R. J. *Hydrolases in organic synthesis*; Wiley-VCH: Weinheim, 1999.
(32) EPI Suite Software version 3.11; Environmental Protection Agency, http://www.epa.gov/opptintr/exposure/docs/episuite.htm, 2000.
(33) Reprinted from *J. Inorganic Biochemistry*, *96*, Bassindale, A. R.; Brandstadt, K. F.; Lane, T. H.; Taylor, P. G., *Enzyme-catalysed siloxane bond formation*, 401-406, **2003**, with permission from Elsevier.
(34) Birk, Y. *Int. J. Peptide Protein Res.* **1985**, *25*, 113-131.
(35) Kassell, B. In *Proteolytic enzymes*; Academic Press: New York, 1970; Vol. 19, pp 853-862.
(36) Schwert, G. W.; Takenaka, Y. *Biochim. Biophys. Acta* **1955**, *16*, 570-575.
(37) Segal, B. G. *Chemistry experiment and theory*; John Wiley & Sons: New York, 1985.

Chapter 13

# "Sweet Silicones": Biocatalytic Reactions to Form Organosilicon Carbohydrate Macromers

Bishwabhusan Sahoo[1], Kurt F. Brandstadt[2],
Thomas H. Lane[2], and Richard A. Gross[1,*]

[1]NSF I/UCRC for Biocatalysis and Bioprocessing of Macromolecules,
Polytechnic University, 6 Metrotech Center, Brooklyn, NY 11201
[2]Dow Corning Corporation, Midland, MI 48686
*Corresponding author: http://chem.poly.edu/gross

An enzymatic route for the synthesis of organosilicon carbohydrates was discovered and reduced to practice. Immobilized lipase B from *Candida antarctica* (Novozyme® 435) catalyzed the formation of regioselective ester bonds between carboxylic acid-endblocked organosilicones and a C1-O-alkylated sugar under mild reaction conditions (i.e. low temperature, neutral pH, solventless). Specifically, the acid-endblocked organosilicones reacted with the primary hydroxyl group at the C6 position of α,β-ethyl glucoside during the regioselective esterification in a one-step reaction without performing any protection-deprotection steps. It was observed that the lipase-catalyzed reactions did not require activation of the acid groups. In comparison to organic materials, the hydrophobic organosilicon molecules were acceptable substrates. Given the ability to perform selective reactions and maintain the integrity of the siloxane bonds, lipases appear to be useful catalysts in the efficient synthesis of structurally defined organosilicon carbohydrate bioconjugates.

Historically, reaction conditions have inhibited the synthesis of structurally defined organosilicon carbohydrates. Typically, esters and polyesters are synthesized with an acid or base catalyst in a one-step process at high temperatures for long periods[1]. Although these reaction conditions favor the equilibrium of condensation, they also promote uncontrolled side reactions. While reasonable reaction rates and conversions may be achieved through acid or base catalysis, these catalysts may induce the decomposition of potentially useful functional groups (e.g. epoxy) and bonds (e.g. siloxane). In addition, the usual acid and base catalysts are not regioselective and may catalyze esterification at all reactive groups on a multifunctional monomer. Consequently, the ability to control the structure of the material is essentially lost.

Given the ability to self-assemble, amphiphilic organosilicon carbohydrates were documented to have unusual properties in solution or as neat materials[2,3]. The physical properties of these "sweet silicones"[2] are dependent on the structure of the attached carbohydrate. They may be used as surfactants, adhesion promoters, or chiral templates. Braunmuhl et al.[2,3] synthesized polydimethylsiloxanes with pendent maltoheptaoside or maltoheptaonamide groups by hydrosilylation and/or amidation. Subsequently, potato phosphorylase was used to enzymatically catalyze the formation of poly(dimethylsiloxane-graft-($\alpha$,1$\rightarrow$4)-D-glucopyranose molecules with $\alpha$-D-glucose-1-phosphate in a citrate buffer at 37°C. The amylose side chains were determined to have helical structure. The amylose side chains were determined to have a helical structure. This represents an elegant strategy for the synthesis of an amphiphilic organosilicon carbohydrate. However, the need for multiple steps and activation chemistry would be problematic at larger reaction scales.

The ability to conduct regioselective enzymatic reactions may circumvent the necessity to perform protection-deprotection steps during the synthesis of organosilicon carbohydrate bioconjugates. Various immobilization techniques (e.g. solvent engineering, micelle, entrapment, physical adsorption) have been used to optimize the handling and activity of a broad selection of biocatalysts[4-7]. In this chapter, substrates were chosen to investigate the attributes of biocatalytic reactions (e.g. selectivity, mild reaction conditions) during the preparation of organosilicon carbohydrates.

## Experimental

**Materials.** The reagents were purchased at the highest available purity and used as received. $\alpha,\beta$–Ethylglucoside was prepared and purified according to literature methods[8,9]. Glucose (CAS #28905-12-6) was purchased from Aldrich (Milwaukee, WI). 1,3-Bis(3-carboxypropyl)tetramethyldisiloxane (diacid disiloxane, Product #SIB1027.0, CAS #3353-68-2) and 10-15 cst. diacid-endblocked polydimethylsiloxane (diacid PDMS, Product #XG-0886 Bulk) were purchased from Gelest, Inc. (Tullytown, PA). Novozyme® 435 (N435) was a gift from Novozyme (Bagsvaerd, Denmark). N435 (Product #537322) may also be purchased from Sigma (St. Louis, MO). N435 is a commercial source of *Candida antarctica* lipase B immobilized on acrylic resin beads.

**Lipase-Catalyzed Organosilicon Carbohydrate Esterification Reaction.** The reactions were conducted in vacuo with constant stirring in a two-neck round-bottom flask in a heated oil bath. The neat (solventless) reactions were formulated with an organosilicon:organic mole ratio equal to 1:2. The enzyme:monomer weight ratio was 1:10. Prior to reaction, the enzyme was transferred to an oven-dried vial and dried in vacuo (0.1 mmHg) at $25°C$ for 48 hours. α,β–Ethyl glycoside was added to the organosilicon reactant at $70°C$. After homogenization, dried N435 was added to the reaction mixture and heated for 36 hours under vacuum. After the reaction, the mixture was filtered to remove the enzyme and the product was purified by silica gel column chromatography with a chloroform:methanol eluent (9:2) for the diacid disiloxane (x=0, n=3) and (9.5:0.5) ratio of chloroform:methanol for the diacid disiloxane (x=7, 65; n=3).

## Results and Discussion

Lipase (N435) was used to selectively react 1,3-bis(3-carboxypropyl)tetramethyldisiloxane (diacid disiloxane, Scheme 1, x = 0, n = 3) and a diacid-endblocked polydimethylsiloxane (diacid PDMS, Scheme 1, x = 7, n = 3), independently, with the primary hydroxyl group of α,β–ethyl glucoside. The reactions were monitored by thin layer chromatography (TLC). After 36 hours, the reactions were filtered and the products were purified by silica gel column chromatography. The organosilicon carbohydrate products (i.e. tan fluids) were characterized by spectroscopic ($^{13}C$ DEPT NMR, ESI MS, FTIR), chromatographic (GPC-RI), and thermal (TGA, DSC) analytical techniques. Based on the characterization results, lipase was observed to catalyze regioselective esterifications with α,β-ethyl glucoside during the formation of mono- and di-ester products. Specifically, the acid-endblocked organosilicones reacted with the primary hydroxyl group at the C6 position of α,β-ethyl glucoside during the regioselective esterifications. This simplification of an otherwise tedious reaction was a result of the inherent regioselectivity of the lipase. The lipase-catalyzed reactions did not require activation of the acid groups. In comparison to other organic substances, the hydrophobic organosilicon molecules were acceptable substrates in organic media.

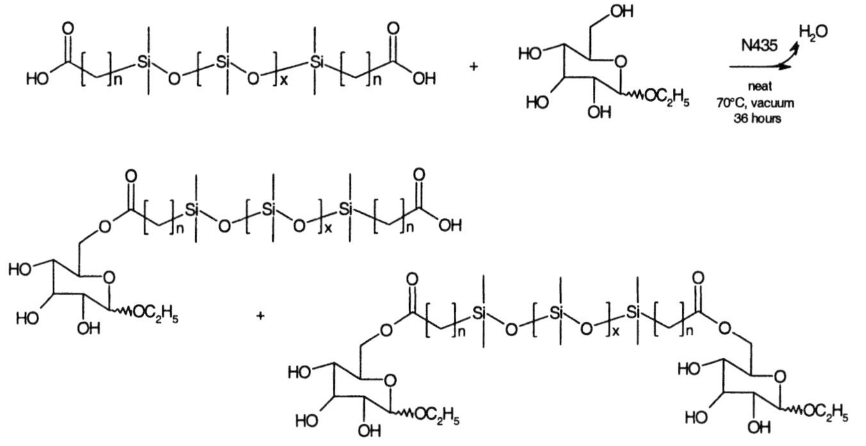

**Scheme 1.** *Lipase-catalyzed esterification of diacid-endblocked siloxanes and α,β–ethylglucoside during the synthesis of organosilicon carbohydrates (x = 0, 7 or 65; n = 3).*

The regioselectivity of the esterification reactions were confirmed by $^{13}$C distortionless enhancement by polarization transfer nuclear magnetic resonance spectroscopy ($^{13}$C DEPT NMR) experiments. Figure 1 details the $^{13}$C DEPT NMR experiment of 1,3-bis(3-carboxypropyl)tetramethyldisiloxane, α,β-ethyl glucoside, and 1,3-bis(1'-ethylglycosyl-6'-propionate) tetramethyldisiloxane. In the $^{13}$C DEPT NMR spectrum of α,β-ethylglucoside, the chemical shift of the 6α,β-carbons was 61.4 ppm. After esterification, the 6α,β-carbons were shifted downfield by 2.1 ppm. Concurrently, the 5α- ☐and 5β-carbons shifted upfield by approximately 3.4 ppm. Since the chemical shifts of the other α,β-ethylglucoside carbons remained nearly unchanged, the esterification was determined to occur with > 98% regioselectivity for the primary hydroxyl position (C6). These observations for the $^{13}$C NMR signals of α,β-ethyl glucoside were identical for the mono- and di-esterified products of both the diacid- disiloxane and PDMS products. Comparatively, similar spectra were reported during the regioselective esterification of α,β-ethyl glucoside and poly(e-caprolactone) at the C6 carbon[10].

Furthermore, the products were characterized by electrospray ionization mass spectrometry (ESI MS). The ESI MS analyses of the organosilicon carbohydrate products (Figures 2-3) support the selective formation of an ester bond between one hydroxyl group of the α,β-ethyl glucoside molecules and the diacid siloxanes. For clarification in Figure 2, the non-reactive cyano functionality was an impurity in the commercial diacid disiloxane starting material. In Figure 3, the distribution of mono- (Δ) di- (O) ester products detail the polydispersity of the diacid PDMS starting material.

Following the esterification of diacid PDMS (x=7, n=3) and α,β-ethylglucoside, the number average molecular weight (Mn) of the organosilicon carbohydrate was calculated to be 1323 (PDI 1.4) based on a gel permeation chromatography (GPC) analysis. The Mn of the diacid PDMS starting material was determined to be 902 (PDI 1.5). Comparatively, a theoretical regioselective esterification with the primary hydroxyl group at the C6 position of α,β-ethyl glucoside would result in a product with an Mn equal to 1310. In the similar way the diacid end blocked PDMS (x=65, n=3) also observe a small increase in the Mw from 12400 to 12675. The GPC analysis supports the lipase-catalyzed regioselective esterification. Based on the molecular weight distribution or polydispersity (PDI), the telechelic organosilicon carbohydrate is not crosslinked. Experimentally, product purification by flash column chromatography resulted in a smaller polydispersity due to the removal of small fraction of low molecular weight fractions.

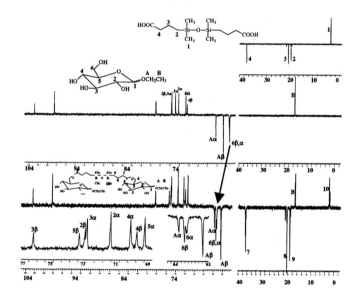

**Figure 1.** $^{13}C$ *DEPT NMR spectra of 1,3-bis(3-carboxypropyl)-tetramethyldisiloxane (top), a,β-ethylglucoside (middle), and 1,3-bis(1'-ethylglycosyl-6'-propionate) tetramethyldisiloxane (bottom).*

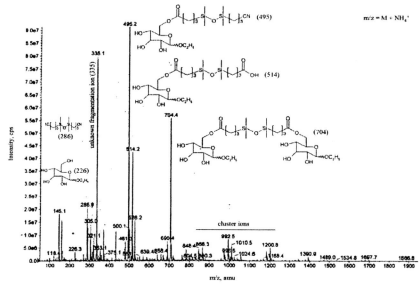

**Figure 2.** ESI MS analysis of the lipase-catalyzed regioselective esterification of 1,3-bis(3-carboxypropyl)tetramethyldisiloxane and α,β-ethyl glucoside.

Based on thermal gravimetric analyses (TGA) and differential scanning calorimetry (DSC) results (Tables I-II), 1,3-bis(1'-ethylglycosyl-6'-propionate) tetramethyldisiloxane experienced a critical mass loss at 184 °C versus 164°C (i.e. diacid disiloxane reactant). Comparatively, the weight loss occurred at two temperatures (281°C, 395°C) in the product versus one temperature (226°C) in the starting material. The $T_g$ of the product and the diacid disiloxane were -26°C and -76°C, respectively. Hence, more energy was required to achieve molecular motion in the ester derivative. During the DSC heating cycles, a crystalline phase (Tm) was observed at 13°C (1.7 J/g) in the product. In contrast, a cold crystallization ($T_{cc}$ = -15°C, 50.4 J/g) and two crystalline phases ($T_m$ = 30°C, 45 J/g; 46 °C, 8.6 J/g) were detected in the diacid disiloxane. Similarly, for the product x=7, n=3 experienced a critical mass loss at 201 °C versus 197°C (i.e. diacid disiloxane reactant, x=7, n=3). Comparatively, the weight loss occurred at three temperatures (218°C, 268°C and 417°C) in the product versus two temperature (249°C and 563°C) in the starting material. The $T_g$ of the product (x=7, n=3) and the diacid disiloxane were -111°C and -34°C, respectively. During the DSC heating cycles, two melting endotherms with peak at 20°C and 26°C was observed in the product. Given the small energy value, the region of order within the product is either small or the molecular interactions are weak.

**Table I.** TGA analysis of organosilicon carbohydrates.

| Product (Ave.DP[1]) | °C | 100 | 200 | 300 | 400 | 500 | 600 | 700 |
|---|---|---|---|---|---|---|---|---|
| $x = 0$[1] | % | 98 | 92 | 49 | 5 | 1 | 1 | 2 |
| $x = 7$[1] | % | 98 | 94 | 65 | 32 | 4 | 2 | 1 |
| $x = 65$[1] | % | 100 | 99 | 95 | 87 | 71 | 4 | 1 |

[1] Ave. DP = the diacid siloxane average degree of polymerization = x + 2.

**Table II.** DSC analysis of organosilicon carbohydrates.

| Product (Ave. DP[1]) | DSC Results |
|---|---|
| $x = 0$[1] | Tg -26°C; Tm 13-17°C<br>Peak Maximum at 13°C, 1.7 J/g area |
| $x = 7$[1] | Tg -34°C; Tm 18-27°C<br>Peak Maximum at 20°C and 26°C |
| $x = 65$[1] | Tg -123°C; Tm -54 to -32°C<br>Peak Maximum at -47°C and -38°C, 30 J/g area |

[1] Ave. DP = the diacid siloxane average degree of polymerization = x + 2.

When the diacid end blocked PDMS is extended to x=65, n=3, it experienced a critical mass loss at 264°C versus 172°C (i.e. diacid disiloxane reactant, x=65, n=3). Comparatively, the weight loss occurred at three temperatures (324°C, 397°C and 529°C) in the product versus two temperature (216°C and 591°C) in the starting material. The $T_g$ of the both product (x=65, n=3) and the diacid disiloxane were -123°C. During the DSC heating cycles, a cold crystallization (Tcc -79oC, 29 J/g) and two melting endotherms with peaks at -47°C, -38oC (30J/g) were observed in the product. This, it indicate the fluxing character dominate due to the PDMS chain lengths in the product.

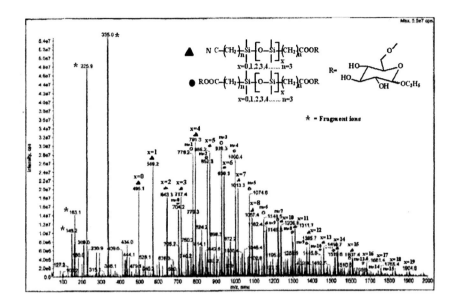

*Figure 3. ESI-MS analysis of the lipase-catalyzed esterification of diacid-endblocked polydimethylsiloxane and α,β-ethylglucoside.*

## Conclusions

An enzyme-catalyzed regioselective reaction of an organosilicon material with a carbohydrate was discovered and reduced to practice under mild reaction conditions (i.e. low temperature, neutral pH, solventless). In particular, pure organosilicon-sugar conjugates were prepared in a one-step reaction, without protection-deprotection steps. This simplification of an otherwise tedious reaction was a result of the inherent regioselectivity of lipase. The lipase-catalyzed reactions did not require activation of the acid groups. In comparison to organic materials, the hydrophobic organosilicones were acceptable substrates. Given the ability to perform a selective reaction and maintain the integrity of the siloxane bonds with lipase, the ability to synthesize structurally defined organosilicon carbohydrates with a diversified set of functional groups may be used to create new materials such as fibers, films, coatings, gels, and surfactants with novel properties.

## Acknowledgements

We are grateful to the Dow Corning Corporation (Midland, MI) for their financial support, encouragement, and scientific collaboration during the course of this work.

## References

1. Hood, J. D.; Blount, W. W.; Sade, W. T. *Journal of Coatings Technology* **1986**, *58*, 49-52.
2. Braunmuhl, V. v.; Jonas, G.; Stadler, R. *Macromolecules* **1995**, *28*, 17-24.
3. Braunmuhl, V. v.; Stadler, R. *Macromol. Symp.* **1996**, *103*, 141-148.
4. Klibanov, A. M. *TIBS* **1989**, *14*, 141-144.
5. Luisi, P. L. *Angew. Chem. Int. Ed. Engl.* **1985**, *24*, 439-450.
6. Reetz, Manfred T.; Zonta, A.; Simpelkamp, J. *Biotechnol. Bioeng.* **1996**, *49*, 527-534.
7. Gill, I.; Pastor, E.; Ballesteros, A. *J. Am. Chem. Soc.* **1999**, *121(41)*, 9487-9496.
8. Bjorkling, F.; Godtfredsen, S. E.; Kirk, O. *Journal of American Chemical Society Communication* **1989**, 934-935.
9. Adelhorst, K.; Bjorkling, F.; Godtfredsen, S. E.; Kirk, O. *Synthesis* **1990**, 112-115.
10. Bisht K. S.; Deng F.; Gross R. A.; Kaplan D.L.; and Swift G.; *J. Am. Chem. Soc.* **1998**, *120*, 1363-1367.

# Polysaccharides

## Chapter 14

# Enzymatic Synthesis of Complex Bacterial Carbohydrate Polymers

Hanfen Li[1], Hesheng Zhang[1], Wen Yi[1], Jun Shao[2], and Peng George Wang[1]

[1]Department of Biochemistry, The Ohio State University, Columbus, OH 43210
[2]Symyx Technologies, Inc., 3100 Central Expressway, Santa Clara, CA 95051

Natural polysaccharides have complex structure including many kinds of monosaccharide units, which are very difficult and costly to generate by chemical synthetic approaches. Over the years, enzymatic approaches have been gaining popularity for the synthesis of oligosaccharides and polysaccharides, and it is becoming increasingly feasible to produce complex carbohydrate polymers in large scale following enzymatic biosynthetic pathways.

Recently glycobiology has emerged as a new and challenging research area at the interface of biology and chemistry. Carbohydrates which constitute one of the most abundant types of biomolecules perform a much broader role in biological science *(1)*. In fact, carbohydrate complexes have been widely used as potential pharmaceuticals for the prevention of infection, the neutralization of toxins, and the immunotherapy of cancer. Bacterial lipopolysaccharides (LPS) consist of a hydrophobic domain known as lipid A, a core oligosaccharide, and a distal polysaccharide (O-antigen) that has been identified as a dominant antigen capable of inducing human protective immune responses. Vaccines comprising carbohydrates coupled to a protein carrier have been proven effective for the prevention of invasive bacterial disease *(2)*. However, these glycoconjugates prepared by using purified polysaccharides or degraded oligosaccharides are sometimes contaminated with other bacterial components. In addition, the molecular size of carbohydrate antigens is another major concern in designing glycoconjugate vaccines. The exploration of these well-structured vaccines has been hindered largely by the technical difficulties in chemical preparation of oligosaccharides. Compared to chemical synthesis, enzymatic approach utilizing glycolsyltransferases has been identified to be more efficient in the production of complex carbohydrates. Considering the extremely narrow substrate specificity of the most known glycosyltransferases, molecular characterization of the biosynthetic pathway of O antigen repeat units is undoubtedly a crucial step towards large-scale synthesis of an immuno-dominant oligosaccharide.

Unlike the linear nature of the nucleic acid and protein molecules, complex carbohydrates exhibit highly branched structures and multiplicity of different types of linkages, known as glycosidic bonds. Glycosyltransferases use activated sugar molecules (donors) to attach saccharide residues to a variety of substrates (acceptor). Bacterial glycosyltransferases are enzymes responsible for the assembly of bacterial cell walls and lipopolysaccharides or polysaccharide structures attached to the lipids of the outer membrane of Gram negative bacterial cells. With increasing availability of recombinant glycosyltransferases in recent years, it can be forseen that more researchers will use these enzymes to construct different glycoconjugates. Now biocatalytic approaches employing enzymes or genetically engineered whole cells are powerful and complementary alternatives to chemical methods *(3)*.

## Carbohydrate Biosynthesis Associated Enzymes

Among the numerous enzymes associated with carbohydrate processing in cells, those used in glycoconjugate synthesis belong to two categories, viz., glycosidases and glycosyltransferases.

## Glycosidases

Glycosidases are enzymes that cleave oligosaccarides and polysaccharides *in vivo*. They can form glycosidic linkages under *in vitro* conditions in which a carbohydrate hydroxyl moiety acts as a more efficient nucleophile than water itself. They have been of tremendous benefit in the enzymatic synthesis of oligosaccharides due to their availability, stability, organic solvent compatibility, and low cost *(4-6)*. Nevertheless, traditional glycosidase catalyzed transglycosylations still suffer from low yields and poor unpredictable regioselectivities. Based on the structure-function relationship information and the mechanisms of glycosidase-catalyzed reactions, modern mutagenesis technology has played an important role in enhancing the enzymatic activity towards the synthesis of oligosaccharides.

## Glycosyltransferases

Glycosyltransferases are enzymes that can transfer a sugar moiety to a defined acceptor so as to construct a specific glycosidic linkage. This "one enzyme-one linkage" concept makes glycosyltransferases useful and important in the construction of glycosidic linkages in carbohydrates *(7-9)*. Glycosyltransferases can be further divided into two groups, viz., the transferases of the Leloir pathway and those of non-Leloir pathways. The Leloir pathway enzymes require sugar nucleotides as glycosylation donors, while Non-Leloir glycosyltransferases typically utilize glycosylphosphate or glycosides. The Leloir transferases are responsible for the synthesis of most glycoconjugates in cells, especially in mammalian systems *(10-12)*.

They can be expressed at high levels in mammalian systems, such as the Chinese hamster ovary (CHO) cells. However, this expression procedure is too tedious and expensive to be applied in practical transferase production. Efforts have been made in expressing mammalian enzymes in insect, plant, yeast, and bacterial cells, but high-level expression remains difficult. Fortunately, glycosyltransferases from bacterial sources can be easily cloned and expressed in *E. coli*, in large quantities *(13)*. They also have a broader range of substrates when compared with mammalian glycosytransferases. Furthermore, some bacterial transferases were found to produce mammalian-like oligosaccharide structures, which makes these enzymes quite promising in synthesis of biologically important oligosaccharides *(14-16)*. The recent expansion in genomic sequencing has allowed many glycosyltransferases to be characterized and expressed in recombinant form.

Fucosyltransferase is considered to be a key enzyme in the biosynthesis of many oligosaccharides that are involved in physiological as well as in pathophysiological processes. Fucosyltransferases are the enzymes responsible for transferring fucose from GDP-Fuc to Gal in an alpha1,2-linkage or to

GlcNAc in alpha1,3-, alpha1,4-,or alpha1,6-linkages. Since all fucosyltransferases utilize the same nucleotide sugar, their specificity will probably reside in the recognition of the acceptor and in the type of linkage formed (17). More than thirty fucosyltransferase genes from prokaryotes and eukaryotes were cloned and there are three main families of α-2, α-3, and α-6-fucosyltransferases gene families. Six conserved peptide motifs have been identified in DNA and protein databanks. Two of these motifs are specific of α-3 fucosyltransferases (18), one is specific of α-2-fucosyltranferases, another is specific of α-6-fucosyltransferases, and two are shared by both α-2- and α-6-fucosyltransferases. The eukaryotic fucosyltransferases in general have the typical topology of type II membrane proteins with the transmembrane hydrophobic domain in their N-terminus, whereas the prokaryotic enzymes apparently lack such domain (18).

## Structures Common to Glycans

Glycan chain modifications found at outer or terminal positions generally result from the actions of glycosyltransferases, which modify glycan precursors (acceptors). These precursors, which are typically expressed by all cell types, include the multiantennary N-linked glycan acceptors, the linear or biantennary O-linked glycan acceptors, and linear lipid-linked acceptors. Although the terminal structures can be unique to various core subtypes, many are found on more than one class of glycans and normally establish the functions of glycoconjugate.

### *N*-acetyllactosamine structures

*N*-acetyllactosamines are a major family of bioactive glycans found on the core structures of N-glycans (19,20), O-glycans (21), and lacto (type 1)- and neolacto (type 2)- glycolipids (22). *N*-acetyllactosamines consist of disaccharide units, Galβ1,4GlcNAc (type 2) and Galβ1,3GlcNAc (type 1), which can be linked to form polylactosamines with β1,3- and/or β1,6-linkages. They are generated by the action of β-1,4-galactosyltransferase or β-1,3-galactosyltransferase, respectively (Scheme 1). These oligosaccharides are often modified with terminal structures such as α2,3- or α2,6-linked sialic acids, ABO-blood group antigens, α1,3-linked galactose, and α1,3/4-linked fucoses.

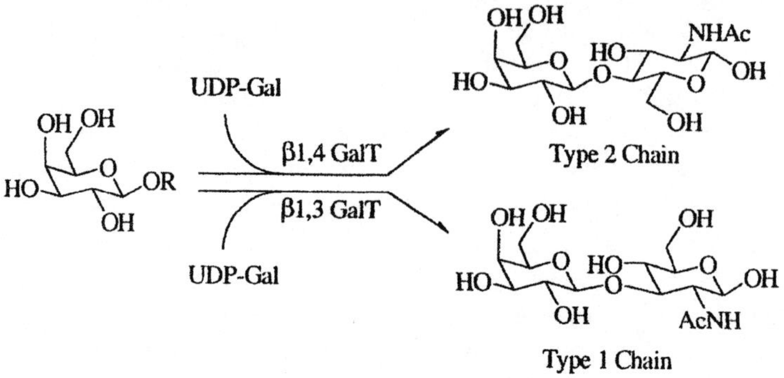

*Scheme 1 Modification of exposed GlcNAc moieties by galactosylation*

## The ABO blood group system

The ABO blood group plays an extremely important role in medical processes such as transfusion and transplantation. ABO blood groups are determined by the presence of two distinct antigens, viz., the A and B antigens, which are oligosaccharide structures (Scheme 2) found on the red blood cell membranes. The immunodominant parts of A / B antigens are synthesized from O antigen by α-1,3-*N*-acetylgalactosaminoltransferase and α-1,3-galactosyltransferase, respectively. Naturally occurring antibodies against α-Gal (Galα1,3Gal) structures (anti-Gal antibodies) are the primary effectors of human hyperacute rejection (HAR) of non-human tissue. Unlike most mammals, humans lack a functional α1,3-galactosyltransferase gene and produce abundant anti-Gal antibodies. It is now accepted that enteric exposure to gram-negative bacteria expressing cell wall or lipopolysaccharide Galα 1,3Gal structures induces human anti-Gal antibody production, similar to the development of human antibodies against ABO blood group antigens *(23)*. α-Gal is structurally similar to the human blood group B antigen and was first reported as a "B-like" antigen on rabbit erythrocytes *(24)*. Human anti-Gal and anti-ABO antibodies are actually ubiquitous and not initially induced by classical peptide antigens. In addition, all these antibodies are considered to result from humeral responses to polysaccharide antigens and to be comprised mainly of low-affinity, cold-agglutinating IgM *(25)*.

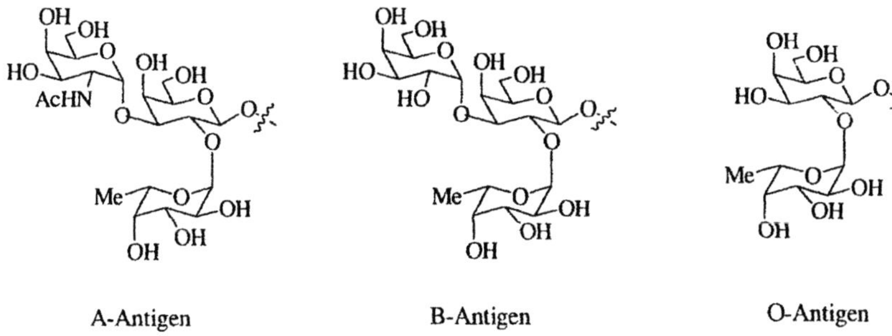

*Scheme 2 Structures of A, B and O oligosaccharide antigens*

### The Galα1,3Gal structure

Carbohydrate structures Galα1,3Galβ1,4GlcNAc-R (or Galα1,3Galβ1,4-Glc-R) and Galα1,3Galβ1,4GlcNAcβ1,3Galβ1,4Glc-R), namely α-Gal epitopes, are abundantly expressed by New World primates and many non-primate mammals but are largely absent from the cells and tissues of Old World primates which include *Homo sapiens (26-32)*. Instead, humans naturally produce large quantities of antibodies against α-Gal epitopes (anti-Gal antibody), which consist of 1-3% of all circulating immunoglobulins produced by about 1% of all B cells (Scheme 3). It has been hypothesized that continuous production of anti-Gal is due to constant antigenic stimulation by α-Gal-like epitopes found on the surface of normal gastrointestinal bacteria *(23)* or trypanosomatidae infection by α-Gal-containing epitopes present on the parasite *(28,33)*. Anti-Gal is also abundant in apes and Old World monkeys (monkeys of Asia and Africa), which, like humans, lack α-Gal epitopes. These anti-Gal antibodies present a major barrier to the use of porcine and other non-primate organs for xenotransplantation in humans because they bind to Galα1,3Gal epitopes on the vascular endothelium in such xenotransplants, and mediate hyperacute graft rejection through complement-mediated endothelial cell cytotoxicity *(34,35)*.

## Biocatalytic Synthesis of Oligosaccharides

In contrast to nucleic acid and protein, the biosynthesis of carbohydrate is not template-driven, but rather defined by the cooperation of glycosyltransferase

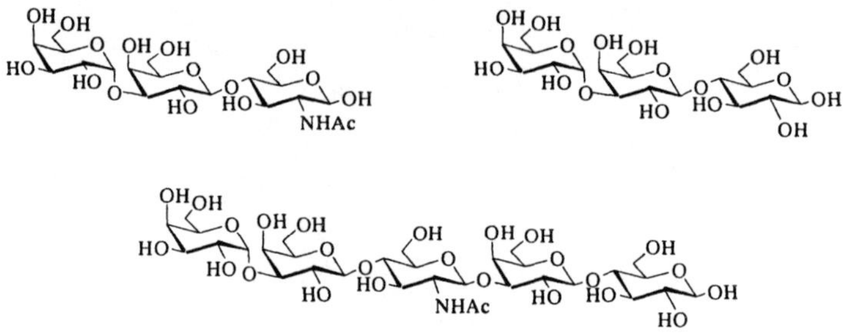

Scheme 3 The structures of α-Gal epitopes.

machinery and their cofactor, nucleotide sugars. Chemical methods often require multiple protection-deprotection steps and long synthetic routes. As could be anticipated, the chemical synthesis of oligosaccharides is not an attractive option to industrial and scientific communities (36). Biocatalytic approaches employing enzymes or genetically engineered whole cells are powerful and complementary alternatives to chemical methods.

**Enzyme-based oligosaccharide synthesis**

Enzymes have been used extensively to simplify the synthesis of complex oligosaccharides and glycoconjugates. Glycosyltransferases and glycosidases are valuable catalysts for the formation of specific glycosidic linkages (3,37). Other enzymes such as aldolases and sulfotransferases can also be exploited for the synthesis of distinct structures that is critical to oligosaccharide functions. Glycosidases are responsible for glycan processing reactions that take place during glycoprotein synthesis. The physiological function of these enzymes is the cleavage of glycosidic linkages. However, under controlled conditions, glycosidases can be used to synthesize glycosidic bonds rather than cleavage of them (Scheme 4). Therefore, these enzymes have been employed as catalysts in oligosaccharide synthesis. Glycosidases are widely available, robust, and require only inexpensive donor substrates. Although glycosidases are generally stereospecific, they only have weak regiospecificity, which may result in the formation of multiple products.

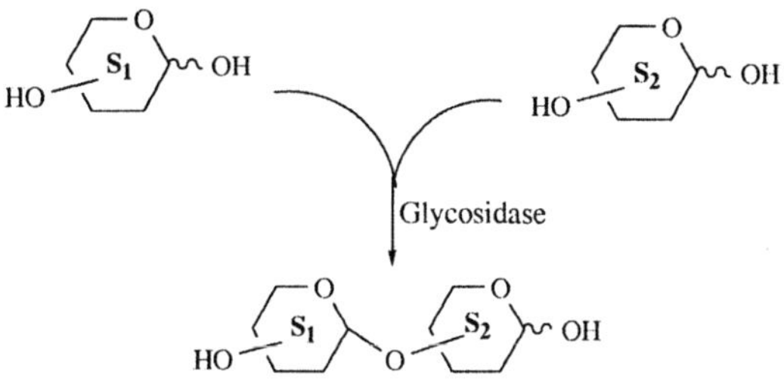

*Scheme 4 Protocol for glycosidase-based
synthesis of oligosaccharide*

Glycosyltransferases of the Leloir pathway are responsible for the synthesis of most cell-surface glycoforms in mammalian systems. These enzymes transfer a given carbohydrate from the corresponding sugar nucleotide donor substrate to a specific hydroxyl group of the acceptor sugar (Scheme 5). These enzymes exhibit very strict stereospecificity and regiospecificity. Moreover, they can transfer with either retention or inversion of configuration at the anomeric carbon of the sugar residue. A large number of eukaryotic glycosyltransferases *(3,38,39)*have been cloned to date and used in large-scale syntheses of oligosaccharides and, in general, exhibit exquisite linkage and substrate specificity. It should be noted that such a considerable number of mammalian enzymes have converged on only nine common sugar nucleotides as donor substrates. Glucosyl-, galactosyl-, and xylosyltransferases employ substrates activated with uridine diphosphate as the anomeric leaving group (UDP-Glc, UDP-GlcNAc, UDP-GlcA, UDP-Gal, UDP-GalNAc, and UDP-Xyl), whereas fucosyl- and mannosyltransferases utilize guanosine diphosphate (GDP-Fuc and GDP-Man). Sialyltransferases are unique in a sense that the glycosyl donor is activated by cytidine monophosphate (CMP-Neu5Ac).

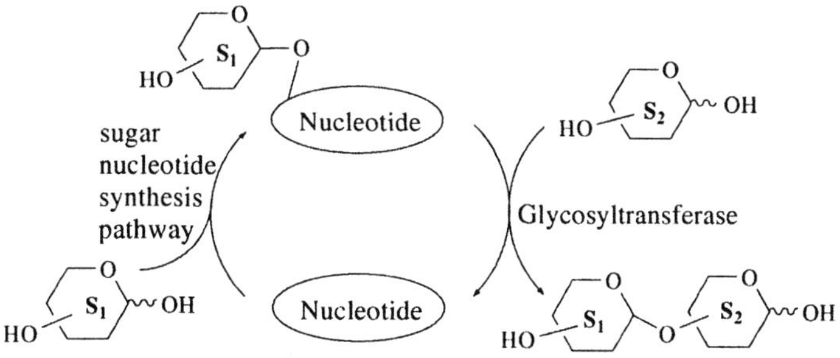

*Scheme 5 Biosynthesis of oligosaccharide by glycosyltransferase*

Recent advances in the area of enzymatic oligosaccharide synthesis are emerging from the identification and cloning of a large number of bacterial glycosyltransferases with many different donor, acceptor and linkage specificities *(13,14,40-42)*. Most eukaryotic glycosyltransferases are not active within prokaryotic expression system because of the absence of post-translation modifications including glycosylation. Bacterial glycosyltransferases, on the other hand, are normally not glycosylated proteins. It has been demonstrated that these enzymes are more easily expressed as soluble and active form in prokaryotic expression system such as *E. coli*. In addition, bacterial glycosyltransferases seem to have relatively broader acceptor substrate specificities, thereby offering tremendous advantages over mammalian enzymes in the chemoenzymatic synthesis of oligosaccharides and their analogues for the development of anti-adhesion therapies for infectious diseases *(43,44)*.

## Oligosaccharide synthesis by biotransformation

The availability of new glycosyltransferases has increased the demand for sugar nucleotides, which have been a problem in the production of oligosaccharides and glycoconjugates. One way to overcome this hurdle is to use a multiple enzymatic system with *in situ* UDP-GalNAc regeneration from inexpensive starting materials. Since the pioneering work by Wong *et al.* on *in vitro* enzymatic synthesis of *N*-acetyllactosamine with the regeneration of UDP-galactose, several glycosylation cycles with regeneration of sugar nucleotides have been developed using either native or recombinant enzymes *(45-49)*.

The isolation of recombinant enzymes is generally a rather laborious operation. A rapidly emerging method for the large-scale synthesis of complex carbohydrates is the use of metabolically engineered microorganisms. So far three strategies have been used in developing the whole-cell biocatalyst, and they are as follow:

### (1) The "bacterial coupling" technology.

This technology was developed by Kyowa Hakko Kogyo Co. Ltd. in Japan *(50-54)*. The key to Kyowa Hakko's technology for the large-scale production of oligosaccharides was a *C. ammoniagenes* bacterial strain engineered to efficiently convert inexpensive orotic acid to UTP (Scheme 6). When combined with an *E. coli* strain engineered to overexpress UDP-Gal biosynthetic genes including *galK* (galactokinase), *galT* (galactose-1-phosphate uridyltransferase), *galU* (glucose-1-phosphate uridyltransferase), and *ppa* (pyrophosphatase), UDP-Gal was accumulated in the reaction solution (72 mM / 21 h). Combining these two strains with another recombinant *E. coli* strain, over-expression of the α-1,4-galactosyltransferase gene of *Neisseria gonorrhoeae* produced a high concentration of globotriose.

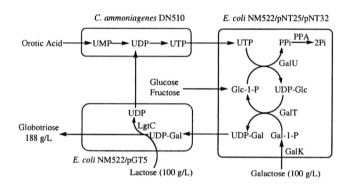

*Scheme 6 Large-scale production of oligosaccharides through coupling of engineered bacteria.*

### (2) The "Superbeads" and "Superbug" technology developed by Wang's Group

For the production of larger oligosaccharides, Wang's group has developed cell-free "superbeads" technology. This approach involves immobilization of all the enzymes along the biosynthetic pathway onto beads. The beads are used as catalysts to produce larger oligosaccharides in a cell-free system. For example, in order to make α-Gal trisaccharide, we have immobilized all the necessary enzymes including GalK, GalT, GalPUT, PykF and galactosyltransferase for *in situ* regeneration of donors onto Ni-containing agarose resins (beads). These agarose beads function as stable and versatile synthetic reagents, which can be used and regenerated to synthesize a variety of oligosaccharides and glycoconjugates (Scheme 7) *(55-58)*. This methodology of enzyme immobilization has all the advantages of solid phase organic synthesis, such as easy separation, increased stability, reusability, and improved kinetics. Larger oligosaccharide synthesis will involve more corresponding glycosyltransferases together with necessary sugar nucleotide regenerating beads. Using this approach, we can produce oligosaccharides containing more than 8 sugar units.

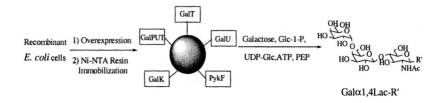

*Scheme 7  globotriose production with superbeads.*

Also the Wang's "superbug" approach can make use of engineered bacteria through fermentation to provide all the necessary enzymes along the biosynthetic pathway starting from monosaccharide through oligosaccharides. This approach relies on a single microbial strain transformed with a single artificial gene cluster of all the biosynthetic genes and uses the metabolism of the engineered bacteria to provide the necessary bioenergetics (ATP or PEP) to drive a glycosylation cycle. The advantage is that whole cells are used as bio-catalysts in the reaction system without any laborious enzyme purification. Only catalytic amount of ATP is needed for this whole cell synthesis. Obviously, this makes the superbug production of oligosaccharide the most cost-effective method *(59-61)*. In fact this biotechnology for mass production of glycoconjugates has matured to such a level that it can provide a variety of products at a fraction of current commercial prices. Currently, our "Superbug" technology can produce oligosaccharides of less than four sugar units efficiently (scheme 8). Since we have incorporated most common sugar nucleotide biosynthetic cycles into the superbug, construction of new superbug for new oligosaccharide simply involves replacing/inserting corresponding new glycosyltransferase gene(s) into the plasmid. Thus, the superbug can be constructed to produce a variety of oligosaccharides depending on the availability of the glycosyltransferases.

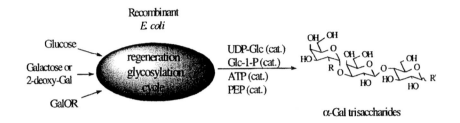

*Scheme 8  α-Gal superbug*

(3) **The "living factory" technology.**

This technology was developed by Dr. Samain's group in France. This technology makes use of the bacteria host cell's own ability to produce nucleotide sugars, while the bacteria are simply engineered to incorporate the required glycosyltransferases (Scheme 9). In high-cell-density cultures, the oligosaccharide products accumulate intracellularly and have been shown to reach levels on the gram / liter scale *(62-67)*.

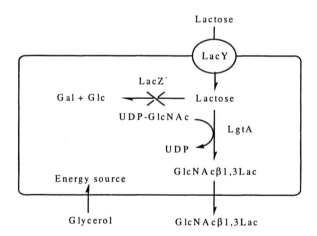

*Scheme 9  Production of trisaccharide GlcNAcβ1,3Lac by E. coli JM109 expressing the lgtA gene that encodes a β-1,3-N-acetylglucosaminyltransferase*

# Bacterial Lipopolysaccharide

Bacterial cells express an enormous variety of polysaccharide structures. These entities are usually found protruding from the outer membrane and include exopolysaccharide (EPS) and lipopolysaccharide (LPS). The polysaccharide components have important structural and functional roles in the life of a bacterial cell. They represent the first line of defense against complement and bacteriophages. They also contain the major antigen determinants that distinguish various serotypes of bacteria, which are sometimes correlated with disease.

## Structure of lipopolysaccharide

LPS is a key component of the outer membrane of Gram-negative bacteria. Lipopolysaccharide typically consists of three distinct regions: (i) a hydrophobic domain known as lipid A (endotoxin), (ii) a nonrepeating "core" oligosaccharide, and (iii) a distal polysaccharide (O-antigen). Lipid A, the hydrophobic anchor of lipopolysaccharide (LPS), is a glucosamine-based phospholipid which makes up the outer monolayer of the outer membranes of most Gram-negative bacteria *(68,69)*. There are $\sim 10^6$ lipid A residues and $\sim 10^7$ glycerophospholipids in a single cell of *Escherichia coli (70)*. Lipid A is the most conserved part of LPS. In many species within the enterobacteriaceae, it has a common structure with a β-1,6-linked disaccharide backbone of amino sugar *(71)*. Variation lies mostly in the number and length of fatty acyl chain, the degree to which the molecule is phosphorylated, the presence or absence of acyloxyacyl substituents, and the presence of additional polar constituents. The core oligosaccharide can be divided into two structurally distinct regions, viz., the lipid A proximal inner core and the outer core (Figure 1). The inner core typically consists of 3-deoxy-D-*manno*-oct-2-ulosonic acid (Kdo) and L-*glycero*-D-*manno*heptose (L,D-Hep) residues. One of the Kdo residues links the core to lipid A. The inner core tends to be well conserved within a genus or family *(72)*. In contrast, the outer core shows more structural diversity, as might be expected for a region with more exposure to the selective pressures of host responses, bacteriophages, and environmental stresses *(73)*. In *E. coli* there are five outer core types that share a basic structure of a three–hexose backbone and two side-chain residues *(74)*. The outer core region provides an attachment site for O polysaccharides.

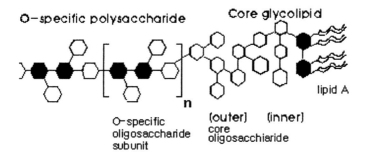

*Figure 1 Bacterial LPS architecture.*

The O antigen is one of the most variable cell constituents, where more than 60 monosaccharides and 30 different noncarbohydrate components have been recognized. The O polysaccharide repeat unit structures can differ in the monomer glycoses, the position and stereochemistry of the *O*-glycosidic linkages, and the presence or absence of noncarbohydrate substituents *(75)*. The structure of the O polysaccharide defines the O-antigen serological specificity in an organism. However, the numbers of unique O antigens within a species may vary considerably, for example, more than 170 O serotypes have so far been identified in *E. coli*. The location of O antigen at the cell surface places it at the interface between the bacterium and its environment. The primary role of the O polysaccharides appears to be protective. In animal pathogens, O polysaccharides may contribute to bacterial evasion of host immune responses, particularly the alternative complement cascade *(73)*.

**O antigen genes in *E. coli***

The differences between various forms of O antigen are due mostly to genetic variation in O antigen gene cluster. Nearly all the genes specific for O antigen biosynthesis are normally clustered in a locus, which is between *galF* and *gnd* genes in *E. coli (76)*. A 39-bp JUMPStart sequence is present upstream of the O antigen biosynthesis gene cluster *(77)*. It has been shown that expression of all genes in the gene cluster is enhanced by RfaH acting on JUMPStart *(78)*.

The O antigen gene cluster includes nucleotide sugar biosynthetic gene, glycosyltransferase genes, and O unit processing genes such as *wzx, wzy*, and *wzz*. Some O antigens include O acetyl groups or other modification groups, and the transferase genes for these residues may be in the cluster. However, genes for the early steps in pathways that are also involved in housekeeping functions are generally not duplicated in the O antigen gene cluster. In general, the genes in the O antigen biosynthesis cluster have a relatively low G+C content, mostly ranging from 30% to 45% for individual gene. This is in contrast to the highly conserved *galF* and *gnd* genes adjacent to the clusters, which have G+C contents typical of most genes of the host chromosome. LPS biosynthesis genes were originally given *rf*** names, but this system cannot copy with the large number of genes recently identified. A new system called Bacterial Polysaccharide Gene Nomenclature (BPGN) system was then set up by researchers in the area (www.microbio.usyd.edu.au/BPGD/default.htm) *(79)*. This system allows each functionally distinctive gene to have a unique name, genes in different clusters, but those with the same function having the same name. All the names generally start with letter "*w*".

**Biosynthesis of O antigen**

The repeating units O polysaccharides are assembled on the membrane-bound carrier, undecaprenyl phosphate (und-P) at the inner face of the cytoplasmic membrane. This is the same $C_{55}$-isoprenoid alcohol derivative used for the synthesis of peptidoglycan and capsular polysaccharides. The repeating units are synthesized from sugar nucleotides by glycosyltransferases that are often soluble or peripheral membrane proteins. However, O polysaccharides are transferred from the carrier lipid and ligated to lipid A-core at the outer face of the cytoplasmic membrane *(80)*. As a result, the assembly processes must include a mechanism whereby either lipid-linked O polysaccharides or lipid-linked O units are delivered to the periplasm. Three O polysaccharide biosynthesis pathways, viz., Wzy-dependent, ABC-transporter dependent, and synthase-dependent, have currently been distinguished by their respective export mechanisms. Despite the export differences, the pathways have similar initiation reactions and are completed by the same ligation process.

Wzy-dependent pathway is most widespread in O polysaccharide biosynthesis. Following the assembly of O repeating units, the individual und-PP-linked O units are exported to the site of polymerization at the periplasmic face of the plasma membrane (Figure 2). This process requires a Wzx protein, the O unit transporter (flippase), which is highly hydrophobic with 12 potential transmembrane domains. Although the Wzx proteins share little primary sequence similarity, they do share structural features with bacterial permeases

*(81)*. At the periplasmic face of the cytoplasmic membrane, und-PP-linked O units are polymerized by O-polysaccharide polymerase, Wzy. The reaction involves transfer of nascent polymer from its und-PP carrier to the nonreducing end of the new und-PP-linked O repeat *(82)*. The released und-PP must be recycled to the active monophosphoryl form. Wzy proteins are all predicted to be integral membrane proteins with 11-13 transmembrane domains, and like the Wzx proteins, they exhibit little primary sequence similarity *(83,84)*. The final component of the Wzy-dependent pathway is the Wzz protein, the chain length determinant. Wzz was hypothesized to act as a timing clock, interacting with the Wzy polymerase, then modulating its activity between two states that favor either elongation or transfer to the ligase (i.e., chain termination).

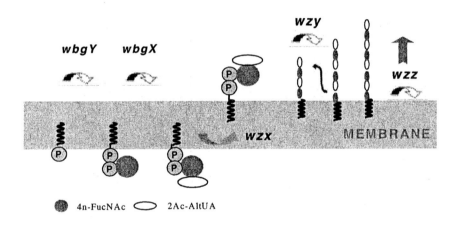

Figure 2  Model of O-unit assembly

### *E. coli* O128 antigen biosynthesis

The primary structure of LPS O-specific polysaccharide (Scheme 10) from *E. coli* O128 has been established *(85)*. The immuno-dominant part of *E. coli* O128 polysaccharide was determined as β-D-GalNAc-(1→6)-[α-L-Fuc-(1→2)]-β-D-Gal trisaccharide by ELISA-inhibition study *(86)*. A total of 19,013 bp *E. coli* O128 antigen biosynthesis gene cluster, which covers a continuous region from the end of *galF* gene to the middle of *hisI*, was amplified by long-range PCR from the chromosomal DNA of *E. coli* O128 strain and completely sequenced (Figure 3), The *E. coli* O128 antigen repeat units consist of one L-fucose, two galactose and two *N*-acetylgalactosamine residues. It can be anticipated that the cluster would contain genes responsible for GDP-L-fucose synthesis, genes coding glycosyltransferases to assemble the O unit and genes for O-antigen process, including O-antigen flippase (*wzx*) for translocation of the completed repeat unit across the cell membrane, polymerase (*wzy*) for the linkage of repeat units and chain length determinant protein (*wzz*).

It has been well established that GDP-L-fucose is synthesized from GDP-D-mannose *via* a three-step pathway catalyzed by two enzymes, GDP-mannose 4,6-dehydratase (*gmd*) and GDP-fucose synthetase (*fcl*), common to both prokaryotic and eukaryotic cells *(87)*. Three more enzymes, phosphomannose isomerase (*manA*), phosphomannomutase (*manB*), and GDP-mannose pyrophosphorylase (*manC*), are needed to generate GDP-mannose from fructose 6-phosphate (Scheme 11). The *E. coli* O polysaccharides are normally synthesized in a Wzy-dependent pathway, which consists of the assembly of O-repeat united on the inner face of cytoplasmic membrane, the exportation by O-antigen flippase Wzx, and the polymerization by polymerase Wzy on the periplasmic face of cytoplasmic membrane *(73)*. Five sugar residues are present in *E. coli* O128 antigen repeat unit. The first step in O-polysaccharides biosynthesis involves the formation of an und-PP-linked sugar by transferring a sugar 1-phosphate residue to undP *(73)*. It has been shown that the *wecA* gene, responsible for adding the first sugar to lipid carrier undP and initiating the O unit biosynthesis, resides outside the *E. coli* O-antigen gene clusters *(88)*. Therefore, we expect to find only four glycosyltransferases within O128 antigen biosynthesis gene cluster.

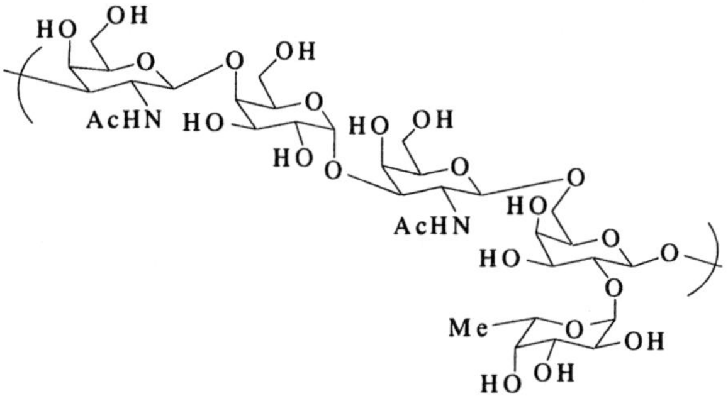

*Scheme 10. E. coli O128 antigen repeat unit*

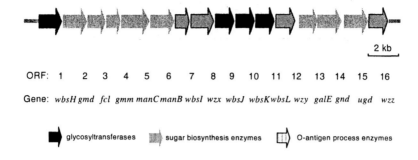

ORF: 1　2　3　4　5　6　7　8　9　10　11　12　13　14　15　16

Gene: *wbsH gmd fcl gmm manC manB wbsI wzx wbsJ wbsK wbsL wzy galE gnd ugd wzz*

■ glycosyltransferases　▨ sugar biosynthesis enzymes　▢ O-antigen process enzymes

*Figure 3. E. coli O128 antigen biosynthesis gene cluster*

To investigate the function of *wbsJ* gene, we cloned this gene into pGEX-4T-1 plasmid *(89)*. The WbsJ was expressed in *E. coli* BL21 (DE3) strain as a GST fusion protein and purified by a one-step GST affinity column. The fusion protein has an apparent molecular weight of 59 kDa as estimated by SDS–PAGE. By using purified enzyme, a mg-scale synthesis was performed with GDP-fucose and methyl β-galactose as donor and acceptor, respectively. A total 4.8 mg of disaccharide product was purified by Bio-Gel P2 gel filtration, and subsequently used for $^1$H and $^{13}$C NMR spectroscopic analysis. Signals were found for a newly introduced fucose residue in the $^1$H NMR spectrum (Figure 4) of the disaccharide product. Therefore, the structure of the disaccharide product was confirmed to be Fucα1,2GalOMe. It can be envisioned that the efficient synthesis of the oligosaccharide could pave the way for the development of a well-structured glycoconjugate vaccine against infantile diarrhea *E. coli* O128.

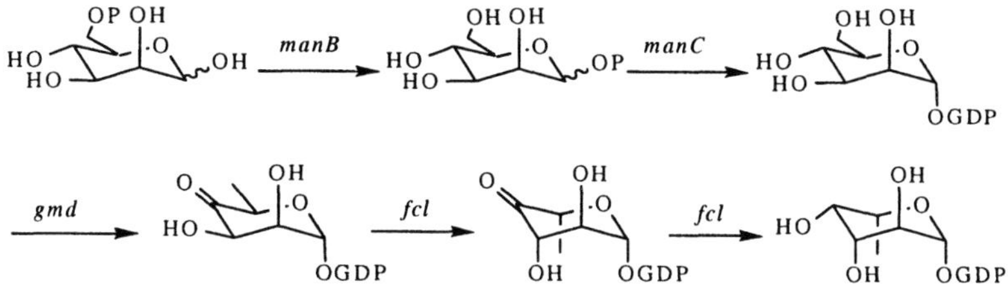

*Scheme 11. Biosynthetic pathway of GDP-fucose*

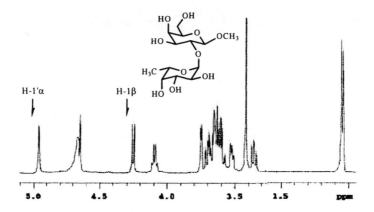

Figure 4  500 MHz proton NMR spectrum of Fucα1,2GalOMe

## *Escherichia coli* O86 antigen Biosynthesis

Several microorganisms possess human blood-group activity and this specificity is often associated with the bacterial O-antigen *(90)*. Based on chemical and serological investigations, Springer *et al. (91,92)* and Kochibe *et al.(93)* have shown that the lipopolysaccharide from *E. coli* O86 contains the blood group B trisaccharide partial structure. The complete structure (Scheme 12) of the LPS O-specific side chains from *E. coli* O86 determined by Addersson *et al. (94)* confirmed that *E. coli* O86 and human blood B type cells share a similar oligosaccharide antigen epitope.

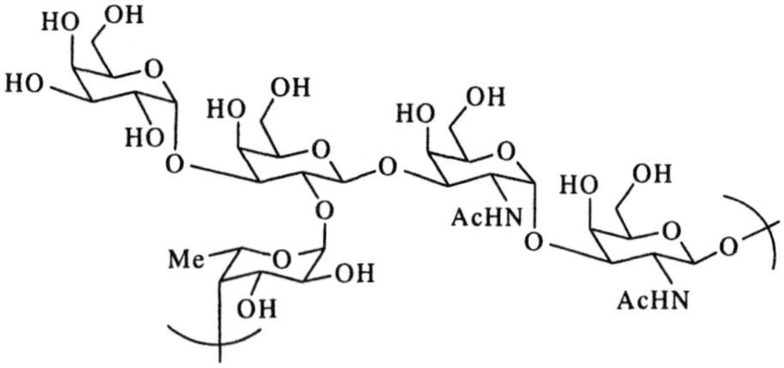

Scheme 12. Structure of *E. coli* O86 antigen

Similar to *E. coli* O128, the *E. coli* O86 antigen units also consist of one L-fucose, two galactose and two *N*-acetylgalactosamine residues. Therefore the cluster would contain genes responsible for GDP-L-fucose synthesis, genes coding glycosyltransferases to assemble the O unit, and genes for O-antigen process, including flippase (*wzx*), polymerase (*wzy*) and chain length determinant protein (*wzz*). Since there is a fucose residue in O86 repeat unit, the GDP-fucose biosynthetic enzymes is expected to be present within its gene cluster, including *manA, manB, manC, gmd and fcl*. All the enzymes, except the first one *manA*, were identified in the *E. coli* O86 biosynthesis gene cluster. The GDP-mannose mannosyl hydrolase (*gmm*) was also found between *fcl* and *manC*, which catalyzes the hydrolysis of GDP-α-mannose to yield GDP and β-mannose. This enzyme has been suggested to participate in the regulation of cell wall biosynthesis by influencing the concentration of GDP-mannose in the cell *(95)*. A total of five sugar residues are present within O86 antigen repeat unit. GalNAc was known to be the sugar residue that links the O antigen to the core region of lipopolysaccharide from *E. coli* O86 strain by the *wecA* gene *(94)*. Therefore, we expect to find only four glycosyltransferases within O86 antigen biosynthesis gene cluster. The WecA *(95)* was proposed to transfer the first GalNAc phosphate to UndP on the cytoplasmic face of the cell membrane (Scheme 13).

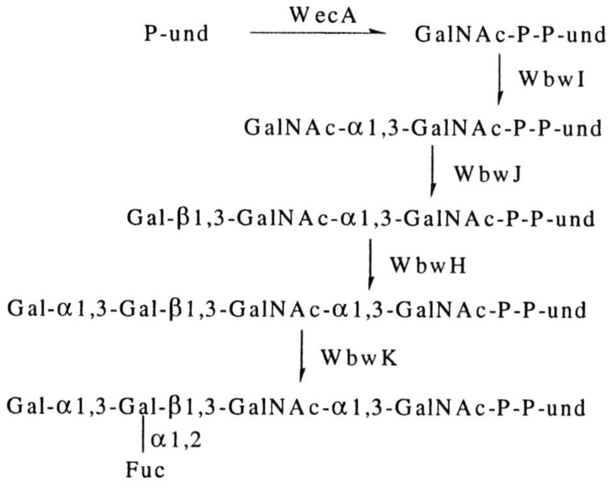

$$\text{P-und} \xrightarrow{\text{WecA}} \text{GalNAc-P-P-und}$$
$$\downarrow \text{WbwI}$$
$$\text{GalNAc-}\alpha\text{1,3-GalNAc-P-P-und}$$
$$\downarrow \text{WbwJ}$$
$$\text{Gal-}\beta\text{1,3-GalNAc-}\alpha\text{1,3-GalNAc-P-P-und}$$
$$\downarrow \text{WbwH}$$
$$\text{Gal-}\alpha\text{1,3-Gal-}\beta\text{1,3-GalNAc-}\alpha\text{1,3-GalNAc-P-P-und}$$
$$\downarrow \text{WbwK}$$
$$\text{Gal-}\alpha\text{1,3-Gal-}\beta\text{1,3-GalNAc-}\alpha\text{1,3-GalNAc-P-P-und}$$
$$|\alpha\text{1,2}$$
$$\text{Fuc}$$

*Scheme 13 Proposed assembly of E. coli O86 repeat unit*

*WbwH* gene shares moderate similarity with some putative bacterial α-galactosyltransferases such as α(1-3) *WbgM* from *E. coli* O55 and α(1-6) CpsF from *S. thermophilus*. Therefore, *WbwH* gene was assigned tentatively as an α-1,3-galactosyltransferase that makes the Galα1,3Gal linkage. Based on the similar gene sequence analysis and deduction, *WbwI* gene encodes an α1,3-*N*-acetylgalactosaminyl transferase that forms the GalNAcα1,3GalNAc linkage of O unit. *WbwJ* gene encodes a galactosyltransferases, which transfers a Gal residue to GalNAc in β(1→3) linkage. *WbwK* gene encodes the fucosyltransferase that links a fucose to the O86 antigen repeat unit.

Currently, four putative glycosyltransferase genes in *E. Coli* O86 O-antigen biosynthetic pathway have been successfully cloned and overexpressed in BL21(DE3), three of which, viz., *WbwH*, *WbwJ*, and *WbwK*, have been proven to be active. Trisaccharide α1,2-Fucβ1,3-Galβ1,3-GalNAcOMe has been synthesized in step-wise manner using *WbwJ* and *WbwK* sequentially (unpublished data). A new dimension will be to characterize each glycosyltransferase biochemically, then use them to synthesize the complete repeating unit of O-antigen, which will then be used as substrate to make the O-antigen polysaccharide.

## Conclusion and Outlook

Significant progress in the study of the enzymatic biosynthesis of complex carbohydrate has been made with the development in protein purification, molecular genetics, and new methods of enzymological analysis. Bioinformatics provide a large amount of putative candidates for carbohydrate active enzymes. The combined enzymatic and genetic approach has overcome formidable obstacles ubiquitous to the study of oligosaccharide and polysaccharide, and has begun to yield new information which definitively addresses basic enzymological issues relevant to these complex carbohydrates. The future direction will be large-scale preparation of complex oligosaccharide and polysaccharides with whole-cell systems. Further development of enzymatic methods will allow synthetic biochemists to create important molecular tools for biochemical, biophysical, and medical applications.

## References

1. Dwek, R. A. *Chem Rev* **1996**, *96*, 683-720.
2. Jennings, H. *J Infect Dis* **1992**, *165 Suppl 1*, S156-9.
3. Palcic, M. M. *Curr Opin Biotechnol* **1999**, *10*, 616-24.
4. Crout, D. H. G.; Critchley, P.; Muller, D.; Scigelova, M.; Singh, S.; Vic, G. *Special Publication - Royal Society of Chemistry* **1999**, *246*, 15-23.

5.  van Rantwijk, F.; Woudenberg-van Oosterom, M.; Sheldon, R.A. *Journal of Molecular Catalysis B: Enzymetic* **1999**, *6*, 511-532.
6.  Vocadlo, D. J.; Withers, S. G. *Carbohydrates in Chemistry and Biology* **2000**, *2*, 723-844.
7.  Roseman, S. *Chem Phys Lipids* **1970**, *5*, 270-97.
8.  Watkins, W. M. *Carbohydr Res* **1986**, *149*, 1-12.
9.  Hehre, E. J. *Carbohydr Res* **2001**, *331*, 347-68.
10. Unligil, U. M.; Rini, J. M. *Curr Opin Struct Biol* **2000**, *10*, 510-7.
11. Davies, G. J.; Henrissat, B. *Biochem Soc Trans* **2002**, *30*, 291-7.
12. Kaneco, M.; Nishihara, S.; Narimatsu, H.; Saitou,N. **2001**, *13*, 147-155.
13. Johnson, K. F. *Glycoconj J* **1999**, *16*, 141-6.
14. Blixt, O.; van Die, I.; Norberg, T.; van den Eijnden, D. H. *Glycobiology* **1999**, *9*, 1061-71.
15. Izumi, M.; Shen, G. J.; Wacowich-Sgarbi, S.; Nakatani, T.; Plettenburg, O.; Wong, C. H. *J Am Chem Soc* **2001**, *123*, 10909-18.
16. DeAngelis, P. L. *Glycobiology* **2002**, *12*, 9R-16R.
17. Breton, C.; Oriol, R.; Imberty, A. *Glycobiology* **1998**, *8*, 87-94.
18. Oriol, R.; Mollicone, R.; Cailleau, A.; Balanzino, L.; Breton, C. *Glycobiology* **1999**, *9*, 323-34.
19. Krusius, T.; Finne, J.; Rauvala, H. *Eur J Biochem* **1978**, *92*, 289-300.
20. Kobata, A. *Eur J Biochem* **1992**, *209*, 483-501.
21. Van den Steen, P.; Rudd, P. M.; Dwek, R. A.; Opdenakker, G. *Crit Rev Biochem Mol Biol* **1998**, *33*, 151-208.
22. Hakomori, S. *Biochem Soc Trans* **1993**, *21 ( Pt 3)*, 583-95.
23. Galili, U.; Mandrell, R. E.; Hamadeh, R. M.; Shohet, S. B.; Griffiss, J. M. *Infect Immun* **1988**, *56*, 1730-7.
24. Galili, U.; Buehler, J.; Shohet, S. B.; Macher, B. A. *J Exp Med* **1987**, *165*, 693-704.
25. Parker, W.; Lundberg-Swanson, K.; Holzknecht, Z. E.; Lateef, J.; Washburn, S. A.; Braedehoeft, S. J.; Platt, J. L. *Hum Immunol* **1996**, *45*, 94-104.
26. Galili, U.; Shohet, S. B.; Kobrin, E.; Stults, C. L.; Macher, B. A. *J Biol Chem* **1988**, *263*, 17755-62.
27. Galili, U.; Macher, B. A. *J Natl Cancer Inst* **1989**, *81*, 178-9.
28. Avila, J. L.; Rojas, M.; Galili, U. *J Immunol* **1989**, *142*, 2828-34.
29. Galili, U. *Immunol Ser* **1991**, *55*, 355-73.
30. Galili, U. *Immunol Today* **1993**, *14*, 480-2.
31. Thall, A.; Etienne-Decerf, J.; Winand, R. J.; Galili, U. *Acta Endocrinol (Copenh)* **1991**, *124*, 692-9.
32. Galili, U. *Biochimie* **2001**, *83*, 557-63.
33. Avila, J. L.; Rojas, M.; Garcia, L. *J Clin Microbiol* **1988**, *26*, 1842-7.
34. Galili, U. *Transplant Proc* **1999**, *31*, 940-1.
35. Galili, U.; Wang, L.; LaTemple, D. C.; Radic, M. Z. *Subcell Biochem* **1999**, *32*, 79-106.

36. Koeller, K. M.; Wong, C. H. *Chem Rev* **2000**, *100*, 4465-94.
37. Crout, D. H.; Vic, G. *Curr Opin Chem Biol* **1998**, *2*, 98-111.
38. Wymer, N.; Toone, E. J. *Curr Opin Chem Biol* **2000**, *4*, 110-9.
39. Ichikawa, Y.; Wang, R.; Wong, C. H. *Methods Enzymol* **1994**, *247*, 107-27.
40. Shao, J.; Zhang, J.; Kowal, P.; Lu, Y.; Wang, P. G. *Biochem Biophys Res Commun* **2002**, *295*, 1-8.
41. Shao, J.; Zhang, J.; Kowal, P.; Wang, P. G. *Appl Environ Microbiol* **2002**, *68*, 5634-40.
42. Zhang, J.; Kowal, P.; Fang, J.; Andreana, P.; Wang, P. G. *Carbohydr Res* **2002**, *337*, 969-76.
43. Sharon, N.; Ofek, I. *Glycoconj J* **2000**, *17*, 659-64.
44. Sharon, N.; Ofek, I. *Crit Rev Food Sci Nutr* **2002**, *42*, 267-72.
45. Fang, J.; Li, J.; Chen, X.; Zhang, Y.; Wang, J.; Guo, Z.; Zhang, W.; Yu, L.; Wang, P. G. *J Am Chem Soc* **1998**, *120*, 6635-6638.
46. Haynie, S. L.; Whitesides, G. M. *Appl Biochem Biotechnol* **1990**, *23*, 155-70.
47. Hokke, C. H.; Zervosen, A.; Elling, L.; Joziasse, D. H.; van den Eijnden, D. H. *Glycoconj J* **1996**, *13*, 687-92.
48. Ichikawa, Y.; Liu, L. ; SHen, G.J.; Wong, C. H. *J. Am. Chem. Soc.* **1991**, *113*, 6300-2.
49. Zervosen, A.; Elling, L. *J. Am. Chem. Soc.* **1996**, *118*, 1836-1840.
50. Koizumi, S.; Endo, T.; Tabata, K.; Ozaki, A. *Nat Biotechnol* **1998**, *16*, 847-50.
51. Endo, T.; Koizumi, S.; Tabata, K.; Kakita, S.; Ozaki, A. *Carbohydr Res* **1999**, *316*, 179-83.
52. Endo, T.; Koizumi, S.; Tabata, K.; Ozaki, A. *Appl Microbiol Biotechnol* **2000**, *53*, 257-61.
53. Endo, T.; Koizumi, S. *Curr Opin Struct Biol* **2000**, *10*, 536-41.
54. Endo, T.; Koizumi, S.; Tabata, K.; Kakita, S.; Ozaki, A. *Carbohydr Res* **2001**, *330*, 439-43.
55. Chen, X.; Fang, J.; Zhang, J.; Liu, Z.; Shao, J.; Kowal, P.; Andreana, P.; Wang, P. G. *J Am Chem Soc* **2001**, *123*, 2081-2.
56. Nahaka, J.; Liu, Z.; Gemeiner, P.; Wang, P. G. *Biotechnology Letters* **2002**, *24*, 925-930.
57. Nahaka, J.; Liu, Z.; Chen, X.; Wang, P. G. *chemistry -- A European Journal* **2003**, *9*, 372-377.
58. Zhang, J.; Wu, B.; Zhang, Y.; Kowal, P.; Wang, P. G. *Org Lett* **2003**, *5*, 2583-6.
59. Shao, J.; Zhang, J.; Kowal, P.; Lu, Y.; Wang, P. G. *Chem Commun (Camb)* **2003**, 1422-3.
60. Chen, X.; Liu, Z.; Zhang, J.; Zhang, W.; Kowal, P.; Wang, P. G. *ChemBioChem* **2002**, *3*, 47-53.

61. Wang, P. G.; Chen, X.; Zhang, J.; Kowal, P.; Andreana, P. *222nd ACS National Meeting, Chicago, IL, United States, August 26-30.* **2001**, MEDI -126.
62. Antoine, T.; Priem, B.; Heyraud, A.; Greffe, L.; Gilbert, M.; Wakarchuk, W. W.; Lam, J. S.; Samain, E. *Chembiochem* **2003**, *4*, 406-12.
63. Dumon, C.; Priem, B.; Martin, S. L.; Heyraud, A.; Bosso, C.; Samain, E. *Glycoconj J* **2001**, *18*, 465-74.
64. Priem, B.; Gilbert, M.; Wakarchuk, W. W.; Heyraud, A.; Samain, E. *Glycobiology* **2002**, *12*, 235-40.
65. Bettler, E.; Samain, E.; Chazalet, V.; Bosso, C.; Heyraud, A.; Joziasse, D. H.; Wakarchuk, W. W.andImberty, A.; Geremia, A. R. *Glycoconj J* **1999**, *16*, 205-12.
66. Samain, E.; Chazalet, V.; Geremia, R. A. *J Biotechnol* **1999**, *72*, 33-47.
67. Samain, E.; Drouillard, S.; Heyraud, A.; Driguez, H.; Geremia, R. A. *Carbohydr Res* **1997**, *302*, 35-42.
68. Raetz, C. R. *Annu Rev Biochem* **1990**, *59*, 129-70.
69. Raetz, C. R.; Dowhan, W. *J Biol Chem* **1990**, *265*, 1235-8.
70. Galloway, S. M.; Raetz, C. R. *J Biol Chem* **1990**, *265*, 6394-402.
71. Takayama, K.; Qureshi, N. *In Bacterial endotoxin lipopolysaccharides; CRC press: Boca Raton, Florida.* **1992**.
72. Rietschel, E.T. *In Bacterial endotoxin lipopolysaccharides; CRC press: Boca Raton, Florida.* **1992**.
73. Raetz, C. R.; Whitfield, C. *Annu Rev Biochem* **2002**, *71*, 635-700.
74. Amor, K.; Heinrichs, D. E.; Frirdich, E.; Ziebell, K.; Johnson, R. P.; Whitfield, C. *Infect Immun* **2000**, *68*, 1116-24.
75. Knirel, Y.A.; Kochetkov, N.K. *Biochemistry - Moscow* **1994**, 1325-1383.
76. Reeves, P. R. *In Bacterial cell wall; Elsevier, Amsterdam.* **1994**.
77. Hobbs, M.; Reeves, P. R. *Mol Microbiol* **1994**, *12*, 855-6.
78. Wang, L.; Jensen, S.; Hallman, R.; Reeves, P. R. *FEMS Microbiol Lett* **1998**, *165*, 201-6.
79. Reeves, P. R.; Hobbs, M.; Valvano, M. A.; Skurnik, M.; Whitfield, C.; Coplin, D.; Kido, N.; Klena, J.; Maskell, D.; Raetz, C. R.; Rick, P. D. *Trends Microbiol* **1996**, *4*, 495-503.
80. McGrath, B. C.; Osborn, M. J. *J Bacteriol* **1991**, *173*, 649-54.
81. Macpherson, D. F.; Manning, P. A.; Morona, R. *Gene* **1995**, *155*, 9-17.
82. Robbins, P. W.; Bray, D.; Dankert, B. M.; Wright, A. *Science* **1967**, *158*, 1536-42.
83. Morona, R.; Mavris, M.; Fallarino, A.; Manning, P. A. *J Bacteriol* **1994**, *176*, 733-47.
84. Daniels, C.; Vindurampulle, C.; Morona, R. *Mol Microbiol* **1998**, *28*, 1211-22.

85. Sengupta, P.; Bhattacharyya, T.; Shashkov, A. S.; Kochanowski, H.; Basu, S. *Carbohydr Res* **1995**, *277*, 283-90.
86. Sengupta, P.; Bhattacharyya, T.; Majumder, M.; Chatterjee, B. P. *FEMS Immunol Med Microbiol* **2000**, *28*, 133-7.
87. Ginsburg, V. *J Biol Chem* **1961**, *236*, 2389-93.
88. Alexander, D. C.; Valvano, M. A. *J Bacteriol* **1994**, *176*, 7079-84.
89. Shao, J.; Li, M.; Jia, Q.; Lu, Y.; Wang, P. G. *FEBS Lett* **2003**, *553*, 99-103.
90. Springer, G. F.; Williamson, P.; Readler, B. L. *Ann N Y Acad Sci* **1962**, *97*, 104-10.
91. Springer, G. F.; Wang, E. T.; Nichols, J. H.; Shear, J. M. *Ann N Y Acad Sci* **1966**, *133*, 566-79.
92. Springer, G. F. *Ann N Y Acad Sci* **1970**, *169*, 134-52.
93. Kochibe, N.; Iseki, S. *Jpn J Microbiol* **1968**, *12*, 403-11.
94. Andersson, M.; Carlin, N.; Leontein, K.; Lindquist, U.; Slettengren, K. *Carbohydr Res* **1989**, *185*, 211-23.
95. Frick, D. N.; Townsend, B. D.; Bessman, M. J. *J Biol Chem* **1995**, *270*, 24086-91.

Chapter 15

# Enzymatic Polymerization: In Vitro Synthesis of Glycosaminoglycans and Their Derivatives

Shiro Kobayashi, Shun-ichi Fujikawa, Ryosuke Itoh, Hidekazu Morii, Hirofumi Ochiai, Tomonori Mori, and Masashi Ohmae

Department of Materials Chemistry, Graduate School of Engineering, Kyoto University, Kyoto 615-8510, Japan

Hyaluronan and chondroitin were prepared as representative molecules of glycosaminoglycans by hyaluronidase-catalyzed polymerizations. $N$-Acetyl-hyalobiuronate oxazoline (**1**) and $N$-acetyl-chondrosine oxazoline (**2**) were designed and synthesized as transition state analogue substrate monomers for hyaluronidase catalysis. Monomer **1** was effectively catalyzed by the enzyme at pH 7.1 and 30°C, giving rise to synthetic hyaluronan (**3**). Monomer **2** was also polymerized by the enzyme at pH 7.5 and 30 °C, affording synthetic chondroitin (**4**). Unnatural chondroitins (**6**) were obtained from monomers (**5**) in a similar manner. All these polymerizations proceeded through perfect regio-selective and stereo-controlled ring-opening polyaddition.

Glycosaminoglycans (GAGs) are one of the naturally occurring linear heteropolysaccharides, which include six biomacromolecules of hyaluronan (hyaluronic acid, HA), heparin/heparan sulfate, chondroitin (Ch) and chondroitin sulfate (ChS), dermatan sulfate and keratan sulfate (*1*). These exist widely in living systems as components of extracellular matrices (ECMs) (*2*) and on cell surface (*3*). In particular, HA, Ch and ChS are their main constituents of ECMs in dermis and cartilage, in which large molecular complexes are formed by association of these molecules (*2*). GAGs play crucial roles in differentiation and proliferation of cells, tissue morphogenesis and wound healing via signaling by interactions with growth factors and morphogens (*2, 4-11*). GAGs having a single repeating structural motif of a disaccharide composed of uronic acid and hexosamine, show a great deal of structural diversity causing discrete structural forms generated by complex patterns of deacetylation, sulfation and epimerization, which are found in particular tissues and influenced by disease and aging (*12, 13*). GAGs containing glucosamines as a hexosamine constituent are called as glucosaminoglycans, and those containing galactosamines as galactosaminoglycans (*1*). HA and Ch are the most well-known glucosaminoglycan and galactosaminoglycan, composed of β(1→4) linked β-D-glucuronyl-(1→3)-*N*-acetyl-D-glucosamine (GlcAβ(1→3)GlcNAc; *N*-acetyl-hyalobiuronate) and β-D-glucuronyl-(1→3)-*N*-acetyl-D-galactosamine (GlcAβ(1→3)GalNAc; *N*-acetyl-chondrosine) repeating units, respectively.

The multi-functional HA is biologically synthesized in cell surface membrane by hyaluronan synthase with two types of sugar nucleotides as substrates, UDP-GlcNAc and UDP-GlcA (*14*). Ch exists as a carbohydrate part of proteoglycans in *C. elegans* (*15*) or in mammalians as a precursor of ChS, mainly existing in cartilage, cornea, and brain matrices (*16*). A number of reports have been published which described the biological functions of Ch and ChS, for example, maintaining cartilage elasticity (*17*), and promotion of neurite outgrowth (*18*) and neuronal migration (*19*). Biosynthesis of Ch is performed in the Golgi apparatus by the catalysis of chondroitin synthase (*20, 21*) or other glycosyltransferases (*22, 23*), followed by selective sulfation by several kinds of specific sulfotransferases (*24*). Detailed mechanisms for the production of Ch and ChS have been unclear.

Thus HA, Ch and ChS play crucial roles in living systems; structurally well-defined samples are essential to elucidate their molecular functions for vital activities in living system. Their chemical or biochemical synthesis is challenging (*25*), and a facile and efficient method to prepare these biomacromolecules has been an important problem.

We have achieved the synthesis of structurally well-defined natural and unnatural oligo- and polysaccharides via enzymatic polymerization utilizing natural glycosyl hydrolases as catalysts (*26-31*); cellulose and xylan prepared by cellulase (*32-34*), an amylose oligomer by amylase (*35*), chitin by chitinase (*36, 37*), alternatingly 6-*O*-methylated cellulose by cellulase (*38*), and a cellulose-xylan hybrid polysaccharide by xylanase (*39*). All these reactions proceeded in perfectly regio-selective and stereo-controlled manners. Mutant glycosidases are also effective as catalysts for the glycosylation reactions (*40-43*). The polymerization employs a substrate monomer activated at the anomeric carbon, which reduces activation energy for the reaction, resulting in a polymer via repeated regio-selective and stereo-controlled glycosylations. For instance, a β-D-xylopyranosyl-(1→4)-β-D-glucopyranosyl fluoride was used as an activated substrate monomer for xylanase catalysis, giving rise to a cellulose-xylan hybrid polysaccharide with an alternating structure of (1→4)-β-D-xylopyranoside and (1→4)-β-D-glucopyranoside (*39*). Synthesis of chitin was achieved by employing an *N,N'*-di-acetyl-chitobiose oxazoline derivative as an activated substrate monomer for chitinase from *Bacillus* sp., which belongs to the glycoside hydrolase family 18 (*44, 45*). The structure of the oxazoline monomer is close to an oxazolinium transition state in the hydrolysis of chitin by the enzyme. Therefore, the high affinity of the monomer to the enzyme drives the reaction to repeated glycosidic bond formation, resulting in production of chitin polymer.

The present paper focuses on precision synthesis of natural HA (*46*) and Ch (*47*) with well-defined structures of biological importance, via hyaluronidase-catalyzed polymerization of sugar oxazoline derivatives (Scheme 1). Similarly, unnatural Chs (**6**) were prepared using the same enzyme (*47*) (Scheme 2). These reactions provide a facile and efficient approach to synthesis of GAGs with well-defined structures.

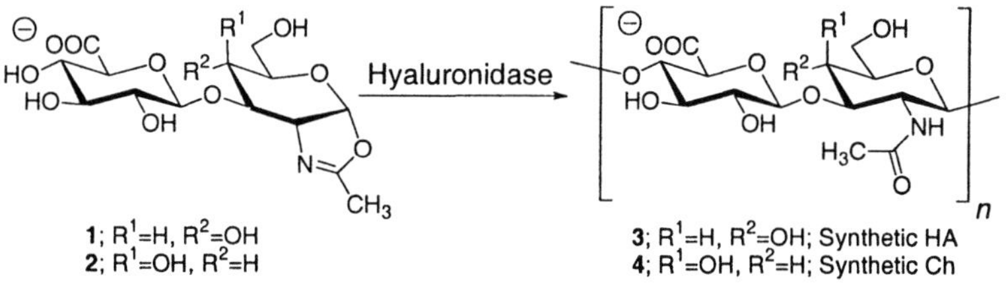

**1**; $R^1$=H, $R^2$=OH
**2**; $R^1$=OH, $R^2$=H

**3**; $R^1$=H, $R^2$=OH; Synthetic HA
**4**; $R^1$=OH, $R^2$=H; Synthetic Ch

*Scheme 1*

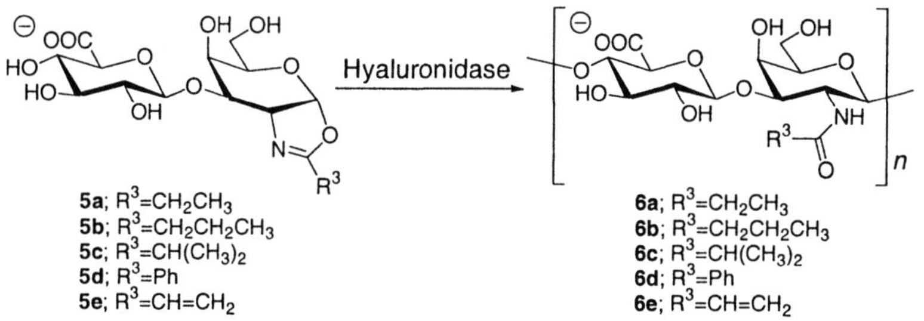

**5a**; $R^3$=CH$_2$CH$_3$
**5b**; $R^3$=CH$_2$CH$_2$CH$_3$
**5c**; $R^3$=CH(CH$_3$)$_2$
**5d**; $R^3$=Ph
**5e**; $R^3$=CH=CH$_2$

**6a**; $R^3$=CH$_2$CH$_3$
**6b**; $R^3$=CH$_2$CH$_2$CH$_3$
**6c**; $R^3$=CH(CH$_3$)$_2$
**6d**; $R^3$=Ph
**6e**; $R^3$=CH=CH$_2$

*Scheme 2*

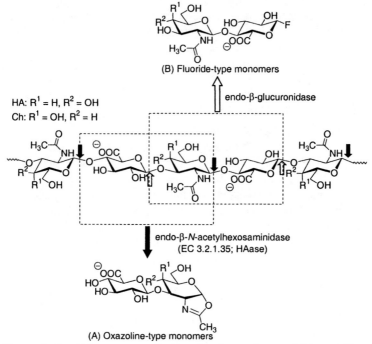

*Figure 1. Possible monomer designs for the synthesis of HA and Ch. Allows show potential bonds for enzymatic cleavage by HAase (black arrows) and endo-β-glucuronidase (white arrows).*

## Monomer Designs

Structures of HA and Ch are illustrated in Figure 1. There are two kinds of glycoside hydrolases responsible for their catabolism. One is endo-β-*N*-acetylhexosaminidase usually called hyaluronidase (HAase; EC 3.2.1.35), which can hydrolyze β(1→4)-*N*-acetyl-hexosaminide linkages (*48*). The other is endo-β-glucuronidase catalyzing hydrolysis of β(1→3)-glucuronide linkages (*48, 49*). These two distinct enzymes are classified to the "retaining" enzymes (*50*), by which the stereochemistry of the anomeric carbon retains the β-form after hydrolysis. According to their catalysis mechanisms, monomers for the synthesis of HA and Ch can be considered via two modes of molecular designs, that is, (A) oxazoline-type monomers for HAases, and (B) fluoride-type monomers for β-glucuronidases. The former is feasible due to availability of the commercial enzymes, on the other hand the latter is not practical because the enzymes are found only in leeches and small marine crustaceans (*48*) or in rabbit liver (*49*). Therefore, the oxazoline-type monomers were selected.

## Monomer Synthesis

Monomer **1** for HA synthesis was prepared according to the reactions outlined in Scheme 3A (*46*): Methyl (2,3,4-tri-*O*-acetyl-α-D-glucopyranosyl bromide)uronate **7** (*51*) was glycosidated with benzyl 2-acetamido-2-deoxy-4,6-*O*-isopropylidene-β-D-glucopyranoside **8** (*52*) by silver triflate to give a disaccharide derivative **9**. The 4,6-*O*-isopropylidene acetal and the 1-*O*-benzyl group of **9** were removed together by hydrogenation, and the generated hydroxyl groups were acetylated to provide **10**. Formation of the oxazoline ring was performed by the action of trimethylsilyl triflate (TMSOTf) to produce **11**. All *O*-acetyl protections of **11** were removed by catalytic amount of sodium methoxide and then methyl ester was hydrolyzed in carbonate buffer at pH 10.5, giving rise to the target monomer **1**.

Monomer **2** for Ch synthesis was also synthesized following the reactions shown in Scheme 3B (*47*): Methyl (2,3,4-tri-*O*-acetyl-α-D-glucopyranosyl trichloroacetimidate)uronate **13** (*53*) was coupled to benzyl 2-azido-4,6-*O*-benzylidene-2-deoxy-β-D-galactopyranoside **14** (*54*) via glycosidation activated by TMSOTf to generate a disaccharide derivative **15**. The 2-azido group of **15** was converted to 2-acetamido group by the action of thioacetic acid and then the 4,6-*O*-benzylidene acetal was cleaved by acid hydrolysis followed by acetylation of the formed hydroxyl groups to give **16**. The 1-*O*-benzyl group of **16** was removed by hydrogenation and the generated hydroxyl group was acetylated. Formation of the oxazoline ring was carried out by using TMSOTf to afford **17**. All *O*-acetyl protecting groups of **17** were then removed by catalytic amount of sodium methoxide in methanol followed by hydrolysis of methyl ester in carbonate buffer at pH 10.5, generating the target monomer **2**.

Monomers **5a-5e** for the synthesis of unnatural Chs were prepared from compound **15** via 8 step chemical reactions (*47*). The 4,6-*O*-benzylidene group of **15** was removed by acid hydrolysis followed by acetylation of the formed hydroxyl groups. The 2-azido and the 1-*O*-benzyl groups were hydrogenated together. The generated 2-amino group was acylated through the reaction with the corresponding acid anhydrides (for **5a** and **5c**) or acid chlorides (for **5b**, **5d** and **5e**) in methanol prior to acetylation of the 1-hydroxyl group. Formation of the oxazoline ring was achieved by the action of TMSOTf followed by removal of all *O*-acetyl protecting groups and hydrolysis of the methyl ester to obtain the 2-substituted oxazoline monomers **5a-5e**.

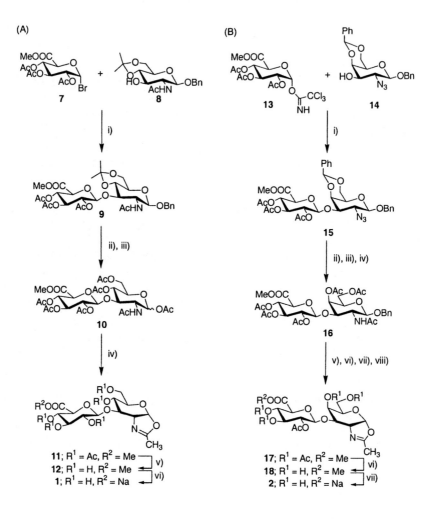

Scheme 3. (A) i) AgOTf, Me₂NC(O)NMe₂ / CH₂Cl₂, -40°C ~ rt, 22h, 29%, ii) Pd-C, H₂ / MeOH, rt, 72h, 98%, iii) Ac₂O / pyridine, rt, 6h, 98%, iv) TMSOTf / ClCH₂CH₂Cl, 50°C, 7h, quant., v) MeONa / MeOH, vi) carbonate buffer (50 mM, pH 10.5 ). (B) i) TMSOTf, MS4A / CH₂Cl₂, 91%, ii) AcSH, 86%, iii) 80% aqAcOH, reflux, iv) Ac₂O / pyridine, 81% (2 steps), v) Pd(OH)₂-C, H₂ / MeOH, vi) Ac₂O / pyridine, quant. (2 steps), vi) TMSOTf / CH₂Cl₂, 89%, vii) MeONa / MeOH, viii) carbonate buffer (50 mM, pH 10.5), 82%.

# Enzymatic Polymerization

**Synthesis of natural HA**

Monomer **1** was subjected to the enzymatic reaction catalyzed by ovine testicular HAase (OTH) and bovine testicular HAase (BTH) at pH 7.1 and 30°C. An NMR system was used for monitoring the reaction progresses by measuring the H-1 proton of **1**, kept at the reaction temperature of 30°C. Without enzymes, **1** gradually disappeared through oxazoline-ring opening, resulting in a disaccharide of *N*-acetyl-hyalobiuronate. After 64 h, 27% of **1** remained in the reaction mixture. Addition of BTH accelerated the consumption of **1** compared to the non-enzymatic reduction. The reaction was finished after 64 h. OTH catalyzed the reaction more effectively, and **1** was completely consumed after 40 h. These results indicate that **1** is catalyzed by HAases leading to oxazoline ring-opening reaction. SEC analysis of the mixture from each enzymatic reaction revealed a polymeric product formation.

$^1$H and $^{13}$C NMR data of the product confirmed the GlcAβ(1→3)GlcNAcβ(1→4) repeating structure of HA; the specific doublet signals at δ 4.55 ($J$ = 7.48 Hz) and 4.47 ($J$ = 6.96 Hz) derived from two kinds of the anomeric protons of the respective β-linked GlcNAc and GlcA on the $^1$H NMR spectrum, and characteristic signals at δ 103.89 (C1 of GlcA), 101.34 (C1 of GlcNAc), 83.43 (C3 of GlcNAc), and 80.79 (C4 of GlcA) on the $^{13}$C NMR spectrum. These results show that **1** was catalyzed by HAases, giving rise to synthetic HA (**3**) via ring-opening polyaddition in a perfect regio-selective and stereo-controlled manner. All spectroscopic data of **3** were in good agreement with those of naturally occurring HA. Yields and molecular weight values of the synthetic HAs were 39% and $M_n$ 1.74×10$^4$ ($M_w$ 6.69×10$^4$; OTH), and 52% and $M_n$ 1.35×10$^4$ ($M_w$ 4.12×10$^4$; BTH).

**Synthesis of natural Ch**

Figure 2 shows the reaction-time courses of **2** with H-OTH (Δ) and without enzyme (O). Monitoring the reaction progresses was performed by HPLC analyses. Consumption of **2** with H-OTH was significantly faster than that without enzyme, indicating the catalysis of the enzyme with **2** involving oxazoline ring-opening. A polymeric product was detected on the size exclusion chromatograms of the reaction mixture with the enzyme. Non-enzymatic reaction only afforded the hydrolysis compound from **2** of *N*-acetyl-chondrosine.

NMR analysis revealed that the polymeric product was synthetic Ch (**4**); from $^1$H NMR analysis, specific doublet signals were observed at δ 4.37 ($J$ = 7.53 Hz) and 4.30 ($J$ = 7.53 Hz) derived from two kinds of the anomeric protons of GalNAc and GlcA, respectively, and from $^{13}$C NMR, characteristic signals were detected at 105.07 (C1 of GlcA), 101.58 (C1 of GalNAc), 81.11 (C3 of

GalNAc), and 80.43 ppm (C4 of GlcA), respectively. These data show the formation of GlcAβ(1→3)GalNAcβ(1→4) repeating structure via perfect regioselective and stereo-controlled ring-opening polyaddition of **2**. All chemical shift values of **4** matched with those of natural Ch.

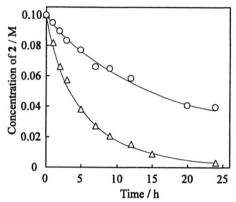

*Figure 2. Reaction-time courses of 2 with H-OTH (Δ) and without enzyme (O).*

Polymerization of **2** was optimized by varying the reaction condition (Table 1). Using HAases from various origins (entries 1-5), H-OTH produced **4** most effectively in a 50% yield (entry 3). Bee venom containing HAase showed a slight activity for the polymerization of **2**. Yields were affected by pH value of the reaction with H-OTH (entries 3, 6, 7, 10-12). The enzyme marked an optimum yield at pH 7.5 (entry 3). The reaction did not occur at pH 9.0 (entry 12). Shorter reaction time generated **4** having larger molecular weight of $M_n$ 4800 ($M_w$ 7100) (entry 8), which is close to that of naturally occurring Ch. Thus the reaction of **2** with H-OTH effectively provided **4** around at pH 7.5.

## Synthesis of unnatural Chs

Polymerization reaction of monomers **5a-5e** with H-OTH was performed under the conditions of pH 7.5 and 30°C. All of the reactions performed in the absence of H-OTH gave the disaccharides through hydrolysis of the corresponding monomers **5a-5e**. Interestingly, monomers of 2-ethyl (**5a**), 2-*n*-propyl (**5b**), 2-isopropyl (**5c**), and 2-vinyl (**5e**) oxazolines were significantly catalyzed by the enzyme, resulting in ring-opening polyaddition of the monomers. It is to be noted that the 2-ethyl oxazoline monomer (**5a**) was consumed at an almost identical rate as the 2-methyl derivative (**2**).

Table 2 indicates the polymerization results of monomers **5a-5e**. Unnatural *N*-propionyl (**6a**) and *N*-acryloyl (**6e**) derivatives of Ch were obtained from **5a** and **5e** in a 46% yield with molecular weight of $M_n$ 2700 ($M_w$ 3600; mainly 12-14 saccharides) and in a 19% yield with that of $M_n$ 3400 ($M_w$ 4600; mainly 16-18 saccharides), respectively (entries 1 and 5). Ring-opening polyaddition of **5a** and **5e** proceeded in a regio-selective and stereo-controlled manner. $^1$H and $^{13}$C NMR confirmed their structures of β(1→4)-linked β-D-glucuronyl-(1→3)-β-*N*-propionyl-D-galactosamine repeating unit (**6a**) and β-D-glucuronyl-(1→3)-β-*N*-acryloyl-D-galactosamine repeating unit (**6e**). Monomers **5b** and **5c** were oligomerized by the enzyme, giving rise to the *N*-butyryl (**6b**) and *N*-isobutyryl (**6c**) derivatives of Ch up to decasaccharide and octasaccharide in small yields (<1%), respectively, as determined by MALDI-TOF/MS (entries 2 and 3). No catalytic actions were observed during the reaction of **5d** with the enzyme, only affording the hydrolyzed disaccharide through oxazoline ring-opening (entry 4). Thus, the natural enzyme effectively acted as a catalyst for production of the unnatural Ch derivative such as *N*-propionyl (**6a**) and *N*-acryloyl (**6e**) derivatives, as well as for that of the natural Ch (**4**). 2-Substituents in the oxazoline ring can be varied in the approximate order 2-methyl ≥ 2-ethyl > 2-vinyl >> 2-*n*-propyl > 2-isopropyl.

Table 1. Enzymatic polymerization of 2 under various conditions.[a]

| Entry | Enzyme | pH | Time / h[b] | Polymer | | |
|---|---|---|---|---|---|---|
| | | | | Yield of 4 / %[c] | $M_n$[d] | $M_w$[d] |
| 1 | OTH | 7.5 | 23 | 35 | 2500 | 3200 |
| 2 | BTH | 7.5 | 40 | 10 | 2800 | 3600 |
| 3 | H-OTH | 7.5 | 23 | 50 | 2100 | 2500 |
| 4 | H-BTH | 7.5 | 40 | 29 | 2600 | 3400 |
| 5 | bee venom | 7.5 | 40 | ~ 1 | - | - |
| 6 | H-OTH | 6.0 | 1 | 28 | 1600 | 1700 |
| 7 | H-OTH | 7.0 | 9 | 47 | 1900 | 2200 |
| 8 | H-OTH | 7.5 | 3[e] | 16 | 4800 | 7100 |
| 9[f] | H-OTH | 7.5 | 3[e] | 30 | 4300 | 6400 |
| 10 | H-OTH | 8.0 | 33 | 49 | 2200 | 2700 |
| 11 | H-OTH | 8.5 | 48 | 32 | 2400 | 3100 |
| 12 | H-OTH | 9.0 | 74 | 0 | - | - |

[a] In a phosphate buffer: 50 mM, [**2**]: 0.1 M, amount of enzyme: 10 wt% for **2**, reaction temperature: 30°C.

[b] Indicating the time for complete consumption of **2**.

[c] Determined by HPLC (over tetrasaccharide).

[d] Determined by SEC using hyaluronan standards.

[e] Reaction was terminated at the indicated time.

[f] Reaction was performed at 20°C.

**Table 2. Enzymatic polymerization of monomers 5a-5e.**[a]

| Entry | Monomer | Enzyme | Time / h | Polymer Structure | Yield[b] / % | $M_n^c$ | $M_w^c$ |
|---|---|---|---|---|---|---|---|
| 1 | 5a | H-OTH | 35 | 6a | 46 | 2700 | 3600 |
| 2 | 5b | H-OTH | 122 | 6b | <1 | - | - |
| 3 | 5c | H-OTH | 168 | 6c | <1 | - | - |
| 4 | 5d | H-OTH | 239 | - | - | - | - |
| 5 | 5e | H-OTH | 24 | 6e | 19 | 3400 | 4600 |

[a] In a phosphate buffer at pH 7.5: 50mM, monomer concentration: 0.1 M, amount of enzyme: 10 wt% for monomer, reaction temperature: 30°C.
[b] Determined by HPLC containing products with molecular weight higher than tetrasaccharides.
[c] Determined by SEC calibrated with hyaluronan standards.

## Mechanisms of Polymerization by Hyaluronidase

Figure 3 illustrates postulated reaction mechanisms of HAase. Catalysis of the enzyme for hydrolysis of Ch proceeds through following steps as illustrated in A and B; the oxygen atom of β(1→4) glycosidic bond is immediately protonated in A after the recognition. Then the carbonyl oxygen atom of GalNAc at the donor site attacks the anomeric carbon atom intramolecularly from α-side as a catalytic nucleophile, which assists the glycosidic bond cleavage, resulting in the oxazolinium transition state in B. Water molecule nucleophilically attacks the anomeric carbon atom in the oxazolinium ring from β-side to achieve the hydrolysis with oxazolinium ring-opening. In the polymerization, protonation occurs at the nitrogen atom of **2** placed in the donor site. The resulting structure is just identical with the oxazolinium ion at the donor site in B. The 4-hydroxyl group of GlcA in the non-reducing end of growing chain or in monomer **2** located in the acceptor site attacks the anomeric carbon of the oxazolinium ion of **2** from β-side to form β(1→4) glycosidic linkage between GalNAc and GlcA as shown in C and D. Thus, **2** can be regarded as a "transition state analogue substrate monomer" that is readily recognized and activated via protonation, lowering the activation energy for the reaction. Sequential repetition of this regio-selective and stereo-controlled glycosylation is a ring-opening polyaddition of **2** catalyzed by the enzyme, giving rise to synthetic Ch.

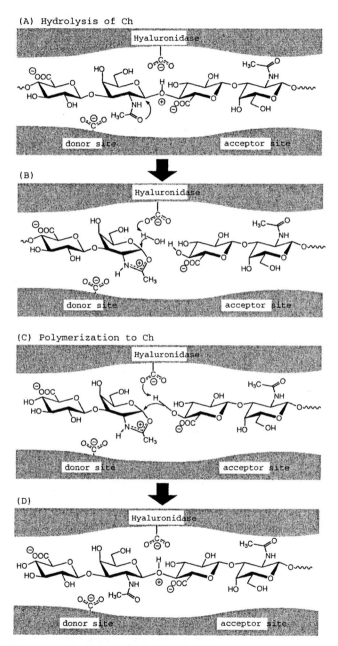

*Figure 3. Postulated reaction mechanisms of HAase in the hydrolysis of Ch (A and B) and the polymerization of 2 (C and D).*

## Conclusion

Enzymatic polymerizations to synthetic HA (**3**) and Ch (**4**) were achieved by the catalysis of HAases with two kinds of transition state analogue substrate monomers, *N*-acetyl-hyalobiuronate oxazoline (**1**) for the HA synthesis and *N*-acetyl-chondrosine oxazoline (**2**) for the Ch synthesis, respectively. These monomers were polymerized by the enzymes under neutral or weak alkaline conditions in a perfect regio-selective and stereo-controlled manner. The synthetic HA with $M_n$ ~20000 can be hardly obtained through chemical and enzymatic degradation of naturally occurring HA as well as the other synthetic methodologies. HA in mitotic cells has a similar molecular weight to the synthetic HA (*55*), which is expected to be a potent tool for investigation of bioactivities of intracellular HA. Molecular weight value of the synthetic Ch was close to that of naturally occurring Ch. Furthermore, the synthetic Ch has a well-defined structure of the β(1→4)-linked GlcAβ(1→3)GalNAc repeating unit, whereas that in living systems has a molecular diversity of small contents of L-iduronic acid (IdoA) that is a C-5 epimer of GlcA (*1*). The clear-cut structure will support the research of biological functions of IdoA in naturally occurring Ch. Synthesis of unnatural *N*-propionyl (**6a**) and *N*-acryloyl (**6e**) derivatives of Ch was successfully demonstrated by enzymatic polymerization with 2-ethyl (**5a**) and 2-vinyl (**5e**) oxazoline monomers by HAase. These unnatural Chs can not be provided through biosynthetic pathways. These have two types of amido groups in all of the galactosamine constituents of Ch with important functionalities for tissue engineering and biomaterial architecture. Particularly, the fact that the vinyl oxazoline monomer of **5e** is a good substrate for the HAase implies possible applications for the production of functionalized Ch with a well-defined structure such as new macromonomers, telechelics and gels utilizing the reactive vinyl group. In addition, their sulfated derivatives are under preparation via enzymatic polymerization. These will open the door to the investigation of correlation between the structures and the bioactivities at a molecular level.

## References

1. Prydz, K.; Dalen, K. T. *J. Cell Sci.* **2000**, *113*, 193-205.
2. Iozzo, R. V. *Ann. Rev. Biochem.* **1998**, *67*, 609-652.
3. Bernfield, M.; Götte, M.; Park, P. W.; Reizes, O.; Fitzgerald, M. L.; Lincecum, J.; Zako, M. *Ann. Rev. Biochem.* **1999**, *68*, 729-777.
4. Bullock, S. L.; Fletcher, J. M.; Beddington, R. S. P.; Wilson, V. A. *Genes Dev.* **1998**, *12*, 1894-1906.
5. Perrimon, N.; Bernfield, M. *Nature* **2000**, *404*, 725-728.

6. Lin, X.; Wei, G.; Shi, Z.; Dryer, L.; Esko, J. D.; Wells, D. E; Matzuk, M. M. *Dev. Biol.* **2000**, *224*, 299-311.
7. Lander, A. D.; Selleck, S. B. *J. Cell Biol.* **2000**, *148*, 227-232.
8. Chudo, H.; Toyoda, H. *Seikagaku* **2001**, *73*, 449-457.
9. Habuchi, O. *Cell Technol.* **2001**, *20*, 204-210.
10. Maeda, N. *Cell Technol.* **2001**, *20*, 1074-1083.
11. Selleck, S. B. *Trend Genet.* **2000**, *16*, 206-212.
12. Maccarana, M.; Sakura, Y.; Tawada, A.; Yoshida, K.; Lindahl, U. *J. Biol. Chem.* **1996**, *271*, 17804-17810.
13. Feyzi, E.; Saldeen, T.; Larsson, E.; Lindhal, U.; Salmivirta, M. *J. Biol. Chem.* **1998**, *273*, 13395-13398.
14. DeAngelis, P. L. *Cell. Mol. Life Sci.* **1999**, *56*, 670-682.
15. Bulik, D.A.; Wei, G.; Toyoda, H.; Kinoshita-Toyoda, A.; Waldrip, W. R.; Esko, J. D.; Robbins, P. W.; Selleck, S. B. *Proc. Natl. Acad. Sci. USA* **2000**, *97*, 10838-10843.
16. Faissner, A.; Clement, A.; Lochter, A.; Streit, A.; Mandl, C.; Schachner, M. *J. Cell Biol.* **1994**, *126*, 783-799.
17. Watanabe, H.; Kimata, K.; Line, S.; Strong, D.; Gao, L.; Christine, A. K.; Yamada, Y. *Nat. Genet.* **1994**, *7*, 154-157.
18. Clement, A. M.; Nadanaka, S.; Masayama, K.; Mandl, C.; Sugahara, K.; Faissner, A. *J. Biol. Chem.* **1998**, *273*, 28444-28453.
19. Maeda, N.; Noda, M. *J. Cell Biol.* **1998**, *142*, 203-216.
20. Kitagawa, H.; Uyama, T.; Sugahara, K. *J. Biol. Chem.* **2001**, *276*, 38721-38726.
21. Kitagawa, H.; Izumikawa, T.; Uyama, T.; Sugahara, K. *J. Biol. Chem.* **2003**, *278*, 23666-23671.
22. Sato, T.; Gotoh, M.; Kiyohara, K.; Akashima, T.; Iwasaki, H.; Kameyama, A.; Mochizuki, H.; Yada, T.; Inaba, N.; Togayachi, A.; Kudo, T.; Asada, M.; Watanabe, H.; Imamura, T.; Kimata, K.; Narimatsu, H. *J. Biol. Chem.* **2003**, *278*, 3063-3071.
23. Uyama, T.; Kitagawa, H.; Tanaka, J.; Tamura, J.; Ogawa, T.; Sugahara, K. *J. Biol. Chem.* **2003**, *278*, 3072-3078.
24. For example, see the following review: Kusche-Gullberg, M.; Kjellén, L. *Curr. Opin. Struc. Biol.* **2003**, *13*, 605-611.
25. For example, see the following review: Karst, N. A.; Linhardt, R. J. *Curr. Med. Chem.* **2003**, *10*, 1993-2031.
26. Kobayashi, S.; Shoda, S.; Uyama, H. *Adv. Polym. Sci.* **1995**, *121*, 1-30.
27. Kobayashi, S.; Shoda, S.; Uyama, H. In *Catalysis in Precision Polymerization*; Kobayashi, S., Ed.; John Wiley & Sons: Chichester, 1997; Chapter 8.
28. Kobayashi, S. *J. Polym. Sci., Polym. Chem. Ed.* **1999**, *37*, 3041-3056.
29. Kobayashi, S.; Uyama, H.; Kimura, S. *Chem. Rev.* **2001**, *101*, 3793-3818.

30. Kobayashi, S.; Sakamoto, J.; Kimura, S. *Prog. Polym. Sci.* **2001**, *26*, 1525-1560.
31. Kobayashi, S.; Uyama, H.; Ohmae, M. *Bull. Chem. Soc. Jpn.* **2001**, *74*, 613-635.
32. Kobayashi, S.; Kashiwa, K.; Kawasaki, T.; Shoda, S. *J. Am. Chem. Soc.* **1991**, *113*, 3079-3804.
33. Kobayashi, S.; Wen, X.; Shoda, S. *Macromolecules*, **1996**, *29*, 2698-2700.
34. Kobayashi, S.; Hobson, L. J.; Sakamoto, J.; Kimura, S.; Sugiyama, J.; Imai, T.; Itoh, T. *Biomacromolecules* **2000**, *1*, 168-173, 509.
35. Kobayashi, S.; Shimada, J.; Kashiwa, K.; Shoda, S. *Macromolecules* **1992**, *25*, 3237-3241.
36. Kobayashi, S.; Kiyosada, T.; Shoda, S. *J. Am. Chem. Soc.* **1996**, *118*, 13113-13114.
37. Sakamoto, J.; Sugiyama, J.; Kimura, S.; Imai, T.; Ito, T.; Watanabe, T.; Kobayashi, S. *Macromolecules*, **2000**, *33*, 4155-4160, 4982.
38. Okamoto, E.; Kiyosada, T.; Shoda, S.; Kobayashi, S. *Cellulose* **1997**, *4*, 161-172.
39. Fujita, M.; Shoda, S.; Kobayashi, S. *J. Am. Chem. Soc.* **1998**, *120*, 6411-6412.
40. Fort, S.; Boyer, V.; Greffe, L.; Davis, G. J.; Moroz, O.; Christiansen, L; Schülein, M.; Cottaz, S.; Driguez, H. *J. Am Chem. Soc.* **2000**, *122*, 5429-5437.
41. Hrmova, M.; Imai, T.; Rutten, S. J.; Fairweather, J. K.; Pelosi, L.; Bilone, V.; Driguez, H.; Fincher, G. B. *J. Biol. Chem.* **2002**, *277*, 30102-30111.
42. Sakamoto, J. Watanabe, T.; Ariga, Y.; Kobayashi, S. *Chem. Lett.* **2001**, 1180-1181.
43. Shoda, S.; Fujita, M.; Lohavisavapanichi, C.; Misawa, Y.; Ushizaki, K.; Tawada, Y.; Kuriyama, M.; Kohri, M.; Kuwata, H. *Helv. Chim. Acta* **2002**, *85*, 3919-3936.
44. Kiyosada, T.; Shoda, S.; Kobayashi, S. *Polym. Prepr. Jpn.* **1995**, *44*, 1230-1231.
45. Tews, I.; van Scheltinga, A. C. T.; Perrakis, A.; Wilson, K. S.; Dijkstra, B. W. *J. Am. Chem. Soc.* **1997**, *119*, 7954-7959.
46. Kobayashi, S.; Morii, H.; Itoh, R.; Kimura, S.; Ohmae, M. *J. Am. Chem. Soc.* **2001**, *123*, 11825-11826.
47. Kobayashi, S.; Fujikawa, S.-I.; Ohmae, M. *J. Am. Chem. Soc.* **2003**, *125*, 14357-14369.
48. Frost, G.-I.; Csoka, T.; Stern, R. *Trends Glycosci. Glycotechnol.* **1996**, *8*, 419-434.
49. Takagaki, K.; Nakamura, T.; Majima, M.; Endo, M. *J. Biol. Chem.* **1988**, *263*, 7000-7006.
50. Davies, G.; Henrissat, B. *Structure* **1995**, *3*, 853-859.

51. Jeanloz, R. W.; Stoffyn, P. J. In *Methods in Carbohydrate Chemistry*; Whistler, R. L.; Wolfrom, M. L., Eds. Academic Press Inc.: New York, 1962, p 221.
52. Rana, S. S.; Barlow, J. J.; Matta, K. L. *Carbohydr. Res.* **1981**, *96*, 231-239.
53. Schmidt, R. R.; Kinzy. W. *Adv. Carbohydr. Chem. Biochem.* **1994**, *50*, 21-123.
54. Horito, S.; Lorentzen, J. P.; Paulsen, H. *Liebigs Ann. Chem.* **1986**, 1880-1890.
55. Evanko, S. P.; Wight, T. N. *J. Histochem. Cytochem.* **1999**, *47*, 1331-1341.

## Chapter 16

# Sugar Polymer Engineering with Glycosaminoglycan Synthase Enzymes: 5 to 5,000 Sugars and a Dozen Flavors

### Paul L. DeAngelis

Department of Biochemistry and Molecular Biology, Oklahoma Center for Medical Glycobiology, University of Oklahoma Sciences Center, Oklahoma City, OK 73104

Glycosaminoglycans [GAGs] are used currently as therapeutics, but the controlled manipulation and/or synthesis of these sugars has been difficult or impossible in the past. A variety of GAG polysaccharides and oligosaccharides may now be synthesized *in vitro* with the advent of recombinant *Pasteurella multocida* synthases. Narrow size distribution polymers ranging from 5 to 1,500 kDa are produced with synchronized, stoichiometrically-controlled reactions. Alternatively, monodisperse oligosaccharides ranging in size from 5 to 22 monosaccharides are synthesized in a step-wise fashion with immobilized enzyme reactors. Either pure or chimeric sugar chains are possible in both types of synthetic schemes. Overall, these strategies should allow the production of a wide variety of novel polymers with potential utility in the treatment of cancer, wound-healing, thrombosis, immune dysfunction, or inflammation.

## Introduction

**Glycosaminoglycans** [GAGs] are linear heteropolysaccharides composed of repeating disaccharide units containing a derivative of an amino sugar (either glucosamine or galactosamine). **Hyaluronan** [HA], **chondroitin**, and **N-acetylheparosan** (or heparosan) contain glucuronic acid [GlcUA] as the other component of the disaccharide repeat. These sugar polymers are essential for vertebrate life playing many roles as structural elements and recognition or adhesion signals.

Pathogenic bacteria often use a polysaccharide **capsule**, an extracellular sugar polymer coating surrounding the microbial cell, to surmount host defenses (*1*). A few bacteria synthesize GAG capsules that contribute to their virulence because the host and microbial polymers are chemically identical or similar and thus are relatively non-immunogenic (*2*). In addition to this molecular camouflage strategy, another emerging scenario is that these bacteria may also be able to commandeer vertebrate systems that employ GAGs.

We have utilized the enzymes derived from a certain bacterial species, *Pasteurella multocida*, with favorable properties to create biomaterials suitable for a variety of health applications. Our chemoenzymatic methodologies allow the preparation of either **defined oligosaccharides** or **narrow size distribution polysaccharides** with a variety of GAG compositions.

Three of the most prominent capsular types of *P. multocida*, a widespread Gram-negative animal bacterial pathogen, are sources of acidic GAGs. Carter Types A, D, or F produce hyaluronan, heparosan, or chondroitin polysaccharides, respectively (Table 1).

Table 1. *Pasteurella* GAGs and Synthases.

| Polysaccharide | Repeat Structure | Capsule Type | enzyme | ref. |
|---|---|---|---|---|
| Hyaluronan, HA | [β4GlcUA- β3GlcNAc] | A | **pmHAS** | (*3*) |
| Chondroitin (unsulfated) | [β4GlcUA- β3GalNAc] | F | **pmCS** | (*4*) |
| Heparosan (unsulfated) | [β4GlcUA- α4GlcNAc] | D | **pmHS** | (*5*) |

The various *Pasteurella* glycosyltransferases, called **GAG synthases** (2-5), required for the production of the GAG chain transfer two distinct monosaccharides to the growing chain in a repetitive fashion according to the overall reaction:

$$n \text{ UDP-GlcUA} + n \text{ UDP-HexNAc} \rightarrow 2n \text{ UDP} + [\text{GlcUA-HexNAc}]_n \quad (1)$$

where HexNAc = N-acetylglucosamine [GlcNAc] or N-acetylgalactosamine [GalNAc]. Depending on the specific GAG and the particular organism examined, the degree of polymerization, $n$, ranges from $\sim 10^{2-4}$. The bacterial polymers are not further modified in contrast to the vertebrate systems where chondroitin and heparosan are found most often in a sulfated and/or epimerized state.

*Escherichia coli*-derived recombinant pmHAS, pmCS, and pmHS will **elongate exogenously supplied GAG-polymer acceptors** *in vitro* (4-6). The HexNAc-transferase or the GlcUA-transferase activities of the *Pasteurella* enzymes can be assayed separately *in vitro* by supplying the appropriate acceptor oligosaccharide and only one of the UDP-sugar precursors. **Single monosaccharides are added to the growing chain sequentially to the non-reducing reducing terminus of the linear polymer chain** (6). The intrinsic fidelity of each transfer step assures the production of the GAG repeat structure. The reducing end of the acceptor may be derivatized with another molecule (*e.g.* a tag or a drug) or immobilized on a surface (*e.g.* glass, metal, or plastic) (7).

Two **tandemly repeated sequence elements** are present in the pmHAS and pmCS polypeptides (Figure 1). Each element contains a set of two short sequence motifs: a DGS followed by a DXD (X = S or C) about 45 residues downstream. Mutation of the aspartate residue in any one DGS or DXD motif of pmHAS **converts the dual-action synthase into a single-action glycosyltransferase** (8,9). The GlcNAc-transferase and the GlcUA-transferase activities are relatively independent based on kinetic comparisons of the various mutants to the wild-type pmHAS enzyme. The *Pasteurella* chondroitin synthase, pmCS, contains separate GalNAc-transferase (a slightly mutated version of the GlcNAc-site of pmHAS) and GlcUA-transferase sites (4). A further practical improvement of pmHAS or pmCS is the conversion of the native sequence membrane proteins into soluble enzymes by the removal of the carboxyl-terminal membrane association domain (4,8). The *Pasteurella* heparosan synthase, pmHS, also appears to contain at least two domains, but it is not similar at the amino acid level to pmHAS or pmCS (5).

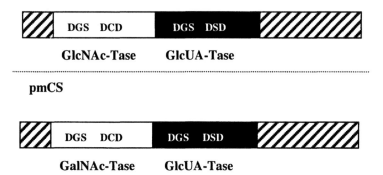

*Figure 1. Schematic Model of Domain Structure of pmHAS and pmCS. These enzymes contain two distinct catalytic sites (important amino acid residues depicted in single letter code) responsible for transfer of the HexNAc or the GlcUA monosaccharide. Certain regions are dispensable for glycosyltransferase activity (hatched area).*

We have created recombinant versions of the *Pasteurella* pmHAS and pmCS enzymes that possess **attractive features for chemoenzymatic synthesis** including: fast reaction rates *in vitro*, good solubility and stability properties, and the lack of strong feedback inhibition by UDP (the byproduct of polymerization) in addition to their essential property of elongating exogenously supplied oligosaccharides. Furthermore, the mutant single-action enzymes that possess only one of the two component transferase activities (*i.e.* either a HexNAc-transferase or a GlcUA-transferase) allow stepwise synthesis strategies as in Equation 2 or 3; the mutant enzymes cannot repetitively polymerize both monosaccharides of the disaccharide repeat.

UDP-HexNAc + [GlcUA-HexNAc]$_x$ →

$$UDP + HexNAc\text{-}[GlcUA\text{-}HexNAc]_x \qquad (2)$$

UDP-GlcUA + [HexNAc-GlcUA]$_x$ →

$$UDP + GlcUA\text{-}[HexNAc\text{-}GlcUA]_x \qquad (3)$$

## Experimental Procedures

**Enzymes and Reactors**

The soluble, truncated dual-action pmHAS$^{1-703}$ or dual-action pmCS$^{1-704}$ enzyme were used for long chain polysaccharide synthesis in synchronized reactions. A pair of single-action enzymes: the GlcNAc-Tase pmHAS$^{1-703}$(D527N,D529N) and the GlcUA-Tase pmHAS$^{1-703}$(D247N,D249N) were used in an immobilized state for stepwise oligosaccharide synthesis. The analogous chondroitin synthase-based GalNAc-Tase mutant, pmCS$^{1-704}$(D520N,D522N), was also immobilized. Mutants with two lesions per motif are preferable to mutants with single lesions when reactions employ high levels of UDP-sugars (*10*). The enzyme reactors (~18 mg protein on 4 ml of packed beads in a small glass column) were catalytically active for months with storage at 4°C in 50 mM Tris, pH 7.2, 1 M ethylene glycol buffer (TEG) buffer.

**Synchronized, Stoichiometrically-controlled Polysaccharide Synthesis**

The synchronized syntheses in general contained a soluble dual-action synthase (pmHAS or pmCS as desired), a GAG-based acceptor, UDP-HexNAc (UDP-GlcNAc or UDP-GalNAc as needed), UDP-GlcUA, and 5 mM MnCl$_2$ in TEG reaction buffer. Reactions are incubated at 30°C for 6 to 48 hrs. The HA tetrasaccharide, the starting acceptor for the synthesis of longer HA chains, was generated by exhaustive degradation of HA polymer with testicular hyaluronidase. Alternative acceptors included various chondroitin polymers (unsulfated or sulfate A, B, or C). The sizes of GAG polysaccharides were analyzed on agarose gels in 1× TAE buffer and detected with Stains-All dye as described previously (*11*). Size exclusion chromatography/multi-angle laser light scattering (SEC/MALLS) analysis was used to determine the absolute molecular weights of the polymers. Polymers were separated on two tandem Toso Biosep TSK-GEL columns (6000PWXL followed by 4000PWXL) eluted in 50 mM sodium phosphate, 150 mM NaCl, pH 7 at 0.5 mL/min. The eluant flowed through an Optilab DSP interferometric refractometer and then a Dawn DSF laser photometer (632.8 nm; Wyatt Technology, Santa Barbara, CA) in the multi-angle mode. The manufacturer's software package was used to determine the absolute average molecular weight using a dn/dC coefficient of 0.153.

**Stepwise Oligosaccharide Synthesis with Enzyme Reactors**

In the typical oligosaccharide synthesis, 90 μmoles of acceptor oligosaccharide and 110-135 μmoles (1.2 to 1.5 equivalents) of UDP-sugar (~15 mM final) in reaction buffer (TEG plus 17 mM $MnCl_2$) were circulated over an enzyme reactor at room temperature. For converting the HA tetrasaccharide starting material (with a GlcUA at the nonreducing terminus) into the pentasaccharide, the GlcNAc from UDP-GlcNAc was transferred with the GlcNAc-Tase reactor (~1 to 2 hours) as in Equation 2.

The next UDP-sugar (in this specific case, UDP-GlcUA) was added to the reaction mixture and then applied to the next reactor (converting pentasaccharide into the hexasaccharide with immobilized GlcUA-Tase) to perform the reaction of Equation 3. This repetitive synthesis was continued by adding the next appropriate UDP-sugar and switching enzyme reactors.

The reactions were monitored by thin layer chromatography (silica TLC plates developed with *n*-butanol/acetic acid/$H_2O$, 1.5:1:1 or 1:1:1) and napthoresorcinol staining (0.2% w/v in 96% ethanol/4% sulfuric acid; 100°C). For mass spectrometry (MALDI-TOF MS), the matrix solution (50 mg/ml 6-aza-2-thiothymine in 50% acetonitrile, 49.9% water, 0.1% trifluoroacetic acid, 10 mM ammonium citrate) was mixed 1:1 with the samples containing ~0.1 μg/μl oligosaccharide in water, spotted onto the target plate, and vacuum dried. The samples were analyzed in the negative ion, reflectron mode on a Voyager Elite DE mass spectrometer.

# Results and Discussion

The recombinant *Pasteurella* synthases produced in the appropriate bacterial host will elongate certain exogenously supplied glycosaminoglycan acceptor chains (*4-6*). Specifically, if the host bacterium does not produce one of the required sugar precursors, UDP-GlcUA (*i.e.* lacks the gene for UDP-glucose dehydrogenase), then the recombinant synthase will not have an endogenous nascent GAG chain upon isolation from the cell. These **virgin catalysts probably have open or available active sites that readily bind and then extend the exogenously supplied acceptor chain.** For example, recombinant pmHAS will readily add sugars onto HA tetrasaccharides to create longer polymer chains in the range of 5 to 2,000 kDa. The recombinant pmHAS and pmCS enzymes appear to function in a non-processive fashion *in vitro*; the GAG chain is bound, extended, and released repeatedly during the polymerization process.

## Model for Monodisperse HA Production

The recombinant pmHAS synthesizes HA chains *in vitro* if supplied with both required UDP-sugars in a suitable reaction buffer. If recombinant pmHAS is supplied a HA-like oligosaccharide *in vitro*, then the overall incorporation rate is elevated up to 20- to 60-fold (6). It was suggested that the rate of initiation of a new HA chain *de novo* was slower than the subsequent elongation (*i.e.* repetitive addition of sugars to a nascent HA molecule). Therefore, the observed **stimulation of synthesis by exogenous acceptor appears to operate by bypassing the kinetically slower initiation step** allowing the reaction in Equation 4 to predominate. We now further hypothesize that the polymerization by pmHAS in the presence of HA acceptor is a **synchronized process** *in vitro* and thus a more uniform defined HA should be obtained; all chains will be elongated in parallel resulting in a more homogenous final population.

$$n \text{ UDP-GlcUA} + n \text{ UDP-HexNAc} + [\text{GlcUA-HexNAc}]_x \xrightarrow{fast}$$

$$2n \text{ UDP} + [\text{GlcUA-HexNAc}]_{x+n} \quad (4)$$

## Model for Controlling Size of HA Products

As noted above, chain initiation appears to be the rate-limiting step for pmHAS-catalyzed HA production. Due to this kinetic phenomenon, the synthase will add all available UDP-sugar precursors to the acceptor termini before much new chain initiation occurs. If the polymerization is indeed a synchronized process, then **the amount of acceptor should affect the final size of the HA product when a limited amount of UDP-sugar is present**. If there are many termini (*i.e.* Eq. 5 where $z$ is large), then the available UDP-sugars will be distributed among many molecules and thus result in many short polymer chain extensions. Conversely, if there are few termini (*i.e.* Eq. 5 where $z$ is small), then the available UDP-sugars will be distributed among a few molecules and thus result in a few long chain extensions.

$$n \text{ UDP-GlcUA} + n \text{ UDP-HexNAc} + z [\text{GlcUA-HexNAc}]_x \xrightarrow{fast}$$

$$2n \text{ UDP} + z [\text{GlcUA-HexNAc}]_{x+(n/z)} \quad (5)$$

To test our speculation, we performed a series of assays utilizing various levels of HA tetrasaccharide with a fixed amount of UDP-sugar and pmHAS. The size and polydispersity of the HA products was analyzed by SEC-MALLS (Table 2). In this example, decreasing amounts of tetrasaccharide (sample #1 > #5) were employed. As predicted, higher acceptor/UDP-sugar ratios resulted in shorter products. With this general strategy, we were able to generate HA from 27 kDa to 1.3 MDa with polydispersity ranging from 1.001 to 1.2 (*12*). For reference, a polydispersity value of 1 corresponds to an ideal monodisperse polymer.

By calculating the **ratio of acceptor to UDP-sugar**, it is possible to **select the final HA size desired**. Overall, the products of synchronized, stoichiometrically-controlled reactions exhibit much more narrow size distributions in comparison to commercially available HA preparations from natural sources including chicken or bacteria (Fig. 2). The utility of HA preparations with a narrow size distribution should be extensive because many biological phenomena (*e.g.* angiogenesis, cell proliferation and signaling) respond differentially depending on the molecular weight of HA (*13*). Our preliminary studies of *in vitro* syntheses with pmCS suggest that a wide variety of monodisperse chondroitin polymers may also be synthesized.

Table 2. SEC-MALLS analysis of HA produced via synchronized, stoichiometrically-controlled reactions.

| Sample | $M_n$ | $M_w$ | polydispersity ($M_w/M_n$) |
|---|---|---|---|
| #1 | 283,000 | 284,000 | 1.001 |
| #2 | 346,000 | 347,000 | 1.002 |
| #3 | 422,000 | 424,000 | 1.004 |
| #4 | 490,000 | 493,000 | 1.006 |
| #5 | 570,000 | 575,000 | 1.010 |

$M_n$ = number average molecular weight
$M_w$ = weight average molecular weight

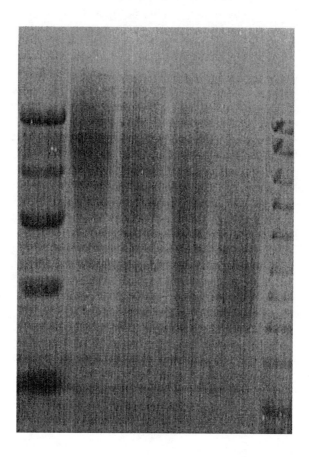

*Figure 2. Agarose gel analysis of various monodisperse products of synchronized, stoichiometrically-controlled reactions.*

**S** = *a mixture of the HA polymer products (1.3, 0.9, 0.6, 0.3, 0.027 MDa from top to bottom) from 5 different synchronized reactions loaded in 1 lane;* **Commercial HA** = *4 different HA preparations;* **D** = *DNA standard ('kilobase ladder' - 12 kb at top).*

## Production of Chimeric GAG Chains

We recently found that the various *Pasteurella* GAG synthases will also use **non-cognate acceptor molecules** (*14*). For example, we added a HA chain onto an existing chondroitin sulfate chain using pmHAS in Figure 3. This particular family of hybrid molecules may be useful as artificial proteoglycans for tissue engineering that do not contain the protein component of the natural cartilage proteoglycans.

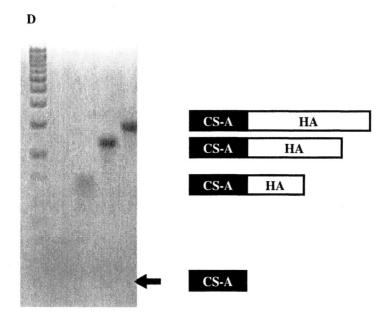

*Figure 3. Agarose gel analysis of HA-chondroitin hybrid polymers. HA chains of increasing length were added onto chondroitin sulfate A (CS-A) using pmHAS resulting in new bands with lower mobility. (**D** = DNA standard ladder. Arrow marks the position of the CS-A starting material)*

## Chemoenzymatic Synthesis of Defined Oligosaccharides

The *Pasteurella* HA synthase, a dual-action polymerizing enzyme that normally elongates HA chains rapidly, possesses two active sites (8,9). Single sugars are added to the growing chain sequentially to the non-reducing terminus allowing the step-wise synthesis of oligosaccharides, if desired. We used mutagenesis to convert the native enzyme into two single-action glycosyltransferases. The resulting GlcUA-transferase and GlcNAc-transferase are appropriate for producing short sugar chains in a controlled, stepwise fashion without purification of the intermediates. For example, in Figure 4, in the synthesis converting the HA tetrasaccharide into a HA tetradecasaccharide, HA11-mer was converted into HA12-mer in one, quantitative step with an immobilized enzyme reactor.

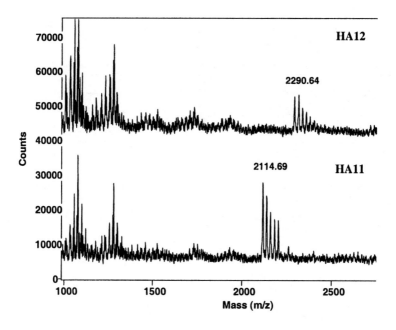

*Figure 4. Matrix-assisted laser desorption ionization time-of-flight spectrum of crude reaction mixtures from the oligosaccharide synthesis reactions. The mass of the H+ form of the target sugar is indicated; the train of higher mass ions are forms with increasing amounts of Na+ (22 additional amu in comparison to H+) on the carboxylates of GlcUA residues.*

We have produced oligosaccharides each with a single length in the size range of 10 to 22 monosaccharides in a few days (*10,15*). Only simple desalting at the end of the reaction is required to purify the target sugar. The experimental masses and the predicted masses of these synthetic oligosaccharides were in excellent agreement. These small HA oligosaccharides have biological activities (*e.g.* inducing cancer cells to undergo apoptosis or stimulating angiogenesis) distinct from long polysaccharides (*13*).

Due to the relaxed acceptor polymer specificity of pmHAS and pmCS, oligosaccharides with a mixed GAG structure may be synthesized by employing different HexNAc-transferase reactors during the synthesis (*15*). For example, Figure 5 depicts the TLC analysis of the HA-chondroitin octasaccharide, [GlcUA-GalNAc]$_2$[GlcUA-GlcNAc]$_2$; this hybrid compound is isobaric with the HA octasaccharide (*i.e.* the epimers GlcNAc and GalNAc have same mass), but normal phase chromatography can distinguish the two sugars.

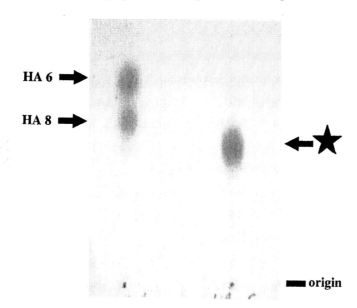

*Figure 5. TLC analysis of HA-chondroitin octasaccharide. The novel sugar (marked with star) migrates slower than the pure HA octasaccharide (HA 8).*

**Current Research Aims**

We have also created **new chimeric enzymes** (*e.g.* a polypeptide composed of segments of both pmHAS and pmCS) that have **reduced selectivity for the sugar transfer reaction** (*11*). Normally, native sequence pmHAS will only transfer UDP-GlcNAc while native sequence pmCS will only transfer UDP-GalNAc; relatively strict donor specificity is observed. On the other hand, certain artificial synthase constructs will transfer either natural UDP-hexosamine or the unnatural substrate UDP-GlcN. Thus, depending on the UDP-sugar precursors supplied, one can produce novel GAGs containing (*i*) a blended mixture of HA and chondroitin or (*ii*) GAGs with unnatural functionalities such as free amino groups. In the former case, the new polymer will have some attributes of both GAGS. In the latter case, the new reactive functionalities on the polymer chain are useful for creating cross-linked viscoelastic gels or polyvalent drug-carriers. The full scope and range of the artificial synthases' catalytic prowess is a subject of current investigation.

Other new catalysts under development include the heparosan synthase enzyme, **pmHS,** and its derivatives. Heparin-like polymers will interact with other proteins in the vertebrate body including coagulation factors and growth factor receptors. We expect that pmHS will possess certain catalytic properties exhibited by pmHAS and pmCS.

## Conclusions

Overall, depending on the choice of the catalyst, the acceptor molecule and the UDP-sugar precursors, a wide variety of potentially useful GAG-based materials based on the hyaluronan, chondroitin, and/or N-acetylheparosan may be created. These new syntheses allow unprecedented control of sugar chain length, size distribution, and composition.

## Acknowledgements

I thank Drs. Wei Jing and F. Michael Haller, Tasha A. Arnett, Bruce A. Baggenstoss, Daniel F. Gay, Leonard C. Oatman, Breca S. Tracy, and Carissa L. White for performing various experiments and laboratory support. This work was supported in part by grants MCB-9876193 from National Science Foundation, OARS program AR02.2-019 from the Oklahoma Center for Advancement of Science and Technology, and GM56497 from National Institutes of Health, and a sponsored research agreement from Hyalose LLC.

## References

1. Roberts, I.S. *Annu. Rev. Microbiol.* **1996**, *50*, 285-315.
2. DeAngelis, P.L. *Glycobiology* **2002,** *12*, 9R-16R.
3. DeAngelis, P.L.; Jing, W.; Drake, R.R.; Achyuthan, A.M. *J. Biol. Chem.* **1998**, *273*, 8454-8458.
4. DeAngelis, P.L.; Padgett-McCue, A.J. *J. Biol. Chem.* **2000**, *275*, 24124-24129.
5. DeAngelis, P.L.; White, C.L. *J. Biol. Chem.* **2002,** *277*, 7209-7213.
6. DeAngelis, P.L. *J. Biol. Chem.* **1999**, *274*, 26557-26562.
7. DeAngelis, P.L. **2002** U.S. Patent 6,444,447.
8. Jing, W.; DeAngelis, P.L. *Glycobiology.* **2000**, *10*, 883-889.
9. Jing, W; DeAngelis, P.L. *Glycobiology* **2003**, *13*, 661-671.
10. DeAngelis, P.L.; Oatman, L.C.; Gay, D.F. *J. Biol. Chem.* **2003**, *278*, 35199-35203.
11. Lee, H.G.; Cowman, M.K. *Anal. Biochem.* **1994**, *219*, 278-287.
12. Jing, W.; DeAngelis, P.L. unpublished.
13. Toole, B.P.; Wight, T.N.; Tammi, M.I. *J. Biol. Chem.* **2002**, *277*, 4593-4596.
14. Tracy, B.S.; DeAngelis, P.L. unpublished
15. DeAngelis, P.L.; Gay, D.F. unpublished

Chapter 17

# Enzyme-Catalyzed Regioselective Modification of Starch Nanoparticles

Soma Chakraborty[1], Bishwabhusan Sahoo[1], Iwao Teraoka[1], Lisa M. Miller[2], and Richard A. Gross[1,*]

[1]NSF Center for Biocatalysis and Bioprocessing of Macromolecules, Othmer Department of Chemical and Biological Science and Engineering, Polytechnic University, 6 Metrotech Center, Brooklyn, NY 11201
[2]National Synchrotron Light Source, Brookhaven National Laboratory, Upton, NY 11973
*Corresponding author: http://chem.poly.edu/gross

This chapter describes the selective esterification of starch nanoparticles catalyzed by Candida antarctica Lipase B (CAL-B) in its immobilized (Novozyme 435) and free (SP-525) forms. The starch nanoparticles were made accessible for acylation reactions by formation of Aerosol-OT (AOT, bis[2-ethylhexyl]sodium sulfosuccinate) stabilized microemulsions. Analysis of these particles by light scattering showed at the high and low ends of the distribution aggregates with many 40-nm nanoparticles and AOT-stabilized micelles with one or two 40 nm starch nanoparticles, respectively. Acylation of the starch nanoparticles in microemulsions at 40°C for 48 h with vinyl stearate, ε-caprolactone and maleic anhydride gave products with degrees of substitution (D.S.) of 0.8, 0.6 and 0.4, respectively. In all cases, substitution occurred regioselectively at the C-6 position of the glucose repeat units. The acylated particles after surfactant removal retained nanodimensions and could be redispersed in DMSO or water. Infrared microspectroscopy (IRMS) showed that AOT coated starch nanoparticles can diffuse into the outer 50 μm shell of catalyst beads. The close proximity of the lipase and substrates promotes the modification reactions.

Starch is an abundant, inexpensive, naturally occurring polysaccharide. It is biocompatible, biodegradable, and nontoxic, so it can be used as a medical material for drug delivery. Literature reports describe the use of chemically modified forms of starch for sustained drug delivery systems. For example, epichlorohydrin cross-linked high amylose starch was used as a matrix for the controlled release of contramid[1]. A complex of amylose, butan-1-ol and an aqueous dispersion of ethylcellulose was used to coat pellets containing salicylic acid to treat colon disorders[2]. Starch has also been used as a carrier for phenethylamines[3], acetylsalicylic acid[4], and estrone[5]. Hydrogels composed of starch/cellulose acetate blends were reported as bone cements[6]. While starch-based biomaterials appear promising, scientific challenges remain to be solved to accelerate their development. Critical to the use of starch matrices for drug delivery is the ability to uniformly modify its structure. This is difficult due to the presence of three hydroxyl groups per glucose residue that are in different chemical environments. Furthermore, to solubilize starch for homogeneous modification, polar aprotic solvents such as dimethylsulfoxide are needed. For example, to modify the primary (6-O) hydroxyl sites of amylose, it was heterogeneously persilylated, the persilylated derivative in carbon tetrachloride was acylated with an anhydride and then the silyl protecting groups were removed[7].

The modification of polysaccharides using enzymes has been studied to attain more uniformly substituted products under mild reaction conditions. Hydroxyethylcellulose (HEC) particles were suspended in dimethylacetamide and acylated with vinyl stearate using *Candida antartica* Lipase B (CAL-B) as the catalyst. After 48 hrs, a product with degree of substitution (D.S.) of 0.1 was formed[8]. HEC in film or powder form was modified using lipase-catalysis and ε-caprolactone (CL) to form low D.S. HEC-g-PCL copolymers[9]. Problems that limit the utility of these reactions include the use of polar aprotic solvents that strip the critical water from enzymes lowering their activities[10] and the use of heterogeneous reaction conditions that restrict the modification of large particles and films to a small fraction of the substrate that resides at the surface.

To overcome the use of polar aprotic solvents, enzymes have been incorporated within reverse micelles using the anionic surfactant Aerosol-OT [AOT, bis(2-ethylhexyl)sodium sulfosuccinate]. AOT forms thermodynamic water droplets surrounded by a surfactant monolayer in oil (isooctane). Water entrapped within the reverse micelles resembles the polar pockets in cells[11]. Incorporation of enzymes within reverse micelles soluble in non-polar media facilitates productive collisions and reactions between enzymes and non-polar substrates. Several types of lipase-catalyzed reactions in AOT/isooctane have been studied(12). Dordick and coworkers incorporated proteases from *Subtilisin carlsberg* and *Bacillus licheniformis* within AOT-coated reverse micelles(13). This enzyme within reverse micelles was active for the acylation of amylose in film and powder form. However, as above, the inability of the enzyme to diffuse into the bulk of these substrates limits the modification of polysaccharide films and powders to surface regions.

Nanoparticles, nanospheres, microspheres and nanogels are legos for the nanoscale construction of sensors, tissues, mechanical devices, and drug delivery systems. For the latter, the sub-cellular dimensions of nanoparticles allow their penetration deep into tissues through fine capillaries. For medical applications, nanoparticles are often constructed from poly(lactic acid), poly(glycolic acid)(14-16) and poly(alkylcyanoacrylate)(17,18). Polymeric nanoparticles formed from the monomers 2-acryloxyethyltrimethylammonium chloride, 2-hydroxyethylacrylate and poly(ethylene glycol)diacrylate have been used as gene and antisense delivery agents(19) and poly(L-lysine)-*g*-polysaccharides(20). Polyvinylpyrrolidone nanoparticles were constructed for the sustained release of antigens and allergens(21). Miller et al reported that polymeric nanoparticles can transport drugs across the blood-brain barrier(22). Starch microspheres were developed and studied for the delivery of insulin via the nasal system(23).

Previous work cited above motivated the current study that used enzymes to selectively modify starch nano-particles. By this strategy, a new family of structurally and dimensionally well-defined nano-particles was prepared from an abundant, biocompatible, and natural building material. Starch nanoparticles were incorporated into reverse micelles stabilized by AOT. The nanodimensions of AOT-coated starch particles and their solubility in non-polar media such as toluene allowed their diffusion within the pores of the macroporous resin that served as the support for CAL-B. The selective modification of starch that occurs throughout the nanoparticles instead of at the surface of large particles is described. Dynamic light scattering probed the dimensions of the modified surfactant-free starch particles.

## Material and Methods

**General Chemicals and Procedure**: Starch nanoparticles (Ecosphere) of average size 40nm were prepared by Ecosynthetix and provided as a gift. The method used to prepare the nanoparticles was described in detail elsewhere(24). All chemicals and solvents were of analytical grade, purchased from Aldrich Chemical Co. Inc., and were used as received unless otherwise noted. ε-Caprolactone (CL) was a gift from Union Carbide Company. Novozyme 435 and SP-525 were gifts from Novozymes. Toluene and CL were dried over calcium hydride and distilled under reduced pressure in a nitrogen atmosphere. AOT was dried in vacuo (16 hrs, 25°C, 1mm Hg) prior to use. Novozyme-435 was dried over $P_2O_5$ in vacuo (16 hrs, 25°C, 0.5 mm Hg) prior to use. Molecular sieves (4Å, 4-8 mesh size) were purchased from Aldrich Chemical Co. Inc and dried for 24 hrs at 160 °C prior to use.

**Instrumental Methods.** $^1H$ and $^{13}C$-DEPT-135 NMR spectra were recorded at 25°C on a DPX300 spectrometer at 300 and 75.13 MHz, respectively (Bruker Instruments, Inc.). The proton ($^1H$) and carbon ($^{13}C$)-DEPT-135 NMR chemical shifts in parts per million (ppm) were referenced relative to tetramethylsilane (TMS). To perform the $^1H$ and $^{13}C$ NMR experiments, 8.0 wt% and 20.0 wt% of samples were dissolved in their respective solvents. FTIR spectra were recorded using a Thermo Nicolet Magna 760 FTIR spectrometer using potassium bromide (KBr) discs prepared from powdered samples mixed with dry KBr in the ratio of 1:100 (sample:KBr). The spectra were recorded in a absorbance mode from 4000 to 400cm$^{-1}$ at a resolution of 8 cm$^{-1}$. Particle size distribution of starch nanoparticles was characterized at 25°C by using a N4 Plus particle sizer (Beckman-Coulter) with a He-Ne laser ($\lambda$ = 632.8 nm). Samples for the measurement were prepared by dissolving nanoparticles at 1% (w/w) in their respective solvents. Unmodified starch nanoparticles were dissolved in DMSO as well as in water. Starch nanoparticles modified with either CL or vinyl stearate were dissolved in DMSO. Maleic anhydride-modified starch nanoparticles were dissolved in water. DMSO and aqueous solution samples were filtered through hydrophobic and hydrophilic PTFE filters, respectively (both 0.2μm pore size; Millipore). Filtered samples were collected into dust-free fluorometer cells (10 x 10 mm). Starch nanoparticles incorporated in the reverse micelles of AOT (AOT:Starch = 2:1 w/w) and dispersed in toluene were measured without filtration. N4 Plus measurements have an autocorrelation function of the light scattering intensity at different scattering angles that converts it into a distribution of the apparent particle diameter $d$, assuming that the light scattering comes from spherical particles independently making diffusional motion. The distribution is weighted by the scattering intensity of each particle of the relevant size at the scattering angle used. Novozyme beads used for the IR microspectroscopic (IRMS) analysis were recovered from the reaction vessel after 24 hrs reaction between starch nanoparticles and vinyl stearate at 40°C in toluene. Beads were washed with toluene and then with water to remove the particles from their surface. Washed beads were dried in vacuo (16 hrs, 25°C, 1mm Hg). These beads were embedded in paraffin wax and microtomed at room temperature into sections with a thickness of 12 μm. Sections were mounted on a $BaF_2$ disk and the IR microspectra were recorded using a Perkin Elmer Spectrum Spotlight infrared microscope equipped with a motorized x-y stage and an MCT-A detector. The IR microscope was coupled to a Spectrum One FTIR spectrometer. Data were collected from 4000 – 700 cm$^{-1}$ in transmission mode, 32 scans/pixel, 4 cm$^{-1}$ spectral resolution, and 6.25 μm spatial resolution. Control spectra of starch, AOT, and Novozyme 435 beads were also collected under the same conditions. Additional details of the general methods and data analysis is described elsewhere(25).

## Experimental

**Incorporation of starch nanoparticles within reverse micelles of AOT/isooctane/$H_2O$ microemulsions**: A concentrated aqueous solution of starch nanoparticles (0.25g/ml) was added dropwise with vigorous stirring to 50ml of 0.1M AOT/anhydrous isooctane solution in a round bottom flask capped with a rubber septum. The dropwise addition was continued. Throughout this process the solution remained clear and phase separation did not occur. This gave reverse micelles of AOT coated starch nanoparticles in isooctane. The isooctane was removed under reduced pressure by using a rotavaporator. The AOT coated micelles were further dried (30°C, 6hrs, 30mm Hg). A similar procedure was adopted to incorporate starch nanoparticles in reverse micelles of CTAB/chloroform and TritonX-100/toluene where, in place of AOT/isooctane, the surfactant/solvent systems used were CTAB/chloroform and TritonX-100/toluene.

**Incorporation of non-immobilized *Candida antartica* Lipase B (CALB) within AOT-coated starch nanoparticles:** Starch nanoparicles (2.2 g) and SP-525(1.1 ml, protein content 0.022g) were added to 7.7ml of deionized water to make the total volume of 8.8 ml. The incorporation of starch nanoparticles within reverse micelles of AOT/isooctane/$H_2O$ microemulsions was performed exactly as above except that SP-525 was present in the aqueous solution.

**Acylation of AOT coated starch nanoparticles with vinyl stearate:** Vinyl stearate was used as the acylating agent for the hydrophobic modification of starch nanoparticles. The starch nanoparticles (2.2 gm, 0.0135 moles of glucose units) coated with surfactant was transferred to a 100 ml round bottomed flask. Dry toluene (50 ml), vinyl stearate (12.63g, 0.0407moles), and Novozyme 435 (0.22 g) were added to the reaction vessel. The molar ratio of acylating agent to glucose residues was 3 to 1. The reactions were performed in the presence of molecular sieves at 6 different temperatures (25, 30, 35, 40, 50, and 60°C) for 48h with magnetic stirring. Control reactions were performed at 40 and 60°C following the above procedure except that Novozyme-435 was not added to the reaction vessel. Reactions were terminated by filtration to remove the enzyme and sieves and then toluene was removed under reduced pressure. To remove AOT, the product was suspended in a mixture of water and methanol (50ml, 1:1 v/v) at 25°C for 30 minutes with magnetic stirring. The suspended nanoparticles were separated by filtration and then, to remove unreacted acylating agent, they were transferred into a beaker containing 50ml of cold chloroform. The suspension was magnetically mixed at 25°C for 30min, separated by filtration, and the residual solvent was removed under reduced pressure. The purified products were characterized by different spectroscopic techniques. For the 48 h reaction with vinyl stearate, the product of DS 0.8 was obtained in 79% yield and the results of NMR studies for this product were as follows: $^1$H NMR in DMSO-d6 (δ in ppm): 5.27 [C1-$H$, m, 1H], 4.22 [C6-$H$, m, 2H], 3.74 [C2-$H$, C4-$H$, m, 2H], 3.42 [C3-$H$, C5-$H$, m, 2H], 2.18 [-O(C=O)C$H_2$-, t, 2H], 1.53 [-O(-C=O)CH$_2$C$H_2$-, q, 2H], 1.32 [-(C$H_2$)$_9$-, s, 18H], 0.72 [-C$H_3$, t, 3H]; $^{13}$C-DEPT in DMSO (showed the carbon signals for the sugar ring) (δ in ppm): 73.68 (C3), 72.54 (C2), 72.18 (C5, unmodified starch unit), 72.13 (C4), 66.74 (C5, modified starch unit), 65.04 (C6, modified starch unit), 61.07 (C6, unmodified starch unit).

**Maleation of AOT coated starch nanoparticles:** Starch nanoparticles (2.2 gm, 0.0135 moles of glucose units) and dry toluene (50 ml) were transferred to a 100ml round bottomed flask. To this was added maleic anhydride (1.323g, 0.0135 moles), Novozyme 435 (0.22g) and hydroquinone (3 mg, free radical inhibitor). The reactions were performed at 40 °C for 48h with magnetic stirring. The procedure for termination and working-up of the reaction was identical to that for the acylation of starch nanospheres with vinyl stearate. The purified product was characterized by different spectroscopic techniques. NMR spectral data of starch maleate nanoparticles (DS 0.4, 48 hr reaction, 67% product yield) were as follows: $^1$H NMR in D$_2$O (δ in ppm): 5.85 [H$_a$, bs, 1H, maleic anhydride unit proton –(O=C-C$H_a$=CH$_b$COOH)], 6.65 [H$_b$, bs, 1H, maleic anhydride unit proton –(O=C-CH$_a$=C$H_b$COOH)], 5.52 [C1-$H$, m, 1H], 4.36 [C6-$H$, m, 2H], 4.04-3.51 [C2-$H$, C3-$H$, C4-$H$, m, 3H], 3.02 [C5-$H$, m, 1H], $^{13}$C-DEPT spectrum (starch signals, in D$_2$O, δ in ppm): δ 145.28 and 120.23(C$_a$ and C$_b$, -C$_a$=C$_b$-), 72.43(C4), 73.88 (C3), 72.91 (C2), 72.83 (C5, unmodified starch unit), 69.56 (C5, modified starch unit), 61.91 (C6, unmodified starch unit), 66.12 (C6, modified starch unit).

**Reaction of AOT coated starch nanoparticles with ε-caprolactone (CL):** The reaction procedure was identical to that for the maleation of starch nanoparticles except that CL (4.40 ml, 4.536g, 0.0405 moles) was used in place of maleic anhydride. The purified product is characterized by different spectroscopic techniques. NMR spectral data of caprolactone modified starch nanoparticles (DS 0.6, 48 hr reaction, 82% product yield) were as follows: $^1$H NMR in DMSO (δ in ppm): 5.52 [C1-$H$, m, 1H], 4.21 [C$H_2$-O-C(O), caprolactone methylene proton, m], 4.13 [C6-$H$, m, 2H], 3.52-3.86 [C2-$H$, C3-$H$, C4-$H$, C5-$H$, m, 4H], 2.21 [-C(O)C$H_2$-, of caprolactone units], 1.58 [m, methylene protons and C$H_2$OH of capralactone units], 1.32[m, other methylene protons of capralactone units]; $^{13}$C-DEPT (starch signals in DMSO, δ in ppm): δ 73.68 (C3), 72.54 (C2), 72.18 (C5, unmodified starch unit), 72.13 (C4), 70.21 (C5, modified starch unit), 65.01 (C6, modified starch unit), 61.07 (C6, unmodified starch unit).

**Acylation of starch nanoparticles using SP-525:** The acylation of AOT-coated starch nanoparticles with vinyl stearate was performed as described above except that the catalyst (CALB) was incorporated within the microspheres instead of as part of heterogeneous Novozyme-435 beads.

## Results and Discussion

**Formation of surfactant coated starch nanoparticles:** The loading of surfactant stabilized reverse micelles with starch nanospheres was studied. The surfactant-solvent systems used were AOT (sodium di-2-ethylhexylsulfosuccinate)/isooctane, CTAB (cetyl trimethyl ammonium bromide)/chloroform and TritonX-100 (polyoxyethylene(10) isooctylphenylether)/toluene. These surfactants were selected due to their use by others for the solubilization of enzymes in reversed micelles(26-28). To load the reverse micelles, a concentrated (0.25g/mL) aqueous solution of starch nanospheres was added to the different surfactant/solvent pairs. The percent incorporation of starch in the nanospheres was determined based on the amount of the starch nanoparticles that could be added to the surfactant/solvent system prior to the formation of a cloudy phase separated mixture.

*Table 1.* Incorporation of starch nanoparticles in different surfactant systems

| Entry | [1]Surfactant | Nature | Solvent | [2]Starch nanoparticles incorporated(%) |
|---|---|---|---|---|
| 1 | AOT | Anionic | Isooctane | 50 |
| 2 | CTAB | Cationic | Chloroform | 40 |
| 3 | TRITON X-100 | Neutral | Toluene | 10 |

[1]The surfactant concentration in the organic solvent was 0.1M, [2]starch nanoparticles/surfactant (w/w) x 100.

As displayed in Table 1, the highest incorporation of starch nanoparticles within reverse micelles was found by using the AOT/isooctane microemulsion system. Based on these results, the AOT/isooctane system was used to prepare surfactant coated starch nanoparticles. To isolate the AOT-coated starch nanoparticles, the isooctane was removed under reduced pressure. Previous studies in our laboratory showed that toluene was a preferred organic medium in which to perform Novozyme 435 catalyzed transesterification reactions(29). Hence, the AOT-coated starch nanospheres were solubilized in toluene and evaluated for modification reactions.

**Lipase-catalyzed acylation of AOT-coated starch nanoparticles:** Novozym 435 is a biocatalyst consisting of lipase B from *Candida antarctica* (CAL-B)

immobilized onto a poly(methacrylate) macroporous resin (Lewatit VP OC 1600). It has been shown that the average pore size of Novozym 435 beads is about 100 nm(25). An initial concern in using this catalyst for modifying AOT-coated starch nanospheres was whether the AOT-coated nanospheres would have access to CAL-B within the beads. This would require the diffusion of the nanospheres within the pores of Novozym 435. The ability of Novozym 435 to modify the starch nanospheres was first tested. Later in this paper the results of studies are given that characterize starch nanoparticle size and their ability to diffuse into the catalyst beads. The acyl acceptors for starch nanoparticle modification included vinyl activated straight chain aliphatic acids (e.g. vinyl stearate), maleic anhydride, and ε-caprolactone (Scheme 1). Fatty acid vinyl esters, maleic anhydride, and ε–caprolactone were selected for this study since they are examples of: i) model fatty acids of variable chain length that can be used to regulate the hydrobicity of the nanoparticles, ii) a 4-carbon cyclic anhydride that introduces crosslinkable unsaturated moieties, and iii) a moderate size lactone that can either react to form single units, oligomers or polyester grafts, respectively.

**Scheme 1.** Reactions between AOT coated starch nanoparticles with different acylating agents[a]

[a]*Conditions: Novozym 435 (1%-protein w/w), toluene, 40°C, 48h (i) starch (glucose residues):vinyl stearate 1:3 mol/mol, (ii) starch:maleic anhydride 1:1 mol/mol, (iii) starch:ε-capralactone 1:3 mol/mol*

The reaction between vinyl stearate and AOT-coated starch nanoparticles was performed for 24hrs, at 40°C, with a 3:1 mol/mol ratio of vinyl stearate to glucose residues. The peak positions and assignments are listed in the experimental section and they are consistent with that expected for stearate acylated starch. Proton NMR signals with peaks at 5.27, 4.22, 3.74, and 3.42 ppm are due to the protons of C1, C6, (C2/C4), and (C3/C5), respectively. Other $^1$H signals at 2.18, 1.53, 1.32, and 0.72 ppm were assigned to protons of the stearate moiety. The determination of the degree of substitution (D.S.) by $^1$H NMR was based on the relative intensities of the peaks at 5.27 ppm (proton on C1) and 1.53 (CH$_2$ of stearate). This analysis was performed using D$_2$O exchange that caused the disappearance of the 2OH and 3OH protons that, otherwise, overlapped with the proton of C1 at 5.27 ppm. Comparison of the DEPT-135 spectra (Figure 1) of the native starch nanoparticles and the vinylstearate modified starch nanoparticles (48 hr reaction, D.S.=0.8) shows that acylation of glucose residues caused: i) the signal at 72.18ppm for C5 of unmodified glucose units to shift upfield to 66.74ppm, and ii) the peak at 61.07ppm for C6 of the unmodified substrate to shift downfield to 65.04 ppm. The other ring carbons at 73.68 (C3), 72.54 (C2), and 72.13ppm (C4) were found at almost identical positions in the starting substrate and the modified product. The peaks at 72.18 and 61.07 ppm in the product spectrum correspond to C5 and C6 of unmodified sugar units. From analysis of the DEPT-135 spectra in Figure 1 we conclude that Novozym 435 catalysis of starch nanosphere acylation under the present reaction conditions is selective for C6. In other words, only the OH attached to the C6 of glucose unit participates in ester bond formation.

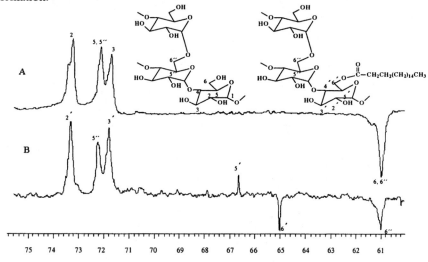

**Figure 1.** *DEPT-135 (75 MHz) spectra recorded in DMSO-d6 of A) native starch nanoparticles(1) B) vinyl stearate modified starch nanoparticles (4) with D.S. =0.8 (showed the carbon signals for sugar ring)*

**The effect of reaction temperature on starch acylation:** Studies were performed to assess how reaction temperature effects Novozym 435 catalyzed acylation with vinyl stearate of AOT-coated starch nanospheres. The results of this work are compiled in Table 2. By increasing the reaction temperature from 25 to 50°C but keeping all other reaction variables constant, the efficiency of the reaction increased from 6.7 to 40.0% and the D.S. increased from 0.2 to 1.2. Further increasing the reaction temperature to 50 and 60 °C resulted in a decrease in the reaction efficiency. Since late in the study 50°C was identified as the preferred temperature, 40°C was used for subsequent experiments described below. Acylation of starch did not occur for reactions performed at 40 and 60°C but without enzyme. Hence, we conclude that the acylation is catalyzed by Novozym 435 (CAL-B).

**The effect of substrate chain length on starch acylation:** To determine how the chain length of the acyl donor effects starch nanosphere modification, vinyl activated fatty acids with 2, 3, 10, and 18 carbons were investigated (see Table 3). There was no reaction when vinyl acetate (2-carbon chain length) was the acyl donor.

*Table 2.* Effect of the reaction temperature on the acylation with vinyl stearate of AOT coated starch nanospheres.[a]

| Entry | Temperature (ºC) | [b]Reaction Efficiency(%) | D.S. |
|---|---|---|---|
| 1 | 25 | 6.7 | 0.2 |
| 2 | 30 | 10.0 | 0.3 |
| 3 | 35 | 16.7 | 0.5 |
| 4 | 40 | 26.7 | [c]0.8±0.06 |
| 5 | 50 | 40.0 | 1.2 |
| 6 | 60 | 30.0 | 0.9 |

*[a]Reaction time: 24h; starch nanoparticles:Novozym 435=10:1(w/w); ratio of acylating agent to glucose unit 3:1 mol/mol; b[vinyl stearate esterified to starch/total vinyl stearate in the reaction] x 100; [c]Reaction performed in triplicate to determine the standard deviation.*

By increase of the fatty acid chain length from 3 to 10 and 10 to 18 carbons, the reaction efficiency increased from 3.3 to 13.3 and 13.3 to 26.7%, respectively. Hence, for acylation of AOT coated starch nanospheres, the reactivity is increased by using fatty acid acyl donors of longer chain length. In an attempt to avoid vinyl ester activation of the free acids, stearic acid was used in place of vinyl stearate. Unfortunately, esterification was not observed with steric acid that lacks the vinyl ester activating group.

**Table 3.** Effect of the acylating agent chain length and activation on its reactivity [a]

| Entry | Acylating agent | Reaction Efficiency(%) | D.S. |
|---|---|---|---|
| 1 | Vinyl Stearate | 26.7 | 0.8 |
| 2 | Vinyl Decanoate | 13.3 | 0.4 |
| 3 | Vinyl Propionate | 3.3 | 0.1 |
| 4 | Vinyl Acetate | No reaction | 0.0 |

[a]Reaction time: 24h; starch nanoparticles:Novozym 435=10:1(w/w); ratio of acyl donor to glucose units was 3:1 mol/mol.

**Acylation of AOT-coated starch nanoparticles with maleic anhydride:** The reaction between maleic anhydride and AOT-coated starch nanoparticles was performed for 24hrs, at 40°C, with a 1:1 mol/mol ratio of maleic anhydride to glucose residues. The peak positions and assignments are listed in the experimental section and they are consistent with that expected for the product. The D.S. was calculated from the $^1$H NMR (300 MHz) by comparing the relative intensity of the peaks at 5.85 (O=C-C$\underline{H}_a$=CH$_b$COOH) and 5.52 (proton at C1) ppm. As above, the only changes in the DEPT-135 spectrum after reaction with maleic anhydride (product D.S. 0.4) were new signals at 69.56 and 66.12ppm for C5 and C6, respectively. The upfield shift of C5 (by 3.27 ppm) and downfield shift for C6 (by 4.21 ppm) in combination with no other changes in the spectrum for carbons of the glucose ring shows that esterification with maleic anhydride also occurred selectively at the C6 position of glucose residues.

That a product with higher D.S. was attained by acylation with vinyl stearate rather than maleic anhydride is consistent with the preference of Novozym 435 for substrates of longer chain length (see Table 3). Furthermore, the activation energy of maleic anhydride ring-opening is likely increased by the development of charge-charge repulsion between the developing carboxylate group and the negatively charged AOT surfactant molecules. Nevertheless, maleation to D.S. 0.4 builds into the nanoparticles a substantial quantity of both unsaturated moieties for crosslinking and pendant carboxylic acids that can be used to regulate particle charge density.

**Acylation of AOT-coated starch nanoparticles with ε-caprolactone (CL):** The reaction between ε-caprolactone (CL) and AOT-coated starch nanoparticles was performed for 24hrs, at 40°C, with a 3:1 mol/mol ratio of CL to glucose residues. The peak positions and assignments are listed in the Experimental Section. Analysis of the DEPT-135 spectrum of products as above showed that the CL units were selectively esterified at the C6 position of glucose units. The D.S. of products was obtained by $^1$H-NMR from the relative intensity of the peaks at 5.52 ppm (proton of C1) and 2.18 ppm (-C(O)-C$\underline{H}_2$). The degree of polymerization (D.P.) was determined from the relative intensity of the caprolactone methylene protons (C$\underline{H}_2$-O-C(O)) at 4.46 ppm and the methylene protons (C$\underline{H}_2$-O-C(O)) at the C6 ring position of esterified glucose residues. Similar regioselective was ohbserved for the ethyl glucoside initiated ring-opening polymerization of CL catalyzed by Novozym 435(30).

**Reaction progress with time:** The effect of reaction time on the D.S. for acylations of AOT-coated starch nanoparticles with vinyl stearate, CL, and maleic anhydride are shown in Figure 2 and discussed below. Acylation with vinyl stearate starts after a 2 h lag period and by 24 h reached D.S. 0.7. Extending the reaction time to 48 h showed at most an increase in the D.S. to 0.8. The lag period observed for stearate acylation is likely due to the immiscibility of hydrophobic vinyl stearate and hydrophilic starch molecules. However, once a low degree of stearate ester formation occurred the solubility of vinyl stearate in the hydrophobically modified starch nanoparticles increases thus accelerating the reaction. In contrast, a lag period was not observed for the reaction with CL. This is consistent with a relatively higher miscibility of CL than vinyl stearate with the starch nanoparticles.

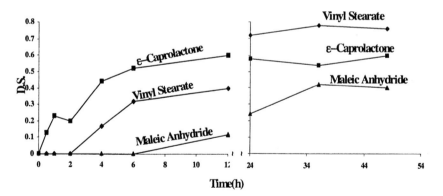

*Figure 2. Effect of reaction time on degree of substitution for the reaction of starch with different acylating agent; starch (glucose residues):vinyl stearate 1:3 mol/mol, starch:maleic anhydride 1:1 mol/mol, starch:ε-capralactone 1:3 mol/mol*

While stearate substitution at 4h was only 0.2, CL substitution had already reached 0.4 and an average degree of polymerization (D.P.) of 1.7. By 12h, the degree of CL and stearate substitution were 0.6 (D.P. = 2.0) and 0.4, respectively. With an increase in the reaction time from 12 to 48 h, neither the D.S. nor the D.P. increased substantially beyond 0.6 and 2.0. However, addition of another 3 equivalents of CL per glucose residue to the reaction system at 24 h resulted in oligo(CL) side chains with an average D.P. of 5 units with no further increase in the D.S. Reports by others that have attempted to form starch-g-PCL by ring-opening polymerizations using chemical catalysts (e.g. stannous octanoate) have similarly reported the formation of starch-g-polycaprolactone. However, unlike the chemical reaction where polycaprolactone emanates randomly from the primary(C6-) and secondary hydroxyl(C2-, C3-) positions, require high temperature like 100°C(31), starch-g-polycaprolactone from lipase-catalysis has grafts that uniformly begin by attachment to the primary hydroxyl groups of glucose residues.

Maleation of starch nanoparticles occurred slowly. In 12h the D.S. was 0.1 and it increased to 0.4 by 48h. The reactions with maleic anhydride were performed in the presence of hydroquinone to avoid the possibility of free-radical side reactions. Furthermore, when succinic anhydride was used in place of maleic anhydride, its reactivity was similarly slow.

**Characterization of unmodified and modified starch nanoparticles by light scattering:** One goal of this work was to determine if the method used to modify the starch nanoparticles gave products that were also nanoparticles. Dynamic light scattering (DLS) experiments were performed to determine the particle size distribution of the starting material and products. Prior to analysis, the modified nanoparticles were treated to remove AOT and other residual reactants (see experimental section). Figure 3 shows the particle size distributions of

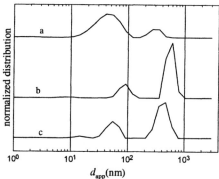

*Figure 3. Distribution of particle diameter d for a) unmodified, b) stearate-modified (D.S. = 0.6), and c) ε-caprolactone-modified (D.S. = 0.4) starch nanoparticles recorded in DMSO at the scattering angle of 28.2°. Each curve is normalized by the total area of the peaks above the baseline.*

unmodified and modified starch nanoparticles in DMSO recorded at the scattering angle of 28.2°.
Also, these measurements were also performed in DMSO at 21.5° and 64.3°. Irregardless of the scattering angle the distributions of the unmodified and modified nanoparticles had peaks at almost the same locations. Unmodified nanoparticles exhibited two prominent peaks at around 45 and 300 nm. The peak area for the smaller size particles was almost three times as large as the peak area for the larger particles. There was a third weak peak at around 2 nm. Since the scattering intensity by suspended particles was roughly proportional to the product of their mass concentration and the molar mass of each particle, the lower peak at 300 nm shows that these large particles are present at negligibly small amounts. The predominant hydrodynamic diameter of unmodified starch nanoparticles in DMSO was 40 nm. This result from light scattering is in excellent agreement with an SEM micrograph given by the supplier of the nanoparticles. Analysis by DLS of the vinyl stearate modified nanoparticles (D.S. = 0.8) showed all the peaks including the third weak peak shifted to larger sizes (Figure 3). Now the peak area for the particles at around 500 nm was three times as large as the area for the particles at around 90 nm. The change in the distribution indicates that the modification caused the equilibrium to shift towards the formation of larger aggregates within each peak and between different peaks. Nevertheless, the modified nanoparticles still remain dispersed in DMSO. For the CL modified nanoparticles (D.S. = 0.6), the shift in the peak positions of the two major peaks was not as large as that for the vinyl stearate modified starch. Furthermore, the increase of the peak area for the larger particles was less. Hence, relative to the vinyl stearate modified particles, modification with CL resulted in particles with a lower attractive interaction. Noteworthy is that all the samples were passed through a 0.2µm filter. Therefore, particles of $d$ 300 nm are dynamic aggregates that are in equilibrium with the smaller particles. Figure 4 shows the size distribution of unmodified and maleic anhydride modified starch nanoparticles dispersed in water, obtained at 30.2°. The results obtained at 23° and 62.6° were almost identical. Unmodified starch nanoparticles showed two prominent peaks at 50 nm and 350 nm, both slightly larger than their counterparts in DMSO. Unlike in DMSO, the larger unmodified particles have a greater peak area than that of the corresponding smaller particles. Thus, the relative population of the larger particles in water is not negligible. A third peak was also observed at around 10 nm. Apparently, relative to DMSO, water has a poorer ability to disperse the starch nanoparticles. Modification with maleic anhydride resulted in a shift of the three peaks to higher particle diameters with growth of the larger particles.

260

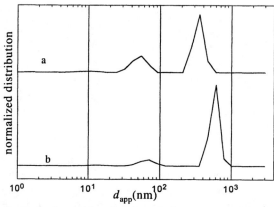

***Figure 4.*** *Distribution of particle diameter d for a) unmodified and b) maleic anhydride-modified starch nanoparticles (D.S.=0.4) recorded at 30° in water. Each curve is normalized by the total area of the peaks above the baseline.*

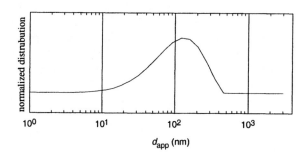

**Figure 5**. Distribution of particle diameter d of AOT coated-starch nanoparticles recorded at 30° in toluene.

In addition to the analysis of modified starch nanoparticles after they were isolated and stripped of AOT, light scattering experiments were also performed to characterize unmodified AOT-coated starch nanoparticles. Light scattering studies in toluene at 30° showed that these particles have a unimodal size distribution and diameters that range from 5 to 350 nm (Figure 5).

Reverse micelles of AOT without starch nanoparticles were not observed. The critical micelle concentration of AOT is far above the concentration used in the sample preparation. Thus, the low end of the distribution in figure 3 represents a suspension that holds only one or two 40-nm starch nanoparticles or their fragments. In contrast, at the high end of the distribution, aggregates that contain many 40-nm nanoparticles were found.

**Diffusion of starch nanoparticles into Novozym 435 beads:** Previous work showed that the average pore size of Novozym 435 beads is about 100 nm[25]. Furthermore, CAL-B is not distributed throughout the entire Novozym 435 bead; it is located in an outer shell with a thickness of 80 to 100 μm for beads with an average diameter of about 500 μm(25). In this work, we questioned whether the AOT-coated nanospheres could diffuse within the Novozym 435 beads. If this was not possible, reactions of the nanospheres would be restricted to the outer regions of Novozym 435 beads, thus limiting the fraction of enzyme available for catalysis of starch acylation. Hence, the extent of penetration by AOT-coated starch nanopartices within the catalyst bead was investigated with infrared microspectroscopy. Novozyme beads used for the IR analysis were withdrawn from the reaction between starch nanoparticles and vinyl stearate at 40°C at 12 h. Beads were embedded in paraffin, sectioned to a thickness of 12 μm with a microtome, and deposited on $BaF_2$ disks. Infrared images were collected from the bead cross-sections. In order to determine the spectral features unique to the starch, CAL-B and Novozym polymer, control infrared spectra of the starch and the unreacted Novozym bead were also obtained (Figure 6A). In the Novozym spectrum, the C=O stretching mode between 1720–1740 $cm^{-1}$ was unique to the Novozym polymer(25), so this spectral range was integrated in each pixel of the infrared image to determine the polymer distribution in the bead (Figure 6B). Similarly, the Amide II protein band from 1500–1575 $cm^{-1}$ was integrated to determine the CAL-B distribution (Figure 6C), and the C-O stretching mode between 1025 – 1058 $cm^{-1}$ was integrated to determine the starch distribution in the bead (Figure 6D). As can be seen, both the CAL-B protein content and the starch nanoparticle content were localized within an outer shell of the bead. Since the starch nanoparticles penetrated ~50 μm into the bead, and CAL-B is located in the outer 80 to 100 μm region of bead, the AOT-coated starch nanoparticles penetrate well within the beads where they encounter much of the available enzyme. This is reasonable since light scattering showed that the average diameter of the AOT-coated starch nanospheres was 5 to 350 nm (Figure 5) and the average pore diameter of the beads is 100 nm(25). Once inside the pore, CAL-B and starch molecules are in intimate contact so that

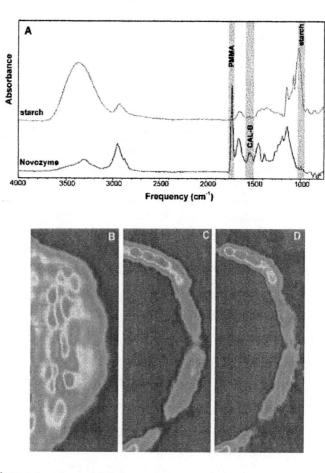

*Figure 6.* Distribution of modified AOT-coated starch nanoparticles within Novozym 435 as observed by Infrared Microscopy (IRMS). (A) Control spectra of pure starch nanoparticles (top) and unreacted Novozym beads (bottom). Spectral peaks unique to the Novozym, CAL-B, and starch are highlighted in gray. (B) Distribution of Novozym polymer (PMMA) in the bead cross-section. (C) Distribution of CAL-B enzyme in the bead cross-section. (D) Distribution of starch nanoparticles in the bead cross-section.

starch hydroxyl groups can react with enzyme activated ester complexes. This observation helps explain why Novozym 435 is so effective at modifying starch nanoparticles.

**Incorporation of CALB within AOT coated starch nanospheres:** As an alternative to the modification of starch nanoparticles by immobilized CAL-B, catalysis by non-immobilized or "free" CAL-B was also studied. Free CAL-B was incorporated into the reversed micelles during its preparation in AOT/isooctane/water (see experimental section). This was accomplished by mixing an aqueous solution of SP-525 (as-received) and starch nanospheres with AOT in isooctane (see above). The resulting AOT-coated starch nanospheres containing CAL-B were reacted with vinyl stearate (3 mol equiv per glucose residue) in toluene at 40°C for 24h to give a product with D.S. 0.5. Under similar reaction conditions, except using Novozym 435 as the catalyst, the product D.S. was 0.7. Furthermore, analysis of the regioselectivity of acylation by DEPT-135 NMR analysis showed that it too was regioselective for the C6 position. Thus, the use of free CAL-B is a reasonable alternative to that of Novozym 435.

## Conclusion

Starch nanoparticles were incorporated into AOT-coated reverse micelles that dissolved in toluene. Light scattering analysis of a suspension of these particles in toluene showed that at the low end of the size distribution there were particles with one or two 40-nm starch nanoparticles. At the high end of the distribution, aggregates that contain many 40-nm nanoparticles were found. By forming dispersions of these starch nanoparticle clusters in the non-polar medium toluene, starch molecules are highly accessible to non-polar substrates and the lipase CAL-B that is immobilized within pores of beads. Furthermore, toluene promotes the activity of the CAL-B for acylation reactions. Previous work that attempted lipase-catalyzed acylations of starch or other polysaccharides used large particles or films where only a small fraction of the polymer on the surface layers was accessible to the enzyme. The other alternative of forming a solution of the polysaccharide in polar aprotic solvents such as DMSO is flawed since DMSO and other suitable solvents that dissolve polysaccharides strip critical water from the enzyme-catalyst rendering it with little or no activity. When AOT-coated dispersions of starch nanoparticles in toluene were exposed to physically immobilized CALB (Novozym 435), acylations of ε-caprolactone (CL), vinyl stearate and maleic anhydride were successfully accomplished. The reactions were performed at 40°C for time periods up to 48h. DEPT-135 spectra of the products showed that starch acylation occurred regioselectively at the C6 position of glucose residues. Using vinyl stearate at 40°C, the product degree of substitution reached 0.8. The acylation with CL occurred most rapidly so that, by 12 h, the degree of CL substitution was 0.6 (D.P. = 2.0). In comparison, by 12 h, the degree of starch substitution with vinyl stearate and maleic anhydride were 0.4 and 0.1, respectively. By increasing the concentration of CL available in the reaction, starch-g-polycaprolactone was formed with grafts that uniformly emanate from the primary hydroxyl groups of glucose residues. Infrared microspectroscopy (IRMS) showed that AOT coated starch nanoparticles can diffuse into the outer 50 μm shell of catalyst beads. Since CAL-B is located in the outer 100 μm shell of the beads the lipase, starch and acylating agent are all in close proximity promoting the occurrence of esterifications. Studies by dynamic light scattering proved that, after isolation of acylated particles with surfactant removal, nanodimension products could be redispersed in DMSO or water.

## Acknowledgements

We are grateful to the members of the NSF I/UCRC for Biocatalysis and Bioprocessing of Macromolecules for their financial support, encouragement, and stimulating discussions. We acknowledge the generous gift of enzyme from Novozymes, and Ecosynthetix for the starch nanoparticles.

## References

1. Lenaerts. V. ; Moussa. I. ; Dumoulin. Y.; Mebsout. F.; Chouinard. F.; Szabo. P., Mateese. M. A.; Cartilier. L., Marchessault. R. *J. Control. Release* **1998**, 53, 225.
2. Vandamme. Th.F.; Lenourry. A.; Charrrueau. C. ; Chaumeil. J-C. *Carbohydrate Polymers*, **2002**, 48, 219.
3. Zikha, A.; Weiner B.Z.; Tahan M. *Journal of Medical Chemistry*, **1972**, 15, 410.
4. Sjoholm, I.; Laakso, T.; Stjarnkvist, P. *Journal of Pharmaceutical Science* **1986**, 76, 134.
5. Won, C.; Chu C.; Yu, T. *Carbohydrate Polymers* **1997**, 32, 239.
6. (6) Pereira, C.S.; Cunha, A.M.; Reis, R.L. *Journal of Material Science: Materials in medicine* **1998**, 9,825.
7. Roesser, D.S.; McCarthy, S.P.; Gross, R.A; Kaplan, D.L. *Macromolecules*, **1996**, 29(1), 1.
8. Cheng, H.N.; Gu, Q-M. *ACS. Sym. Ser.* 840, 203.
9. Li, J.; Xie, W.; Cheng, H.N.; Nickol, R.G.; Wang, P.G. *Macromolecules.* **1999**, 32, 2789
10. Klibanov, A.M. *TIBTECH* **1997**,15, 97.
11. Membrane Mimetic Chemistry: Characterizations and applications of micelles, microemulsions, monolayers, bilayers, vesicles, host-guest systems and polyions, Wiley, 1982, p 60.
12. Schwugner M.J.; Stickdorn K.; Schomäcker, R. *Chem.Rev.* **1995**, 95, 849.
13. Bruno, F. F.; Akkara J.A.; Ayyagari, M.; Kaplan, D.L.; Gross, R.; Swift, G.; Dordick. J. S. *Macromolecules* **1995**, 28, 8881.
14. Murthy, S. K.; Qi, Z.; Remsen, E. E.; Wooley, K.L.; *Polymeric Materials Science and Engineering* **2001**, 84, 1073.
15. Govender, T.; Stolnik, S.; Garnett, M. C.; Illum, L.; Davis, S.S. *Journal of Controlled Release* **1999** 57(2), 171.

16. Perez, C.; Sanchez, A.; Putnam, D.; Ting, D.; Langer, R.; Alonso, M.J.; *Journal of Control Release* **2001,** 75(1-2), 211.
17. Couvreur, P. ; Kante, B. ; Lenaerts, V. ; Scailteur, V. ; Roland, M. ; Speiser, P. *Journal of Pharmaceutical Sciences* **1998**, 69(2), 199.
18. Fontana, G.; Licciardi, M.; Mansueto, S.; Schillaci, D.; Giammona, G. *Biomaterials* **2001,** 22(21), 2857.
19. McAllister, K.; Sazani, P.; Adam, M.; Cho, M.J.; Rubinstein, M.; Samulski, R.J.; DeSimone, J.M. *J. Am. Chem. Soc.* **2002**, 124, 15198.
20. Maruyama, A.; Ishihara, T.; Kim, J.; Sung, W.; Akaike, T. *Colloids and Surfaces A: Physiochemical and Engineering Aspects* **1999**, 153, 439.
21. Madan, T.; Munshi, N.; De, T.K.; Maitra, A.; Sarma, P. U.; Aggarwal, S.S. *International Journal of Pharmaceutics* **1997**, 159, 135
22. Miller, G. *Science* **2002**, 297, 116.
23. Björk, E.; Edman, P. *International Journal of Pharmaceutical* **1990**, 62, 187.
24. Patent no. WO 00/40617, **2000**
25. Mei, Y.; Miller, L.; Gao, W.; Gross, R.A. *Biomacromolecules* **2003**, 4, 70.
26. Bru, R.; Sánchez-Ferrer, A.; García-Carmona, F.; *Biochem. J.* **1995**, 310, 721.
27. Menger, F.M.; Yamada, K. *J. Am. Chem. Soc.* **1979**, 101(22) ,6731.
28. O'Connor, C. J. ; Sun, C.Q. ; Lai, D.T. ; *Surfactant Science Series* **2003**, 109, 171.
29. Kumar A.; Gross R. A. *Biomacromolecules* **2000**, 1, 133.
30. Bisht, K.S. ; Deng, F. ; Gross, R.A. *J.Am. Chem. Soc.* **1998**, 120, 1363.
31. Choi E.J. ; Kim C.H. ; Park J.K. *Macromolecule* **1999**, 32, 7402.

## Chapter 18

# Reactions of Enzymes with Non-Substrate Polymers

### H. N. Cheng[1], Qu-Ming Gu[1,2], and Lei Qiao[1,3]

[1]Hercules Incorporated Research Center, 500 Hercules Road, Wilmington, DE 19808–1599
[2]Current address: National Starch and Chemical Company, 10 Finderne Avenue, Bridgewater, NJ 08807
[3]Current address: 3-Dimensional Pharmaceuticals, Inc., Eagleview Corporate Center, 665 Stockton Drive, Exton, PA 19341

Occasionally enzymes are found to react with polymers that are not the natural substrates of the enzymes. In this work we reviewed several such reactions, e.g., xanthan with cellulase, guar with protease and lipase, pectin with papain and β-galactosidase, and ester formation with protease. The results are variable, depending on the enzyme and the polymer involved. Sometimes we achieve only a modest effect with the enzymatic treatment, and the resulting materials are not likely to lead to commercial products. Sometimes the result is interesting and merits further research and development. In many of these cases, the reactions provide a better fundamental understanding of enzyme-substrate relationships, and the enzymes may be amenable to directed evolution/gene shuffling and process improvements.

It is well known that most enzymes have specific activities for their respective substrates. This substrate specificity is an advantage in enzyme-catalyzed synthesis, and many applications of biocatalysis depend on this property *(1,2)*. However, sometimes reactions between enzymes and non-substrate polymers are encountered. Perhaps these are side reactions discovered by accident. Perhaps these are the outcome of deliberate enzyme screening. Sometimes these reactions have low product yields or entail undesirable byproducts; as such, they are unlikely to have commercial relevance. However, many of these reactions are interesting for a variety of reasons and deserve to be investigated further.

At least four cases whereby an enzyme reacts with a non-substrate polymer may be identified. These cases are not mutually exclusive, and mixed cases are also possible.

First, some enzymes have broad specificity towards a wide range of substrates. These enzymes are potential workhorses and can be used for many polymer reactions. They are also good starting enzymes to use to screen for new reactions. A second case involves enzymes that are specific towards their own substrates but may still have some reactivity, albeit limited, towards other materials. In this case, the product yield may vary widely, depending on individual cases. Similarly, the reaction rate may vary depending on the enzyme-substrate pair, solvent, and other reaction conditions. In the (infrequent) happenstance, an enzyme may be found to have fortuitously good reactivity towards a non-substrate polymer under some specific reaction conditions.

Two other cases are related to compositional or structural heterogeneity. Often a protein sample isolated from an organism consists of a mixture of several enzymes. (Even some commercial enzyme samples contain several components.) Whereas the main enzyme component may carry out one set of reactions, the minor component(s) may have reactivity towards a different set of reactions. A similar situation arises when an enzyme exhibits multiple enzyme forms. Finally, unexpected enzymatic activity may be found when the polymer is heterogeneous. For example, a polymer may contains minor amounts of other materials, and an enzyme may react with the minor materials. A related situation occurs when the polymer contains a structural variation or a substructure that is susceptible to enzymatic attack.

In the literature, several examples of these cases are known and have been reported previously. In this work, we aim to review all four cases and provide illustrations of these types of reactions. In general, a variety of behaviors has been observed and is variously instructive.

## Results and Discussion

**Enzymatic Reactions of Guar**

Guar is a neutral polysaccharide *(3)*, consisting of mannan backbone with galactose side chains; the ratio of mannose versus galactose is approximately 2:1. Several enzymes are known to react with guar (Figure 1): 1) mannanase, which cleaves the backbone polymer and lowers the molecular weight, 2) α-galactosidase, which removes the galactose side chain, and 3) galactose oxidase, which oxidizes the 6-OH on galactose to an aldehyde.

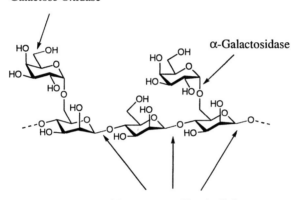

Figure 1. Structure of guar and some known enzymes

In an earlier paper *(4)*, we have described the reactions of guar with four different enzymes: lipase, hemicellulase, pectin methyl esterase, and protease. The molecular weight decreases in all cases, but in very different ways (Table 1). Hemicellulase shows the largest decrease in Brookfield viscosity and molecular weight. Lipase appears only to react with part of the guar. The polydispersity ($M_w/M_n$) of lipase-treated guar solution becomes very large after enzymatic reaction. Pectinase and protease appear to give only small decreases in molecular weights, although the Brookfield results are more pronounced.

As noted in the earlier paper *(4)*, the results of these enzyme treatments are not unexpected. Guar is known to contain > 80% soluble galactomannans, 7% proteins, and 2-3% fatty acids. Presumably the hemicellulase contains some mannanase, which cleaves the mannan backbone of guar. The lipase most likely reacts with the fatty acid portion, and protease reacts with the protein portion of guar, causing the observed changes in viscosity.

We can take a closer look at the protease reaction. The use of different proteases gives somewhat different results. For example, three proteases (Thermolysin, subtilisin Carlsberg, and Alcalase®) all react with cationic guar. However, in an aqueous reaction, Alcalase was found to be the most reactive (Table 2). Up to 95% of the protein could be removed, and there was no significant change in its physical properties. A bonus was a 50% decrease in the turbidity of the guar solution after treatment.

Table 1. Effect of enzymes on guar[a] in solution

| Enzyme Used | $B.V.^b$ | $M_n^c$ | $M_w^c$ | $M_w/M_n$ |
|---|---|---|---|---|
| none | 15,920 | 107,300 | 1,811,000 | 16.9 |
| lipase M (Amano Enzyme) | 1,720 | 37,600 | 1,336,200 | 35.6 |
| hemicellulase (Sigma-Aldrich) | 30 | 12,600 | 161,100 | 12.8 |
| Pectinex 3XL (Novozymes) | 880 | 151,600 | 1,280,500 | 8.5 |
| protease N (Amano Enzyme) | 1,160 | 179,200 | 1,414,700 | 7.9 |

[a] Enzymatic treatment was carried out on 60 g of 1% solutions (w/w) using Supercol® U neutral guar from Hercules Incorporated at pH 6.95; 10 mg of lipase, hemicellulase, protease, and 20 µL Pectinex® enzyme of Novozymes A/S were used.
[b] Brookfield viscosity obtained at room temperature for 1% solution at 3 cpm using spindle 3.
[c] SEC measurement, PEO standard

Table 2. Effect of different proteases on cationic guar

| Sample | Enzyme | Guar Conc. (%) | Rxn Time (hr) | Remaining Protein (%) |
|---|---|---|---|---|
| CG-1 | none | 2.5 | 6, 24 | 100 |
| CG-2 | Flavourzyme | 2.5 | 6 | 85 |
| CG-3 | Savinase | 2.5 | 6 | 46 |
| CG-4 | Alcalase | 2.5 | 6, 24 | 25 |
| CG-5 | Alcalase | 10.0 | 24 | 5 |

## Xanthan and Cellulase

Xanthan is an anionic bacterial polysaccharide (3), comprising a linear polymer of β-(1→4)-D-glucopyranosyl units with a trisaccharide side chain attached to every other glucose unit of the backbone. The trisaccharide side chain consists of a D-glucuronic acid inserted into two modified mannosyl units. The non-terminal D-mannosyl unit is stoichiometrically substituted at O-6 with an acetyl group, and approximately one-half of the terminal mannosyl units contains a pyruvic acid as a 4,6-cyclic acetal. It is believed that xanthan gum exists as a single or double helix with side chains aligned along the polymer backbone. The extraordinary stability of xanthan gum to oxidizers, acids, alkalis, and enzymes is believed to arise from these structural characteristics.

Table 3. Reduction of Xanthan Molecular Weight by Enzymes.

| Enzyme(s) | Units Enz per g xanthan | Xanthan conc. (%) | T (°C) | $\Delta\eta^a$ % | $M_w^b$ $\times 10^6$ |
|---|---|---|---|---|---|
| None | 0 | - | - | 0 | 9.5 |
| Celluclast® 1.5L | 28 | 1.0 | 60 | -34 | 4.8 |
| Celluclast 1.5L | 28 | 2.0 | 60 | -22 | 5.5 |
| Celluclast 1.5L | 28 | 2.5 | 60 | -32 | 5.9 |
| Celluclast 1.5L | 28 | 10.0 | 60 | n.d. | 6.8 |
| Cellulase (Sigma) | 165 | 2.0 | 37 | -33 | 3.3 |
| Carezyme® 1000L | 165 | 2.0 | 37 | -2 | 5.6 |
| Celluzyme® 0.7 T | 165 | 2.0 | 37 | -13 | 3.9 |
| Hemicellulase | 165 | 2.0 | 37 | -12 | 5.3 |
| Celluclast 1.5L + Mannanase | 34 42,500 | 2.0 | 60 | -18 | n.d. |
| Cellulase (Sigma) + Mannanase | 83 42,500 | 2.0 | 37 | -21 | 3.6 |
| Cellulase (Sigma) + Hemicellulase | 165 18 | 2.0 | 37 | -29 | 4.9 |
| Cellulase (Sigma) + Hemicellulase | 165 165 | 2.0 | 37 | -30 | 4.2 |
| Celluclast 1.5L + Mannanase + α-Glucosidase + β-Glucosidase | 34 42,500 10 10 | 2.0 | 37 | -12 | n.d. |

$^a$ Δη is the percentage change in Brookfield viscosity of 1% aqueous solution, relative to unmodified xanthan gum.
$^b$ Data from SEC. n.d.= not determined.

It is known that enzymes such as protease, hemicellulase, pectinase, and amylase have very little effect on xanthan. This resistance to enzymatic attack is presumably due to the steric hindrance caused by the trisaccharide side chains and the sugar composition of the side chains. However, xanthan is completely biodegradable, and "xanthanase" has been reported *(5)*. In addition, xanthan lyase is also known that cleaves the side-chain sugars and may have a role in xanthan degradation *(6)*. The current knowledge of xanthan depolymerization is still incomplete, and xanthanase is not commercially available.

It has been reported previously that cellulase can hydrolyze xanthan in the disordered conformation in solution. The reaction conditions typically involve 50°C in the absence of external salt. A decrease in solution viscosity is usually observed *(7)*.

Since cellulases are commercially available, we looked at this reaction further. A number of several commercial cellulases were screened for their ability to reduce the Brookfield viscosity of xanthan solution. Mannosidase and glucosidase were also used in combination with cellulase with the hope of exposing the cellulose backbone by removing the side chains. All of the cellulases tested were found to be somewhat active toward reducing the molecular weight of xanthan gum (Table 3).

The native xanthan gum has a molecular weight of $8\text{-}10 \times 10^6$, and the enzyme-treated materials have molecular weights ranging from $3.3\text{-}6.8 \times 10^6$. Thus the molecular weight change is modest under the reaction conditions tested. However, if only a slight decrease in molecular weight (or solution viscosity) is desired, this approach is one way to achieve it.

It may be noted that the combined use of several enzymes did not accelerate the backbone cleavage, at least for the enzymes tested. Thus, mannosidase, glucosidase, and hemicellulase appear not to be effective as biocatalysts in this case.

## Pectin and β-Galactosidase

Pectin comprises a linear polymer of D-galacturonic acid, interspersed by L-rhamnose, with some neutral sugar side chains attached *(3)*. A large fraction of the galacturonic acid exists as methyl esters. The degree of esterification affects solubility and gel-forming characteristics of pectin. It has been hypothesized that some of the galacturonic acids are located near the chain ends.

We speculated that a suitable hydrolase might be able to remove these galacturonic acids near the chain ends.

A number of hydrolases were attempted. These enzymes were incubated with pectin, and the reaction was monitored by thin layer chromatography (TLC). Among the enzymes tested, β-D-galactosidase from *Aspergillus oryzae* produced free galacturonic acid rapidly, and no methyl ester of galacturonic acid could be detected. After reaction, the pectin was precipitated out with alcohol. $^{13}$C-NMR analysis showed a modest increase in the degree of esterification of the enzyme-treated pectin (from 67.5% to 75.6-79.4%). However, the molecular weight of the pectin was also reduced (from 175,000 to 39,500). Size exclusion chromatography (SEC) of the commercial β-D-galactosidase showed that the enzyme contains several components. Different components are likely involved with the exo and the endo hydrolytic activities observed.

A search of the literature indicated that β-D-galactosidase is known to be biologically active towards pectin *(8)*. In fact, β-D-galactosidase is one of the enzymes involved in the ripening process of fruits *(8a)*. Thus, the observed β-D-galactosidase activity is not inconsistent with the literature.

**Protease and Ester Formation**

It is known in the literature that some enzymes can catalyze the acylation reaction of polysaccharides *(9,10)*. Not surprisingly, lipases can serve this function *(9)* because lipases are known to mediate ester hydrolysis or synthesis. Yet, some proteases can also catalyze this reaction *(10)*. For example, Alcalase® protease immobilized on Celite® can transfer the acrylic group onto hydroxyethylcellulose (HEC) to form an ester*(9a)* (Figure 2). Thus, in this case Alcalase appears to have broad substrate specificity and can work well with esters.

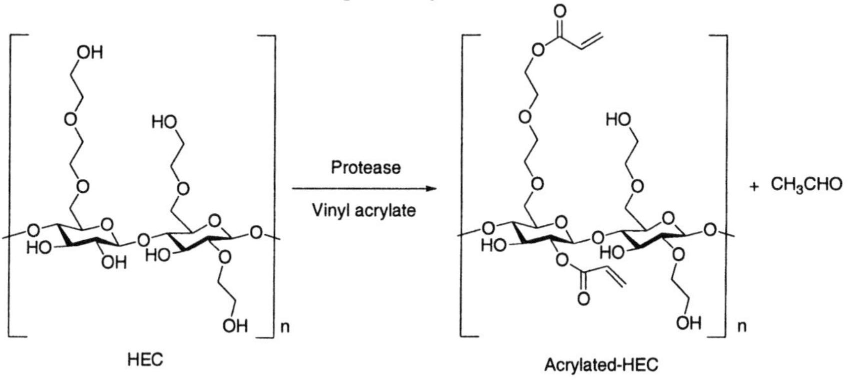

Figure 2. Reaction of vinyl acrylate with HEC, using protease as a catalyst

## Pectin and Amide Formation

Recently we reported the reaction of high methoxy pectin with an amine (or amino acid) in the presence of papaya latex papain *(11)* (Figure 3). Pectin is usually not a substrate of papain, and this reaction is unexpected. This is an example of fortuitous compatibility of an enzyme and a non-substrate.

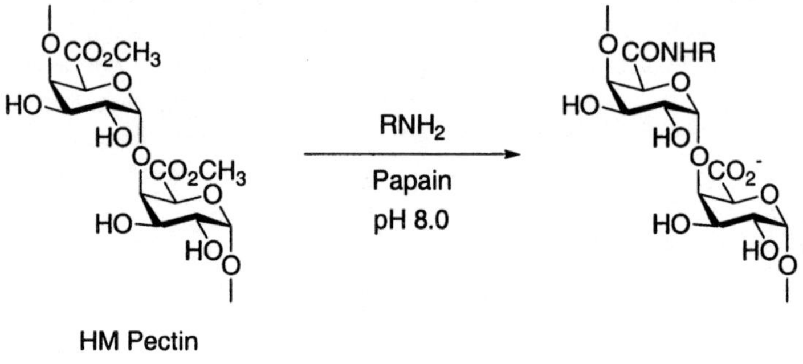

Figure 3. Papain-catalyzed formation of pectin amide

## Potential Uses

Although some of the reactions reviewed here give relatively modest effects, they may still be useful in many ways. For example, one of the recent developments in molecular biology is directed evolution *(12)*, including gene shuffling techniques *(13)*. An enzyme that gives low reactivity or yield may be a good starting point for directed evolution research. Through gene manipulations, it may be possible to modify the amino acid sequence of the enzyme and to improve its reactivity or yield.

Another possibility is to work with the reaction media, temperature, and other process parameters to improve the yield and the reactivity. Whereas this is not always possible, incremental enhancements can sometimes be obtained.

Finally, in selected cases, perhaps these reactions may be useful *as is*. Some natural polymers contain two or more components or substructures. Sometimes selective modification of one of the components may be desirable. In any case, it is beneficial to have knowledge of the different types of reactions possible. When faced with an assignment, a synthetic polymer chemist will then have these reactions (chemical or enzymatic) at his or her disposal.

## Experimental

**Reaction of Protease with Cationic Guar**

Flavourzyme, Savinase and Alcalase® protease all came from Novozymes A/S. Cationic guar powder (from Hercules Incorporated), containing 3.0-3.5% proteins based on BioRad® protein assay, was mixed with tap water at a concentration of 2.5-10%. Without adjusting its pH, the cationic guar was not fully soluble in water but partially swollen. In this way, the enzyme-treated cationic guar could be separated readily from water by centrifugation. Each protease was added at an enzyme: cationic guar ratio of 1:10, and hydrolysis was performed at room temperature for 6-24 hrs. The enzyme-treated cationic guar was recovered by filtration, washed with water to remove the digested proteins, and then dried in an oven. This treatment removed up to 95% of the protein in cationic guar (based on the protein assay results).

**Reaction of Hydrolases with Xanthan.**

The experiments were conducted with 1 - 10 % solutions of xanthan gum, and the Brookfield viscosity was monitored to detect backbone cleavage. After each reaction, the sample was heated at 90°C for 2 hours at pH 11 to deactivate the enzyme and then precipitated in IPA. The solid thus obtained was filtered, washed, and dried. A SEC curve of the reacted xanthan was obtained, using poly(ethylene oxide) (PEO) reference, from which $M_w$ was derived.

The enzymes used included cellulase, α-glucosidase, β-glucosidase, mannanase, and hemicellulase. The following four cellulase samples were studied: Novozymes Carezyme® 1000 L, 1000 Endo Cellulase Units (ECU)/g, from genetically modified *Aspergillus* microorganism; Novozymes Celluclast® 1.5L; Novozymes Celluzyme® 0.7T, 700 Detergent Cellulase Units (DCU)/g, from fermentation of *Humicola* microorganism; and Sigma-Aldrich C-0901, 9.4 U/mg, from *penicillium funiculosun*. The α-glucosidase came from Sigma-Aldrich G-6136, 19 U/mg, and β-Glucosidase also from Sigma-Aldrich G-0395, 5.0 U/mg. Two lots of hemicellulase were used: Sigma-Aldrich H215, lot 68H0422, 0.018 U/mg, and lot 40K1213, 1.57 U/mg, from *Aspergillus niger*. Mannanase came from ChemGen Corp., 855 megaU/liter. The xanthan gum came from Sigma-Aldrich and was used without further purification.

**Reaction of β-Galactosidase with Pectin**

Several hydrolases were incubated with pectin (from Hercules Incorporated) and the reaction was monitored by thin layer chromatography (TLC). After each reaction, we separated the pectin by precipitation with alcohol, which is then analyzed by $^{13}$C-NMR and SEC. The more pronounced results were observed for β-galactosidase (from Sigma-Aldrich).

**Esterification with Protease**

10 g of Natrosol® 250MR hydroxyethylcellulose of Hercules Incorporated was suspended in 20 ml of N,N-dimethylacetamide, followed by the addition of 1 ml vinyl acrylate and 0.5 g Alcalase® immobilized on Celite®. The mixture was thoroughly mixed and then incubated at 50°C for 48 hours. Acetone and isopropanol were then added to precipitate the product and washed with isopropanol. During precipitation and washing, the product flocculated, and the immobilized enzyme was dispersed in the solvent. The solvent was decanted, and the product was washed with IPA for several times until the solvent layer became clear.

# Conclusions

In this work, we have reviewed four cases of enzymatic reactions where the polymers are not the natural substrates of the enzymes.

Case 1. An enzyme has broad specificity and can react with non-substrates under certain reaction conditions. This is the case with Alcalase® protease and esterification.

Case 2. An enzyme has limited reactivity towards the polymer because of the reluctant enzyme/substrate match. An example is xanthan and cellulase. Slight, but real, decrease in molecular weight has been found. In a fortuitous situation an enzyme may show good reactivity towards a non-substrate polymer under a particular set of experimental conditions. The papain-catalyzed reaction of pectin with an amine is such an example.

Case 3. The enzyme is a mixture of several components. It may have one component that can react with a polymer. This is most likely the situation with β-galactosidase and pectin.

Case 4. The polymer is compositionally heterogeneous and contains a mixture of several components. The enzyme can react with either a minor component or a substructural motif. A good example is guar. Protein and fatty acid are minor components in guar that respond to protease and lipase treatment.

## Acknowledgements

The authors wish to acknowledge the technical assistance of G. F. Tozer, S. Mital, and A. J. Walton, and helpful discussions with colleagues at Hercules Incorporated.

## References

1. For example, (a) *Biocatalysis for Fine Chemicals Synthesis*; Roberts, S. M., Ed.; Wiley, Chichester, UK, 1999. (b) *Biotransformations in Organic Chemistry*, 3$^{rd}$ Ed.; Faber, K.; Springer, Berlin, Germany, 1997. (c) *Biocatalysis and Biodegradation*; Wackett, L. P.; Hershberger, C. D.; ASM Press, Washington, DC, 2001.
2. Some recent reviews on polymer biocatalysis include (a) *Biocatalysis in Polymer Science*; Gross, R. A.; Cheng, H. N., Eds.; Amer. Chem. Soc., Washington, DC, 2003. (b) *Enzymes in Polymer Synthesis*; Gross, R. A.; Kaplan, D. L.; Swift, G., Eds.; Amer. Chem. Soc., Washington, DC, 1998. (c) Gross, R. A.; Kumar, A.; Kalra, B. *Chem. Rev.* **2001**, *101*, 2097. (d) Kobayashi, S.; Uyama, H.; Kimura, S. *Chem. Rev.* **2001**, *101*, 3793.
3. For example, (a) *Industrial Gums*; Whistler, R.L.; BeMiller, J. N., Eds., 3$^{rd}$ ed.; Academic Press, New York, 1993. (b) *Food Polysaccharides and Their Applications*; Stephens, A. M., Ed.; Marcel Dekker, New York, 1995.
4. Cheng, H. N.; Qu, Q.-M. in *Glycochemistry*; Wang, P.G.; Bertozzi, C. R., Eds.; Marcel Dekker, New York, 2001, p. 567.
5. For example, (a) Nankai, H.; Hashimoto, W.; Miki, H.; Kawai, S.; Murata, K. *Appl. Environ. Microbiol.* **1999**, *65*, 2520. (a) Ahlgren, J. A. *J. Ind. Microbiol.* **1993**, *12*, 87. (b) Cadmus, M.C.; Slodki, M.E.; Nicholson, J. J. *J. Ind. Microbiol.* **1989**, *4*, 127. (c) Hou, C. T.; Barnabe, N.; Greaney, K. *Appl. Environ. Microbiol.* **1986**, *52*, 37. (b) Sutherland, I.W. *J. Appl. Microbiol.* **1982**, *53*, 385.
6. For example, (b) Ruijssenaars, H.J.; deCont, J. A. M.; Hartman, S. *Appl. Environ. Microbiol.* **1999**, *65*, 2446. (b) Hashimoto, W.; Mimi, H.; Tsuchiya, N.; Nankai, H.; Murat, K. *Appl. Environ. Microbiol.* **1998**, *64*, 3765.
7. For example, (a) Rinaudo, M.; Milas, M. *Int. J. Biol. Macromol.* **1980**, *2*, 45. (b) Sutherland, I. W. *Carbohydr. Res.* **1984**, *131*, 93. (c) Cheetham, N. W. H.; Mashimba, E. N. M. *Carbohydr. Polym.* **1991**, *15*, 195. (d) Christensen, B.E.; Smidsrod, O. *Int. J. Biol. Macromol.* **1996**, *18*, 93.

8. For example, (a) Nakamura, A.; Maeda, H.; Mizuno, M.; Koshi, Y.; Nagamatsu, Y. *Bioscience, Biotechnol., and Biochem.* **2003**, *67*, 68. (b) Esteban, R.; Dopico, B.; Munoz, F. J.; Romo, S.; Martin, I.; Labrador, E. *Plant and Cell Physiology* **2003**, *44*, 718. (c) Ketsa, S.; Chidtragool, S.; Klein, J. D.; Lurie, S. *J. Plant Physiol.* **1998**, *153*, 457.
9. For example, (a) Gu, Q.-M. *ACS Symp. Ser.* **2003**, *840*, 243. (b) Cheng, H. N.; Gu, Q.-M. *ACS Symp. Ser.* **2003**, *840*, 203 (2003).
10. For example, (a) Bruno, F.F.; Akkara, J.A.; Ayyagari, M.; Kaplan, D.L.; Gross, R.A.; Swift, G.; Dordick, J.S. *Macromolecules* **1995**, *28*, 8881. (b) Xie, J.; Hsieh, Y.-L. *ACS Symp. Ser.* **2003**, *840*, 217.
11. Cheng, H. N.; Gu, Q.-M.; Nickol, R. G. *U.S. Patent* 6,159,721, issued on December 12, 2000.
12. For example, (a) Arnold, F. H.; Wintrode, P.; Miyazaki, K. Gershenson, A. *Trends Biochem. Sciences (TIBS)* **2001**, *26*, 100, and references therein. (b) May, O.; Nguyen, P. T.; Arnold, F. H. *Nature Biotechnol.* **2000**, *18*, 317.
13. (a) Stemmer, W. P. C. *Proc. Natl. Acad. Sci. USA.* **1994**, *91*, 10747. (b) Crameri, A.; S-A. Raillard; E. Bermudez; W. P. C. Stemmer. *Nature.* **1998**, *391*, 288. (c) Penning, T. M.; Jez, J. M. *Chem. Rev.* **2001**, *101*, 3027.

# Condensation Polymers: Whole Cell and Related Approaches

Chapter 19

# Preparation, Properties, and Utilization of Biobased Biodegradable *Nodax*™ Copolymers

Isao Noda, Eric B. Bond, Phillip R. Green, David H. Melik, Karunakaran Narasimhan, Lee A. Schechtman, and Michael M. Satkowski

The Procter and Gamble Company, Beckett Ridge Technical Center, 8611 Beckett Road, West Chester, OH 45069

Bio-catalytic preparation, physical and biological properties, and potential utilization of novel $Nodax^{TM}$ copolymers are discussed. $Nodax^{TM}$ is a family of bacterially produced polyesters comprising 3-hydroxybutyrate (3HB) and medium-chain-length 3-hydroxyalkanoate (mcl-3HA) units. The biosynthesis of PHA copolymers involve parallel enzymatic production of 3-hydroxybutyryl CoA and 3-hydroxyacyl CoA units *via* fatty acid biosynthesis and fatty acid oxidation followed by enzymatic copolymerization. PHAs are fully biodegradable under both aerobic and anaerobic conditions. The incorporation of mcl-3HA units effectively lowers the crystallinity and $T_m$ in a manner similar to the effect of alpha olefins in linear low density polyethylene. The lowered $T_m$ well below the thermal decomposition temperature of PHA makes this material easier to process. The reduced crystallinity provides the ductility and toughness required for many applications. Versatile physical properties of $Nodax^{TM}$ PHA copolymers are well suited for the production of broad ranges of articles. Furthermore, ductile PHA can be blended with other degradable polymers like polylactic acid to greatly enhance their properties.

## Introduction

The use of bio-catalysts, such as enzymes, microorganisms, and even higher organisms, to convert non-petroleum-based feed stocks (*e.g.*, fats, oils, sugars, and polysaccharides) to novel and functional polymers is now gaining a strong interest. As opposed to the use of conventional industrial chemical synthesis (which relies primarily on petroleum derivatives as raw materials), such biosynthetic approaches are attractive because of the potential utilization of abundant and renewable bio-based raw materials and the possibility of finding unique bio-catalyzed conversion processes to efficiently produce desirable end-products. Furthermore, in the long run, bio-catalytic production of polymers may offer an intriguing opportunity for establishing the so-called *sustainable supply* of various functional materials. With the needs defined above in mind, the commercial production of plastics by biological means has been investigated.

Polyhydroxyalkanoates (PHAs) are biodegradable high molecular weight aliphatic polyesters produced by many common microorganisms, which accumulate PHAs as an energy storage mechanism in a manner similar to the lipid accumulation in higher organisms (*1*). There are now over 100 different types of known building blocks for PHA copolymers reported (*2*). Poly(3-hydroxybutyrate) (PHB) homopolymer and copolymers of 3-hydroxybutyrate and 3-hydroxyvalerate (PHBV) are the most well known (*3-5*). Recently, copolymers of 3-hydroxybutyrate and 4-hydroxybutyrate (P3HB-4HB) and elastomeric PHA comprising predominantly 3-hydroxyoctanoate (PHO) were also studied (*6,7*).

The properties of PHA copolymers depend strongly on the type, level, and distribution of comonomer units comprising the polymer chains. While PHAs have been investigated by various researchers from academic and industrial laboratories as an interesting class of materials, the successful commercial utilization of PHAs has been slow. The production cost of PHAs by conventional fermentation processes was initially high, making the entry of this material into the commodity plastics market difficult. Limited availability also did not contribute favorably toward the establishment of a robust cost structure and product development. Most importantly, the physical properties of earlier commercial PHAs, like PHBV, were inadequate for many of the applications envisioned for the replacement of commodity plastics.

Because of the remarkable stereo-regularity of the perfectly isotactic chain configuration created by the bio-catalyzed polymerization process, PHB homopolymer has unusually high crystallinity. The high crystallinity results in a rather hard and brittle material that is not very useful for many applications. The melt temperature of PHB is also high (>170°C), relative to the region of its thermal decomposition temperature, making PHB much more difficult to handle using conventional plastic processing equipment. It was hoped the incorporation of a comonomer unit 3-hydroxyvalerate (3HV) could control the excessively high $T_m$ and crystallinity limitations. Unfortunately, the desired effect of 3HV incorporation to regulate the crystallinity and $T_m$ was surprisingly limited due to the isodimorphism phenomenon (4), where 3HV units can be easily included in the crystal lattice of 3HB units and vice versa without the anticipated disruption of crystallinity. An alternative structure of PHA copolymers, therefore, had to be designed to overcome this limitation.

$Nodax^{TM}$ is a registered trademark of a family of newly introduced PHA copolymers comprising 3HB units and a relatively small amount of other medium chain length 3-hydroxyalkanoate (mcl-3HA) comonomer units, which are different from 3HV. These mcl-3HAs are chosen to be larger than 3HV, having the copolymer side groups of at least three carbon units or more (8-10). The simplest form of $Nodax^{TM}$ copolymer is the PHBHx copolymer comprising 3HB and 3-hydroxyhexanoate (3HHx) units. Other 3HA units such as 3-hydroxyoctanoate (3HO) and 3-hydroxydecanoate (3HD) are also available. Unlike 3HV units, larger mcl-3HA units are excluded from the 3HB crystal lattice, so that the incorporation of 3HA into the PHA chain results in a profound lowering of $T_m$ and crystallinity. This effect is analogous to those obtained by the incorporation of medium length alpha olefins into polyethylene to produce linear low density polyethylene (LLDPE). In this report, the biosynthesis of $Nodax^{TM}$ PHA copolymers, some of their basic properties, and potential commercial utility are discussed.

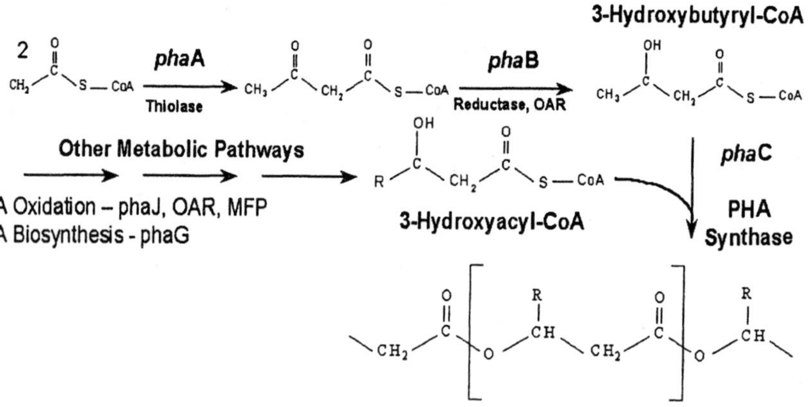

Figure 1. The metabolic pathways to produce PHA copolymers.

The key metabolic pathways utilized in the production of PHA copolymer are shown in Figure 1. Two units of acetyl CoA form acetoacetyl CoA with *pha*A thiolase, which is then converted to 3-hydroxybutyryl CoA with *pha*B reductase. Parallel to these steps are the other metabolic pathways involving fatty acid biosynthesis (*pha*G) and fatty acid oxidation (*pha*J, OAR, MFP), leading to the other larger 3-hydroxyacyl CoA units. Finally, the copolymerization of 3HB CoA and 3HA CoA with *pha*C PHA synthase results in the production of *Nodax*$^{TM}$ PHA copolymers.

Figure 2 shows a high resolution transmission electron micrograph of genetically modified *Ralstonia eutropha* containing numerous inclusion body granules of one of the *Nodax*$^{TM}$ PHA family, poly(3-hydroxybutyrate-*co*-3-hydroxydecanoate) (PHBD). This batch of microbe specimen was examined under the magnification level of 120,550X. Clearly distinguishable white spherical objects in the cells are the inclusion bodies consisting primarily of PHA copolymer, which these microorganisms have accumulated during the fermentation process. Some researchers even started calling the inclusion body granules romantically as *"pearls of microbes."*

The polymer can be harvested directly as individual particles in the form of a latex-like emulsion or extracted in the form of a solution using solvents not containing halogens. Similar accumulation of granules in microbes was observed for many other types of PHA copolymers, like PHBHx. Under the right fermentation conditions, the total yield of PHA copolymers in various microorganisms can reach a surprisingly high level. It was found that well over 80% of the dry cell weight of microorganisms may consist of the PHA polymer. From the industrial point of view, such high level of productivity makes the economic picture of this class of polymeric materials much more attractive than initially anticipated.

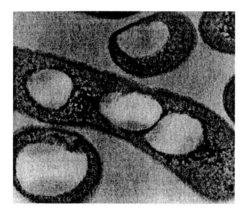

*Figure 2. Ralstonia eutropha containing PHBD granules (120,500 x).*

In this report, we present some of the general physical properties of the $Nodax^{TM}$ family of PHA copolymers and their utilization potentials as alternative forms of truly biodegradable and compostable thermoplastic materials made from renewable resources. We also report the synergy of properties achieved by combining PHA copolymers with other interesting materials like poly(lactic acid) (PLA), which also can be made from renewable biomass.

## Experimental

PHA copolymers were produced by fermentation of wild-type *Aeromonas hydrophila* (*11*) and a species of genetically modified *Ralstonia eutropha*. Additional samples of PHA copolymers were produced by chemical synthesis using the chiral ring opening copolymerization of [R]-β-lactones. The detailed procedure is described elsewhere (*12*).

A standard set of polymer characterization procedures, including molecular weight determination, compositional analysis with NMR and MS, thermal analysis, x-ray diffraction, and mechanical and rheological tests were carried out. The details of test methods have been provided elsewhere (*13*). Blends of PHA and polylactic acid (PLA) were made by melt mixing a PHBHx copolymer (3HHx content of *ca.* 12 mol%) with $T_m$ of 120 °C and a poly(lactic acid) sample (Biomer L9000, Switzerland) with $T_g$ of 60 °C and $T_m$ of 170 °C. Compounding conditions in a single screw Haake extruder zones were set as 170 °C/190 °C/170 °C. Compositions of PLA/PHBHx melt blends were set as 100/0, 90/10, 80/20, 60/40, 40/60 by wt%.

## Results

### Biological Properties

PHA copolymers produced by biological means are fully biodegradable by enzymatic action. Consequently, PHA-based articles should become readily *compostable* under proper conditions. Figures 3 and 4, respectively, show the biodegradation profiles of $^{14}C$-labeled PHA copolymers under aerobic composting conditions and anaerobic sludge digestion (*14*). It can be seen that PHA copolymers have the degradation profile comparable to cellulose, even in the absence of oxygen. Full mineralization into water, carbon dioxide, and methane was achieved for PHA. Such robust biodegradation behavior is known only to this class of thermoplastics.

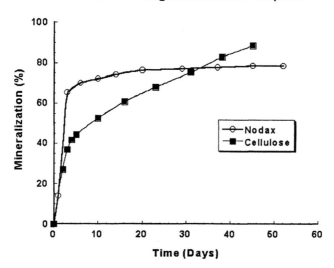

*Figure 3. Biodegradation of PHA copolymer under aerobic composting conditions.*

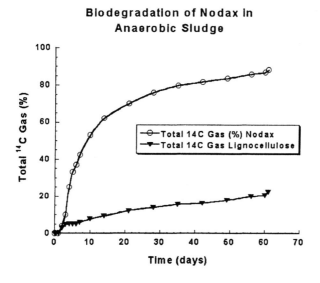

*Figure 4. Biodegradation of PHA copolymer under anaerobic digestion (14).*

The favorable biodegradation profile of PHA copolymers becomes significant when the potential application of this material is in disposable hygiene products. Degradation conditions substantially deprived of oxygen are often found in the under-water environment, including the bottom of rivers, lakes, and ocean floors. Thus, the potential benefit in the environmental waste management is apparent. More importantly, this class of polymer is also expected to undergo very rapid biodegradation in the environment characterized by typical household septic tanks still used in many parts of the world. Thus, products comprising PHA copolymers are not only compostable but also intrinsically *flushable* in many applications.

## Thermo-Mechanical Properties

Figure 5 shows the changes in melt temperature $(T_m)$ of various PHA copolymers with different levels of comonomer contents. The melt temperature of *Nodax*$^{TM}$ PHAs is significantly depressed by the incorporation of mcl-3HAs with side groups having at least three carbon units, such as 3HHx or 3HO. In contrast, PHBV copolymers with short ethyl side groups (3HV) do not depress $T_m$, even at a relatively high level of 3HV incorporation. Figure 5 shows that it is difficult to bring down the $T_m$ of PHBV well below 150 °C. This result has a profound implication in the ease of melt processing for different PHAs.

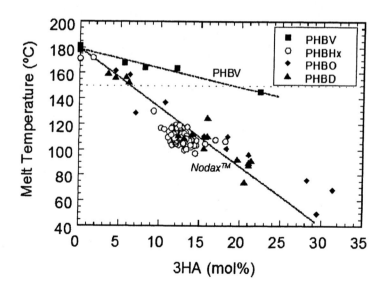

*Figure 5. Melt temperature ($T_m$) vs. 3HA comonomer content of various PHAs.*

PHAs have low to moderate thermal stability because of their propensity to undergo a random chain-scission reaction at an elevated temperature by a β-elimination mechanism. Typical PHAs suffer substantial thermal decomposition at a temperature above 170 °C. This temperature is near the melt temperatures of PHB or PHBV. As one would prefer to keep the melt process temperature of PHAs well below their thermal decomposition temperature, the depression of the $T_m$ of PHAs copolymer to below 150 °C obviously becomes a major advantage compared to PHB homopolymer or PHBV.

The susceptibility of PHA copolymers to thermal decomposition may be used as an advantage. The molecular weight of PHAs can be regulated by the judicious choice of process temperature to intentionally reduce the molecular weight. Typical intended applications for PHAs call for the molecular weight range from 500,000 to 700,000. As biologically produced PHA copolymers can have the initial weight averaged molecular weight well over 1,000,000, the ability to bring down the molecular weight at will to a target range by controlled thermal decomposition provides additional *design space* for the material.

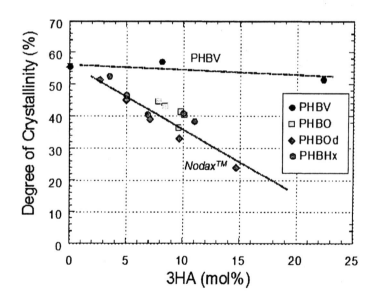

*Figure 6. Crystallinity of various PHA copolymers.*

The crystallinity of PHA copolymers is also greatly affected by the type of comonomer as shown in Figure 6. The incorporation of 3HV units provides very little change in the crystallinity, so PHBV copolymers remain somewhat brittle and fragile even if a large amount of 3HV units are incorporated. On the other hand, a small amount of other mcl-3HA units can effectively depress the crystallinity to make PHAs more ductile and tough. Indeed, the incorporation of mcl-3HA units, as low as 5 mol%, can make the mechanical properties of PHAs comparable to those of ductile polyethylene and greatly expand the potential utility of the material as general purpose plastics. Further incorporation of larger mcl-3HA comonomer units makes PHAs even more soft and flexible.

## Utilization of $Nodax^{TM}$

The relative ease of varying physical properties, through the control of mcl-3HA comonomer content and molecular weight, provides a surprisingly broad range of possible applications for $Nodax^{TM}$ copolymers. Some of the potential uses of $Nodax^{TM}$ copolymers are shown in Figure 7. Applications like fibers call for a relatively hard and stiff grade of PHAs with a low level of mcl-3HA content, while other applications like films may require a softer grade of PHAs with higher mcl-3HA content. Lower melt viscosity desirable, for example, in injection molding is achieved by reducing the weight average molecular weight to an appropriate range, while high molecular weight is often required for other applications like films. Figure 8 shows examples of prototype articles comprising $Nodax^{TM}$, such as molded plastic utensils and a spool of fibers.

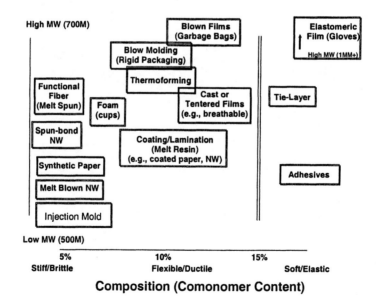

Figure 7. Product design space of $Nodax^{TM}$ copolymers.

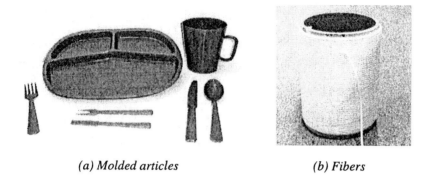

*(a) Molded articles*   *(b) Fibers*

*Figure 8. Examples of prototype products made of Nodax$^{TM}$ copolymers.*

## Polymer Alloys

One of the remarkable properties of PHAs is the ability to mix well with many other polymeris materials to produce highly compatible composites or so-called polymer alloys. Such blends often possess synergistic properties not obtainable by individual components. Of great interest is the compatible blend of *Nodax$^{TM}$* with certain other classes of degradable polymers, especially those made from renewable resources, such as starch and poly(lactic acid) (PLA). For example, hard and somewhat brittle PLA and much more ductile PHA are excellent complementary materials, which can balance the shortcomings of each other when used together. We report here some newly discovered interesting properties of PLA/PHA blends.

Figure 9 shows the remarkable toughness improvement for PLA as a small amount of PHBHx is added. The fortuitous increase of the property of PLA/PHBHx blends is believed to be related to the suppressed crystallization of very finely dispersed PHA particles. It is well known that PHAs do not readily crystallize when they are very finely dispersed as discrete domains (<<5 µm). The retardation of the crystallization is a result of the relatively low frequency of spontaneous nucleation of PHAs. The nucleation rate of PHB at biological temperatures, for example, is estimated to be less than one event per cubic millimeter per second (*15*). Therefore, under very finely dispersed state, the probability of nucleating all the segregated PHA particles becomes extremely low. Indeed this is how bacteria maintain PHA granules in their body in the amorphous state over their lifetime (*16*).

Surprisingly, a fine dispersion of small PHA particles is created in a PLA/PHBHx blend system, as long as the content of PHA is not too high. This is because the interfacial energy between PLA and PHA is rather small. The well dispersed PHA portion of the blend mostly remains in the rubbery amorphous state. The inclusion of such soft particles in a hard PLA matrix will greatly improve the toughness of the blend in a manner similar to rubber-toughened high impact polystyrene (HIPS). Excess amounts of PHA in the PLA blend, however, result in the increase in particle size of the dispersed phase, which then can crystallize rapidly. Once crystallized, PHA loses the ability to effectively toughen the blend. Thus, PLA blends containing much more than 20 wt% PHBHx did not show much improvement in the toughness.

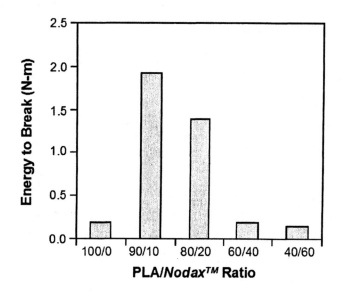

Figure 9. Toughness of PLA/PHA blends as a function of blend composition.

## Conclusions

A new class of bio-based biodegradable PHA copolymers $Nodax^{TM}$ comprising 3HB and medium chain length 3HA units is introduced. The metabolic pathways used in the production of PHA copolymers involve the parallel enzymatic production of 3-hydroxybutyryl CoA and 3-hydroxyacyl CoA units via fatty acid biosynthesis and fatty acid oxidation, followed by the enzymatic copolymerization. $Nodax^{TM}$ PHAs are fully biodegradable under both aerobic composting conditions and anaerobic sludge condition, showing the biodegradation profiles comparable to that of cellulose. By incorporating even a small amount of the 3HA comonomer, the melt temperature and crystallinity of the $Nodax^{TM}$ PHAs can be readily controlled compared to PHBV copolymers. The lowered $T_m$ relative to their thermal decomposition temperature makes this class of materials much easier to melt process. The reduced crystallinity provides the ductility and toughness required for many plastics applications. Versatile physical properties of $Nodax^{TM}$ copolymers are well suited for the production of a broad range of useful articles. Ductile PHAs can be blended with other degradable polymers like poly(lactic acid) to greatly enhance their properties.

## References

1. Limoigne, M. *Compt. Rend. Soc. Biol.* **1926**, *94*, 1291.
2. Steinbüchel, A.; Valentin, H.E. *FEMS Microbiol. Lett.* **1995**, *128*, 219.
3. King, P.P. *J. Chem. Tech. Biotechnol.* **1982**, *32*, 2.
4. Holmes, P. *Phys. Technol.* **1985**, *16*, 32.
5. Bluhm, T.L.; Hamer, G.K.; Machessault, R.H.; Fyfe, C.A.; Veregin, R.P. *Macromolecules* **1986**, *19*, 2871.
6. Doi, Y.; Kunioka, M.; Nakamura, Y.; Soga, K. *Macromolecules* **1988**, *21*, 2722.
7. Lageveen, R.G.; Huisman, G.W.; Preusting, H.; Ketelaar, P.; Eggink, G.; Witholt, B. *Appl. Env. Microbiol.* **1988**, *54*, 2924.
8. Noda, I. United States Patent 5,498,692, **1996**.
9. Shiotani, T.; Kobayashi, G. United States Patent 5,292,860, **1994**.
10. Noda, I. United States Patent 5,990,271, **1999**.
11. Doi, Y.; Kitamura, S.; Abe, H., *Macromolecules* **1995**, *28*, 4822.
12. Schechtman, L.A.; Kemper, J.J. WO 200077072, **2000**.
13. Satkowski, M.M.; Melik, D.H.; Autran, J.-P.; Green, P.R.; Noda, I.; Schechtman, L.A. In *Biopolymers*; Editors Doi, Y.; Steinbüchel, A.; Wiley, Weinheim, **2001**, vol. *3b*, p.231.
14. Federle, T.W.; Barlaz, M.A.; Pettigrew, C.A.; Kerr, K.M.; Kemper, J.J.; Nuck, B.A.; Schechtman, L.A. *Biomacromolecules* **2002**, *3*, 813.
15. Barham, P. J. *J. Mater. Sci.* **1984**, 19, 3826.
16. DeKoning, G. J. M.; Lemstra, P. J. *Polymer* **1992**, *33*, 3292.

## Chapter 20

# Novel Synthesis Routes for Polyhydroxyalkanoic Acids with Unique Properties

Bo Zhang[1], R. Carlson[1], E. N. Pederson[1], B. Witholt[2], and F. Srienc[1]

[1]Department of Chemical Engineering and Materials Science and BioTechnology Institute, University of Minnesota, Minneapolis-St. Paul, MN 55455
[2]Institute of Biotechnology, ETH Zurich, CH-1015, Switzerland

Polyhydroxyalkanoates (PHAs) are naturally occurring, biodegradable and biocompatible polyesters and have attracted considerable interest as a renewable resource based thermoplastic. In this study, we express a PHA polymerase capable of polymerizing medium chain length (R)-3-hydroxy precursor molecules in the cytosol of yeast. We demonstrate that these engineered yeasts are capable of synthesizing PHAs comprised of 6-13 carbon monomers. These monomer units are typically produced via the β-oxidation pathway in specialized organelles known as peroxisomes. Therefore, the presented results indicate that the β-oxidation pathway is not solely restricted to peroxisomes but also appears to be functional in the yeast cytosol. This finding provides a basis for novel metabolic engineering strategies that could make the PHA synthesis process more economical and could yield polymers with unique material properties.

Polyhydroxyalkanoates (PHAs) are polyesters that are synthesized as granular inclusion bodies in numerous species of bacteria. These naturally occurring bioplastics evolved as a microbial storage material for carbon, energy and reducing equivalents *(1)*. PHAs have attracted considerable interest as a natural, biodegradable and biocompatible plastic with the potential to be produced economically by microbial cultivation or by other biological systems *(2)*. Recently, significant research effort has focused on such issues as designing improved synthesis pathways for 'smarter' PHAs which possess more desirable and valuable physical properties.

Physiological data and enzymatic studies *(3)* have shown that there are two distinct classes of PHAs. The two distinct classes are based on the number of carbon atoms in the monomer unit. $PHA_{SCL}$ (short chain length) polymers possess 3-5 carbon monomers (C3-C5), whereas $PHA_{MCL}$ (medium chain length) polymers possess 6-14 carbon monomers (C6-C14) *(2)*. We have previously shown that expression of a bacterial PHA polymerase in the cytosol of *Saccharomyces cerevisiae* leads to the formation of poly (R)-3-hydroxybutyric acid (PHB) *(4, 5)*. We have extended this work by expressing in this yeast a polymerase capable of polymerizing medium chain length (R)-3-hydroxy precursor molecules ($PHA_{MCL}$). We demonstrate that these engineered yeasts are capable of synthesizing $PHA_{MCL}$ consisting of 6-13 carbon monomers (C6-C13) in the cytosol. The metabolites which serve as the $PHA_{MCL}$ monomers are typically produced via the β-oxidation pathway in specialized organelles known as peroxisomes *(6)*. Therefore, the results indicate that the β-oxidation pathway is not restricted to peroxisomes but also appears to be functional in the yeast cytosol. This finding provides a basis for novel metabolic engineering strategies that could make the PHA synthesis process more economical and could yield polymers with unique material properties.

## Experimental

**Strains and media**

All plasmids were maintained and propagated in *Escherichia coli* DH5α. *Saccharomyces cerevisiae* strain BY4743 (Mata/α his3Δ1 leu2Δ0 ura3Δ0) was obtained from Invitrogen. *S. cerevisiae* harboring a PHA synthase plasmid was maintained in SD media (0.67% yeast nitrogen base without amino acids, 2% glucose, and amino acids). For PHA production, a stationary-phase culture was harvested by centrifugation. The cells were washed once in water and resuspended at a 1:10 dilution in fresh SOG1 media containing 0.67% yeast nitrogen base without amino acids, 1% glycerol, 0.4% Tween 80 and fatty acids. Cells were then cultured for an additional 5-6 days before harvesting the cells for PHA analysis. The pH was maintained at 5 with a 5mM citric acid buffer.

**Cloning procedure**

The plasmid p2TG1T-700(H) (Figure 1) was constructed from the plasmid p2TG1T(H) which contains the high copy number $2\mu m$ origin of replication, the HIS3 selection marker, the *S. cerevisiae* TEF1 promoter and the URA3 termination sequence. The *P. oleovorans* PHA polymerase gene *(7)* was isolated from plasmid pPT700 *(8)* using a *ClaI* and *EcoRI* digest. The isolated gene was ligated into a similarly digested p2TG1T(H). A *P. oleovorans* PHA polymerase gene containing a peroxisomal targeting sequence was constructed using PCR-modification of the native gene. The following primers introduce a *ClaI* site at the 5' end of the gene and introduce an *EcoRI* site along with a type 1 peroxisomal targeting sequence (PTS1) to the 3' end: 5'-ATTATCGATGAGTAACAAGAACAACGATGAG-3' and 5'-GGAATTCAACGCTCGTGAACGTAGG-3'. The 3' primer adds a PTS1 triple amino acid peptide to the carboxy terminus of the polymerase enzyme. The PTS1 sequence, -SKL-COOH, has been shown by Elgersma *et al. (9)* to target expression of enzymes to the peroxisome of *S. cerevisiae*. The modified polymerase PCR product was subjected to a *ClaI* and *EcoRI* digest and was ligated into a similarly digested p2TG1T(H) to produce plasmid p2TG1T-755(H). The PHA plasmids were transferred into *S. cerevisiae* using a previously described lithium acetate procedure *(10)*.

**Analysis of PHA**

The cytosolic PHA was studied using gas chromatography-mass spectroscopy analysis of dichloroethane extracts of dried cell material (20-60mg) as described previously *(4)*. All cell samples were washed six times with warm methanol before being subjected to the previously described propanolysis method. Samples were analyzed using a Kratos MS25 GC-MS with a DB-WAX column.

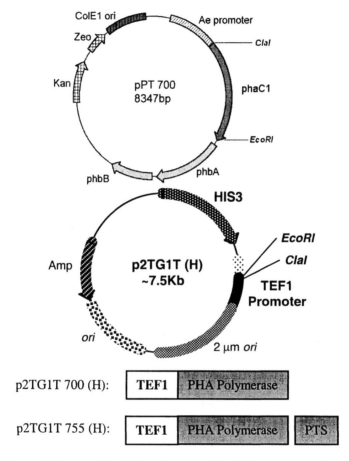

*Figure 1. Vectors for PHA polymerase (phaC1) gene expression*

pPT700 contains the phaC1 gene isolated from P. oleovorans Gpo1 and the phaB and phaA genes from Ralstonia eutropha. The plasmid p2TG1T-700(H) contains the P. oleovorans PHA polymerase under control of the TEF1 promoter and the URA3 termination sequence. The plasmid also contains the high copy number 2μm origin of replication and the HIS3 yeast selection marker. Plasmid p2TG1T-755(H) is identical to p2TG1T-700(H) except the P. oleovorans PHA polymerase contains the peroxisomal targeting sequence(PTS).

# Results

### Expression of the *P. oleovorans* PHA polymerase in the cytosol of yeast

It has been shown previously that yeast can produce medium chain length PHA with the *P. aeruginosa* PHA polymerase when the enzyme is targeted to the peroxisome *(11)*. In the present study, the *P. oleovorans* PHA polymerase is expressed in the cytosol of *S. cerevisiae* BY4743. The plasmid p2TG1T-700(H) is shown in Figure 1. This plasmid contains the high copy number yeast 2μm origin of replication and the HIS3 selection marker. The PHA polymerase is under the control of the constitutive TEF1 promoter and URA3 transcription termination sequence. Plasmid p2TG1T-755(H) is identical to p2TG1T-700(H) except the *P. oleovorans* polymerase is modified to contain the previously described type I peroxisomal targeting sequence.

### Production of Medium Chain Length (MCL)-PHA

The recombinant yeasts were grown as described in the Materials and Methods, and lauric acid (C12) was used as the carbon source. The cytosolic expression of the MCL-PHA polymerase resulted in the production of PHA which accumulated to approximately 0.014% of the total cell dry weight (CDW). Figure 2B shows the GC-MS analysis of *S. cerevisiae* BY4743 harboring plasmid p2TG1T-700(H). The C12 PHA (poly 3-hydroxydodecanoic acid) peak, C10 (poly 3-hydroxydecanoic acid), C8 (poly 3-hydroxyoctanoic acid) and C6 (poly 3-hydroxyhexanoic acid) PHA peaks are all clearly visible. Mass to charge ratios of all peaks were compared to PHA produced by *E. coli* harboring *P. oleovorans* PHA polymerase. The peroxisomally targeted PHA polymerase strain (BY4743/ p2TG1T-755(H)) was used as a positive control. Under the same conditions, this strain accumulated MCL-PHA up to 0.054% of the CDW in the peroxisomes (Figure 2C and Table 1).

### Composition of MCL-PHA produced in the cytosol of yeast

In order to determine the influence of the carbon source on PHA monomer composition, the recombinant yeast were grown in SOG1 media containing one of the following fatty acids: oleic acid, tridecanoic acid (C13), lauric acid (C12) and undecanoic acid (C11). Table 1 and 2 show that the accumulated PHA composition is dependent on the nature of the externally fed fatty acids. When lauric acid (C12) was used as the carbon source, C12 PHA is the major component of the PHA. About 58% of total PHA was comprised of C12

Table 1. PHA content and monomer composition produced by *S. cerevisiae* BY4743, when even-number fatty acids were used as the carbon source.

| Carbon source | Plasmid | PHA content (% of CDW) | Composition of PHA %, w/w | | | |
|---|---|---|---|---|---|---|
| | | | C12 | C10 | C8 | C6 |
| Lauric acid (C12) | p2TG1T-700 | 0.0147±0.0011 | 58.6 | 16.6 | 22.9 | 1.9 |
| Lauric acid (C12) | p2TG1T-755 | 0.0539±0.0041 | 38.7 | 23.8 | 29.9 | 7.6 |
| Oleic acid (C18) | p2TG1T-700 | Not detected | nd | nd | nd | nd |
| Oleic acid (C18) | p2TG1T-755 | 0.0385±0.0076 | 47.1 | 23.2 | 23.9 | 6.7 |

nd: not detected

monomer while no C14 PHA was detected (Table 1). In yeast BY4743 harboring plasmid p2TG1T-755(H), lauric acid was presumably degraded in the peroxisomes and significant amounts of C10-C6 monomers were incorporated into the PHA by the peroxisomally targeted MCL-PHA polymerase.

Similarly, recombinant yeast grown on tridecanoic acid (C13) and undecanoic acid (C11) produced PHA containing odd-chain monomers ranging from C13 to C7 with the major components being C13 and C11 monomers (Table 2 and Figure 3). When the yeast were grown on oleic acid (C18), no PHA was detected in the strain expressing the cytosolic polymerase, however the yeast strain with the PHA$_{MCL}$ polymerase targeted to the peroxisomes accumulated PHA to approximately 0.0385% of its CDW (Table1).

## Discussion

It has been previously shown in *S. cerevisiae* that the expression of a peroxisomally targeted PHA polymerase leads to synthesis of MCL-PHA in the peroxisome *(11)*. In this study, we expressed the *P. oleovorans* MCL-PHA polymerase in the cytosol of yeast. Feeding experiments with different fatty acids led to the cytosolic synthesis of C6-C13 PHA.

The yeast strain cytosolically expressing the PHA polymerase did not produce PHA from oleic acid (C18). However, PHA was produced from oleic acid in the strain which expressed a peroxisomally targeted PHA polymerase. These results suggest that the β-oxidation intermediates do not transverse the peroxisome membrane and that the nontargeted MCL-PHA polymerase is not transported into the peroxisomes.

Based on the observation that the recombinant yeast expressing a cytosolic polymerase accumulate PHA monomers with C-backbones of different lengths than the fed fatty acids, we propose that β-oxidation can occur, at least partially, in the cytosol of *S. cerevisiae* (Figure 4). One possible explanation for this

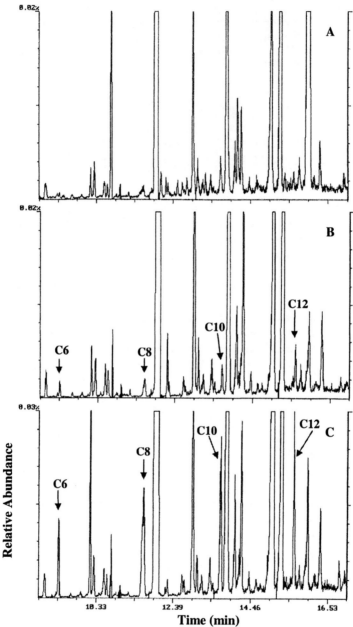

*Figure 2.* GC-MS analysis of PHA produced by *S. cerevisiae* BY4743, when lauric acid (C12) was used as the carbon source. Only peaks, which possess a mass-to-charge ratio value of 131, are shown. A: Wild-type *S. cerevisiae* BY4743; B: *S. cerevisiae* BY4743 harboring plasmid p2TG1T-700; C: *S. cerevisiae* BY4743 harboring plasmid p2TG1T-755. The arrow indicates the position of 6, 8, 10, and 12 carbon PHA monomers.

Table 2. PHA content and PHA monomer composition of polyester produced by *S. cerevisiae* BY4743 harboring plasmid p2TG1T-700(H) when different odd-number fatty acids were used as the carbon source.

| Carbon source | PHA content (% of CDW) | Composition of PHA %, w/w | | | |
|---|---|---|---|---|---|
| | | C13 | C11 | C9 | C7 |
| Tridecanoic acid (C13) | 0.0498±0.0117 | 24.2 | 16.1 | 37.6 | 21.9 |
| Undecanoic acid (C11) | 0.0255±0.0048 | nd | 50.9 | 46.5 | 2.6 |

nd: not detected

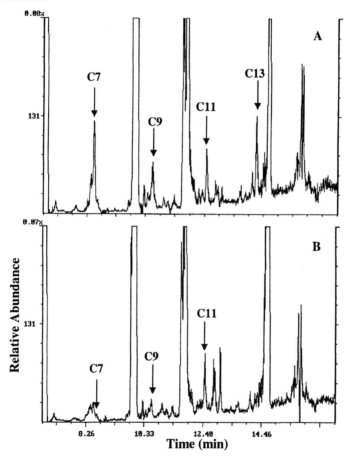

*Figure 3. GC-MS analysis of PHA produced by S. cerevisiae BY4743 harboring plasmid p2TG1T-700(H), when A: tridecanoic acid (C13) and B: undecanoic acid (C11) were used as the carbon source. Only peaks, which possess the mass-to-charge ratio value of 131, are shown. The arrow indicates the position of 7, 9, 11 and 13 carbon PHA monomers.*

observation is that β-oxidation enzymes are synthesized in the cytosol and then transported into the peroxisomes posttranslationally. This creates a temporal window where they could be active in the cytosol. In fact, some studies have shown that 15-25% of β-oxidation enzyme activities can be found in the cytosol of yeast *(12, 13, 14, 15)*. Another potential source of PHA precursors is from fatty acid biosynthesis. Marchesini and Poirier *(16)* have showed that a *fox3* yeast mutant harboring a peroxisomally targeted $PHA_{MCL}$ polymerase can synthesize PHA monomers with C-backbones of different length than the fed fatty acids. They concluded that there is futile cycling operating between fatty acid biosynthesis and fatty acid degradation. So both externally fed fatty acids and fatty acid biosynthesis may contribute to the observed cytosolic $PHA_{MCL}$ synthesis.

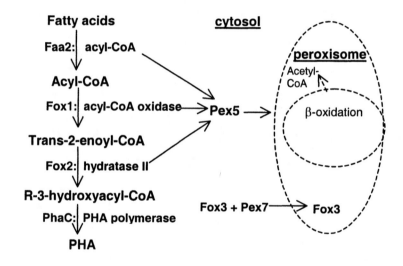

Figure 4. *Expression of PHA synthesis pathway in the cytosol of S. cerevisiae*

## Conclusions

The data indicate that *S. cerevisiae* is capable of synthesizing (R)-3-hydroxy precursor molecules from fatty acids in the cytosol. This study demonstrates that a reaction sequence normally thought confined to peroxisomes, can also function in the cytosol. Because different fatty acids can be offered in the culture medium, it should be possible to synthesize many different PHAs with useful material properties in yeasts expressing the PHA pathway genes in the cytosol based on either externally supplied fatty acids or based on fatty acid synthesis.

## Acknowledgement

We would like to thank the National Science Foundation for the support (BES-0109383).

## References

1. Steinbuchel, A.; Valentin, H. E. *FEMS Microbiol. Lett.* **1995**, 128, 219-28.
2. Lee, S. Y. *Trends in Biotechnology*, **1996**, 14, 431-8.
3. Haywood, G. W.; Anderson, A. J.; Dawes, E. A. *FEMS Microbiol. Letters,* **1989**, 57, 1-6.
4. Leaf, T. A.; Peterson, M. S.; Stoup, S. K.; Somers, D.; Srienc, F. *Microbiology*, **1996**, 142, 1169-80.
5. Carlson, R.; Fell, D.; Srienc, F. *Biotechnol Bioeng.* **2002**, 79:121-34.
6. Hahn, J. J.; Eschenlauer, A. C.; Sleytr, U. B.; Somers, D.; Srienc, F. *Biotechnol. Prog.* **1999**, 15, 1053-7.
7. Huisman, G. W.; Wonink, E.; Meima, R.; Kazemier, B.; Terpstra, P.; Witholt, B. *J. Biol. Chem.* **1991**, 266, 2191-2198.
8. Jackson, J. K. Master Dissertation. University of Minnesota. St. Paul, MN, **1998**.
9. Elgersma, Y.; Vos, A.; van den Berg, M.; van Roermund, C. W.; van der Sluijs, P.; Distel, B.; Tabak, H. F. *J Biol Chem.* **1996**, 271, 26375-82.
10. Soni, R.; Carmichael, J. P.; Murray, J. A. *Curr Genet.* **1993**, 24, 455-9.
11. Poirier, Y.; Erard, N.; Petetot, J. M. *Appl Environ Microbiol.* **2001**, 67, 5254-60.
12. Hettema, E. H.; van Roermund, C. W.; Distel, B.; van den Berg, M.; Vilela, C.; Rodrigues-Pousada, C.; Wanders, R. J.; Tabak, H. F. *EMBO J.* **1996**, 15, 3813-22.
13. van Roermund, C. W.; Drissen, R.; van Den Berg, M.; Ijlst, L.; Hettema, E. H.; Tabak, H. F.; Waterham, H. R.; Wanders, R. J. *Mol Cell Biol.* **2001**, 21, 4321-9.
14. Klein, A. T.; van den Berg, M.; Bottger, G.; Tabak, H. F.; Distel, B. *J Biol Chem.* **2002**, 277, 25011-9.
15. Erdmann, R.; Veenhuis, M.; Mertens, D.; Kunau, W. H. *Proc Natl Acad Sci U S A.* **1989**, 86, 5419-23.
16. Marchesini, S.; Poirier, Y. *J. Biol. Chem.* **2003**, 278, 32596-601.

## Chapter 21

# Kymene® G3-X Wet-Strength Resin: Enzymatic Treatment during Microbial Dehalogenation

### Richard J. Riehle

Hercules Incorporated, Research Center, 500 Hercules Road, Wilmington, DE 19808–1599

Polyaminopolyamide-epichlorohydrin (PAE) resins, typified by Kymene® 557H wet-strength resin, are the predominant commercial products used to manufacture wet-strength paper. Changes in the manufacturing process decreased the level of epichlorohydrin (epi) by-products in PAE resins, typified by Kymene® 557LX. Post-treatment technologies further reduced the level of epi by-products. European regulatory agencies have recently introduced requirements that reduce the allowable level of 3-chloropropanediol (CPD) in food contact paper. A Subtilisin enzyme (Alcalase 2.5L type DX from Novozymes) treatment was combined with a microbial treatment to remove epi by-products to provide a product, Kymene® G3-X wet-strength resin, that allows the papermaker to meet all regulatory requirements.

Polyaminopolyamide-epichlorohydrin (PAE) resins are the predominant commercial products used to manufacture wet-strength paper. PAE resins are typified by the products, Kymene® 557H wet-strength resin, Kymene® 557LX wet-strength resin, and Kymene® SLX2 wet-strength resin, manufactured worldwide by Hercules Incorporated. These resins contain by-products from the reaction of epichlorohydrin with the aqueous polyaminopolyamide (a.k.a, polyamide) (see Figure 1). Two of these by-products are 1,3-dichloropropanol

(DCP) and 3-chloropropanediol (CPD). Due to their toxicity, it is desirable to decrease the levels of DCP and CPD. Changes in chemical labeling requirements for toxic materials prompted the development of a second generation of products. Changes in manufacturing, based on a kinetic understanding of the process, markedly reduced the levels of DCP and CPD and yielded this second generation of products, typified by Kymene® 557LX and Kymene® SLX2 wet-strength resins (*1*). Second generation products have less than 1,000 ppm of DCP and less than 1,000 ppm of CPD on an "as received" basis.

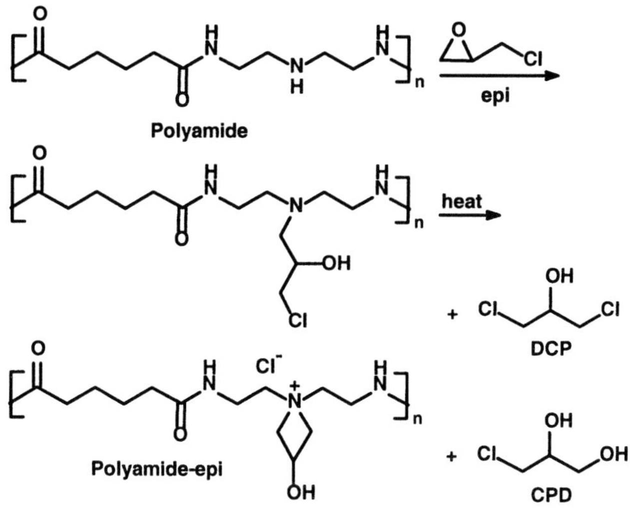

*Figure 1. Manufacture of polyamide-epi (PAE) resins.*

Subsequent regulatory requirements prompted the development of a third generation of products. European regulatory agencies (BfR, formerly the BgVV; MAFF; and SCF) continue to reduce the allowable levels of impurities in food. For example, recent BfR recommendations reduce the allowable levels of DCP and CPD in food contact paper (*2*). Typical regulated food contact papers are tea bag, coffee filter, and milk and juice cartons. Food contact paper using wet-strength resins can contribute DCP and CPD to food.

To meet the requirements for food contact paper, third generation products with lower levels of DCP and CPD are typically manufactured by post-treating the PAE resin. Post-treatment technologies include microbial dehalogenation (*3*), membrane filtration (*4*), and carbon absorption (*5*). Researchers at Hercules Incorporated and the University of Kent developed a third generation PAE resin using a microbial dehalogenation process that converts the two toxic by-products, 1,3-dichloropropanol (DCP) and 3-chloropropanediol (CPD), into carbon dioxide and salt (see Figure 2) (*3*).

The preferred microorganisms used in this process are a consortium of *Arthrobacter histidinolovorans* (HK1) and *Agrobacterium radiobacter* (HK7). These microorganisms are non-toxic, non-pathogenic bacteria isolated from soil. HK1/HK7 utilize DCP and CPD as growth substrates. Microbial growth during the microbial dehalogenation process is the primary source of the biocatalyst.

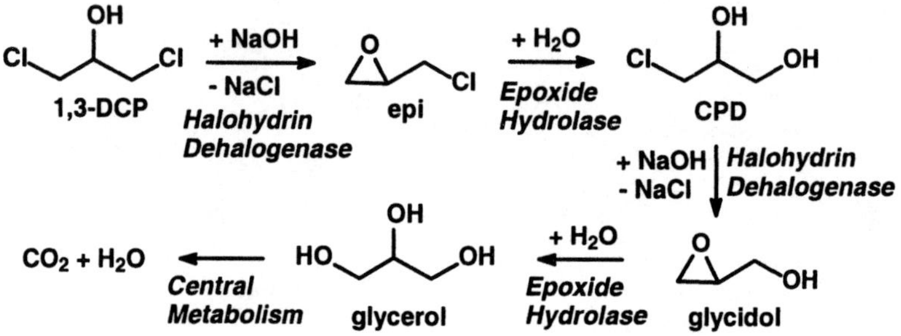

*Figure 2. Chemistry of the microbial dehalogenation process.*

When freshly prepared, third generation PAE resins produced with these technologies have DCP and CPD levels typically less than 10 ppm. However, upon storage, the level of CPD increases, typically to 50 ppm or greater. Researchers at Hercules discovered that this increase is due to "polymer-bound CPD" (PB-CPD) and that both CPD and PB-CPD in PAE resins contribute to CPD in paper. Hercules, therefore, developed and commercialized a new PAE resin, Kymene® G3 wet-strength resin, with greatly reduced levels of CPD and PB-CPD (*6*). A basic ion exchange process can also be used to remove DCP, CPD, and PB-CPD, but it is costly and produces a significant industrial waste stream (*7*).

A key step in the production of Kymene® G3 wet-strength resin used an acid treatment that reduced the performance of the product relative to second generation products. This performance gap was eliminated by combining a Subtilisin protease treatment process with a microbial treatment process that removes DCP, CPD, and PB-CPD from PAE resins. Kymene® G3-X wet-strength resin, the product based on this new manufacturing process, meets all regulatory requirements.

## Experimental

**Materials.** Kymene® E7219 wet-strength resin was obtained from the Hercules Zwijndrecht, Netherlands plant. Alcalase 2.5L type DX (herein referred to as Alcalase) was obtained from Novozymes A/S Bagsvaerd, Denmark. For microbial dehalogenations, Kymene® E7219 was pasteurized at 80 °C for 10 minutes. The nutrient solution consisted of 8,026 ppm of potassium dihydrogen phosphate, 27,480 ppm of urea, 4,160 ppm of magnesium sulfate, and 840 ppm of calcium chloride in tap water. The microorganisms used were: *Arthrobacter histidinolovorans* (HK1) and *Agrobacterium radiobacter* (HK7).

**Typical Alcalase Treatment of PAE Resin (Information for pH 7.3 Reaction).** A 500-mL round-bottom flask was fitted with a condenser, a pH meter, a temperature controlled circulating bath, and a mechanical stirrer. To the flask was added 400 g of 13.3% Kymene® E7219. The temperature was raised to 30 °C, the pH was raised to 7.3 with 6.77 g of 30% aqueous sodium hydroxide, and then 4.92 g of Alcalase was added. The temperature and pH were maintained. Periodically, aliquots of the reaction mixture were removed and analyzed by gas chromatography (GC). After 24 hours, the pH was lowered to 2.9 with 96% sulfuric acid. The resin had a Brookfield viscosity of 20 cps (at 25 °C).

**Typical Microbial Treatment of PAE Resin (Batch 3, 10% Inoculation).** A 500-mL round-bottom flask was fitted with a condenser, a pH meter, a temperature controlled circulating bath, an air sparge tube, and a mechanical stirrer. To the flask was added 350 g of pasteurized 13.5% Kymene® E7219. The pH was raised to 5.9 with 3.85 g of 30% aqueous sodium hydroxide. Then 4.2 g of a nutrient solution and 39 g of inoculum, which contained a blend of microorganisms comprising an inoculum from a biodehalogenated polyaminopolyamide-epichlorohydrin resin (batch 2), were added. The air sparge was started, the temperature was maintained at 30 °C, and the pH was maintained by periodic addition of 30% aqueous sodium hydroxide. The bacterial growth was monitored by Optical Density ($OD_{600}$). Periodically, aliquots of the reaction mixture were removed and analyzed by GC. After 21 hours, inoculum was removed for the next batch. The mixture was then cooled to room temperature and the pH was adjusted to 2.8 with 96% sulfuric acid. The resin had a Brookfield viscosity of 94 cps (at 25 °C).

**Typical Combined Alcalase-Microbial Treatment of PAE Resin (Batch 3, 20% Inoculation).** A 500-mL round-bottom flask was fitted with a condenser, a pH meter, a temperature controlled circulating bath, an air sparge tube, and a mechanical stirrer. To the flask was added 350 g of pasteurized 13.5% Kymene® E7219. The pH was raised to 7.6 with 8.59 g of 30% aqueous sodium hydroxide. Then 3.1 g of nutrient solution, 4.38 g of Alcalase, and 88 g of inoculum, which contained a blend of microorganisms comprising an inoculum from a biodehalogenated polyaminopolyamide-epichlorohydrin resin (batch 2), were added. The air sparge was started and the temperature was maintained at 30 °C. The bacterial growth was monitored by $OD_{600}$ and the biodehalogenation was monitored by GC. The pH of the reaction mixture was maintained by periodic addition of 30% aqueous sodium hydroxide. After 23 hours, the resulting resin was used as inoculum for batch 4. The pH of the remaining resin was adjusted to 2.8 with 96% sulfuric acid. The resin had a Brookfield viscosity of 49 cps (at 25 °C).

## Results and Discussion

The evidence suggests that "CPD-ester" functionality is responsible for polymer-bound CPD (PB-CPD) (6). During the manufacture of polyaminopolyamide-epichlorohydrin (PAE) resins, the acid end-group functionality on the low molecular weight polyaminopolyamide reacts with epichlorohydrin to form CPD-ester functionality (for example, see Figure 3). It is believed that hydrolysis of CPD-ester functionality is responsible for the observed CPD increase during aging in third generation products manufactured using only post-treatment technology. More importantly, the evidence suggests that hydrolysis of CPD-ester functionality significantly contributes to CPD in paper.

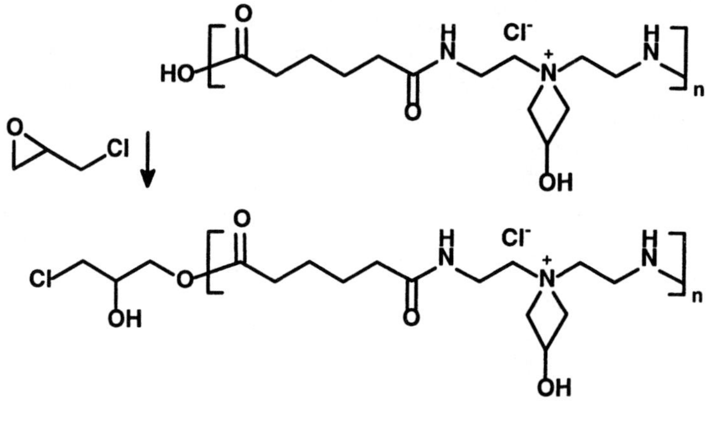

*Figure 3. Formation of PB-CPD during PAE resin manufacture.*

To meet the European regulatory requirements for CPD in paper, the first process the Hercules investigators used to release CPD from PB-CPD was based on acid hydrolysis, resulting in Kymene® G3 wet-strength resin (6). Although this acid treatment was very effective at removing PB-CPD, it also significantly decreased the level of azetidinium functionality, thereby reducing the wet strength efficiency of the resin. To improve wet strength efficiency, we investigated a mild enzymatic process to hydrolyze the PB-CPD in PAE resins. Many enzymes were evaluated. The Subtilisin family of proteases were by far the most effective enzymes. Specifically, the Subtilisin from *Bacillus licheniformis* (a.k.a., Subtilisin Carlsberg) had close to optimal properties. Although a protease, Subtilisin Carlsberg is an effective esterase (*8,9*). Alcalase 2.5L type DX from Novozymes (herein referred to as Alcalase) is an especially effective commercial product. The evidence indicates that the key reaction is hydrolysis of CPD-ester functionality on the PAE resin (for example, see Figure 4).

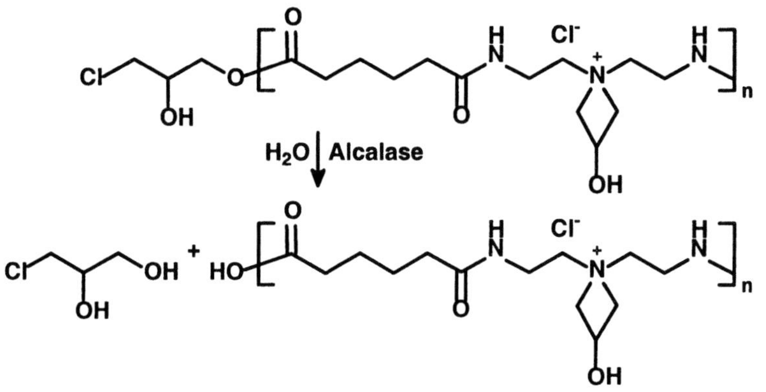

*Figure 4. Alcalase hydrolysis of the PB-CPD on PAE resin.*

## Alcalase Treatment of PAE Resin

The rate of CPD release is dependent upon pH, temperature, time, and the amount of enzyme. Additionally, to obtain a commercially viable PAE resin, the enzymatic treatment conditions must be optimized to manufacture a product with the desired viscosity. At pH greater than 6, the viscosity of PAE resins tends to increase (due to the crosslinking reaction between secondary amine and azetidinium functionality, for example). Using 30 °C and 12.5 g of Alcalase per Kg of 13.5% PAE resin, the amount of CPD released is similar throughout the pH range of 7.3-7.9 (see Figure 5). However, the pH 7.9 process gelled within 23 hours (see Figure 6). Thus, pH 7.1-7.7 is preferred to obtain the desired viscosity.

Under these conditions, Alcalase was reasonably stable, losing about 30-50% of its activity over 23 hours (see Figure 7). The loss of Alcalase activity was pH dependent. As the pH increased, the rate of activity loss increased. A reaction velocity experiment showed that the rate of CPD release was dependent on the Alcalase amount (see Figure 8).

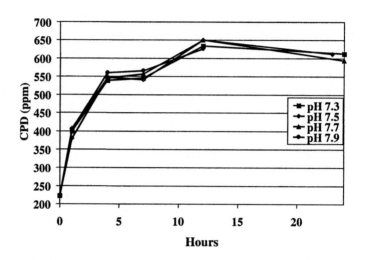

*Figure 5. CPD release by Alcalase hydrolysis of PB-CPD on PAE resin.*

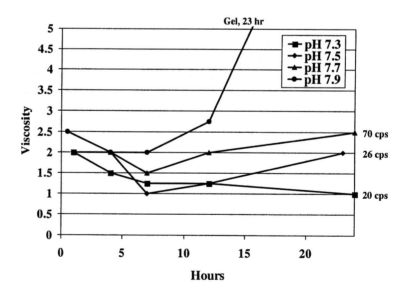

*Figure 6. Viscosity change during Alcalase hydrolysis of PB-CPD on PAE resin.*

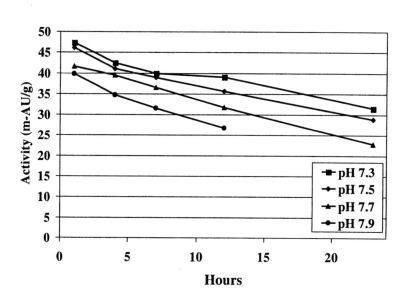

*Figure 7. Loss of Alcalase activity during treatment of PAE resin.*

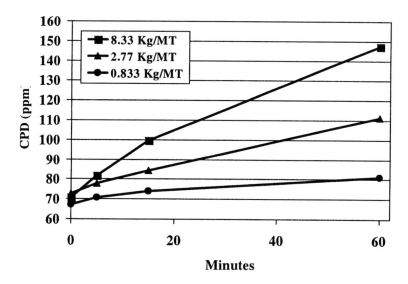

*Figure 8. Release of CPD from PB-CPD during Alcalase treatment of PAE resin.*

## Microbial Dehalogenation of PAE Resin

Microbial dehalogenation of PAE resins removes the DCP and CPD within 24 hours (see Figure 9). This process is typically conducted at pH 5.8-6.0 and at 13.5% PAE solids. The preferred microorganisms used in this process are a consortium of *Arthrobacter histidinolovorans* (HK1) and *Agrobacterium radiobacter* (HK7). These microorganisms utilize DCP and CPD as growth substrates. Microbial growth during the microbial dehalogenation process is the primary source of the biocatalyst, which can be generated in successively larger batchs. In addition to the DCP and CPD in the PAE resin as the carbon source, a nutrient solution is added to provide the other essential elements.

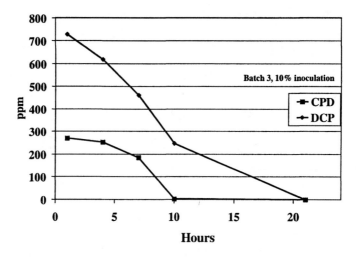

*Figure 9. Microbial dehalogenation of PAE resin.*

## Combined Alcalase and Microbial Dehalogenation Treatment of PAE Resin

A key part of the development effort was to combine the Alcalase treatment with the microbial dehalogenation treatment. The process conditions (pH, temperature, PAE concentration, and the amount of Alcalase) were optimized to provide a PAE resin with the desired properties. With this combined process, DCP, CPD, and PB-CPD were efficiently removed using a consortium of HK1 and HK7 (see Figure 10). The CPD level initially increased partly due to the release of CPD from PB-CPD. Measurement of the microbial optical density indicates the desired exponential growth during the combined Alcalase treatment and microbial dehalogenation process (see Figure 11). Under the mild conditions of this process, the level of azetidinium functionality of the PAE resin is not significantly decreased and therefore, the wet strength efficiency is essentially unchanged by the process.

Additional experimentation shows that the microbial dehalogenation process does not affect the Alcalase treatment. The rate of Alcalase activity loss is only dependent on pH, not on microbial growth (see Figure 12). With this process, a microbial consortium (HK7/HK1) is preferred over a single microbial (HK7) for efficient removal of DCP and CPD (see Figure 13).

Hercules currently manufactures and markets a third generation product, Kymene® G3-X wet-strength resin, based on the combined Alcalase and microbial dehalogenation treatment technology. Relative to second generation products, this product has greatly reduced levels of DCP, CPD, and PB-CPD. Relative to third generation products manufactured using only post-treatment technology, Kymene® G3-X has a greatly reduced level of PB-CPD. It is, therefore, a true third generation product, which herein is defined as having very low levels of DCP, CPD, and PB-CPD and thereby allows complete regulatory compliance in every grade of food contact paper.

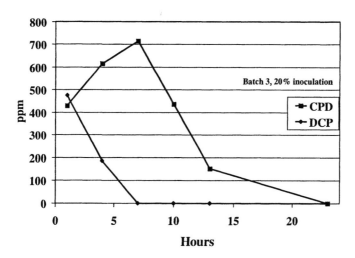

*Figure 10. Combined Alcalase and microbial dehalogenation treatment of PAE resin.*

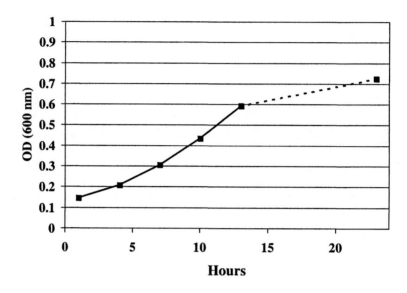

*Figure 11. Rate of microbial growth during combined Alcalase and microbial dehalogenation treatment of PAE resin.*

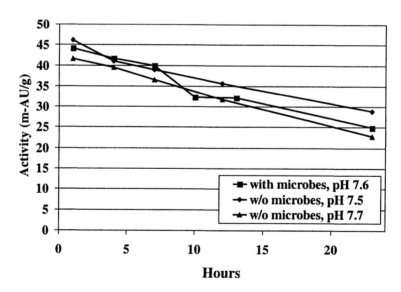

*Figure 12. Loss of Alcalase activity during microbial growth.*

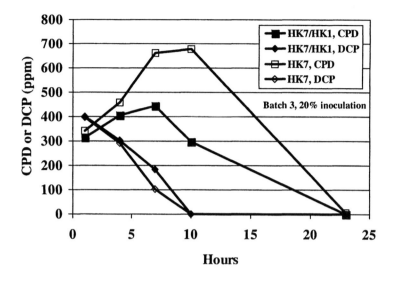

*Figure 13. Microbial preference with the combined Alcalase and microbial dehalogenation treatment of PAE resin.*

## Performance of Alcalase and Microbial Dehalogenation Treated PAE Resin

Wet strength is a key property that PAE resins impart to paper. First generation resins (e.g., Kymene® 557H wet-strength resin) have the highest level of wet strength performance for PAE resins. Kymene® G3-X has only slightly lower wet strength performance than Kymene® 557H (see Figure 14).

The key performance property that distinguishes an optimized third generation resin from other PAE resins is the level of CPD in paper. While Kymene® ULX3 (a third generation product manufactured using microbial dehalogenation post-treatment technology) provides much lower levels of CPD in paper than a first or even a second generation resin, the CPD level in paper is higher than the optimized third generation resin, Kymene® G3-X (see Figure 15).

## Conclusions

Alcalase treatment of PAE resins releases CPD from PB-CPD. An Alcalase treatment combined with a microbial dehalogenation treatment produced a PAE resin, Kymene® G3-X wet-strength resin, which does not contain DCP, CPD, or PB-CPD, and therefore allows papermakers to meet all current European regulatory requirements.

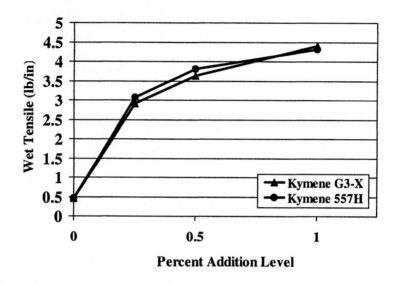

*Figure 14. Wet strength performance of Kymene® wet-strength resins.*

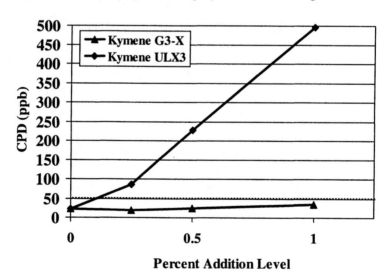

*Figure 15. CPD in paper performance of Kymene® wet-strength resins.*

## Acknowledgements

The author would like to thank the following for their contributions to this research: Tony Allen, Bob Bates, Massimo Berri, Adam Borowski, Ronald Busink, Frank Carlin, Marco Cenisio, H. N. Cheng, Jorg Coolen, Mark Crisp, Jay Dickerson, Alfred Haandrikman, Sybe Hartmans, Michaela Hofbauer, Jack Hoglen, Hal Jabloner, John Lapre, Allison Moore, Joann Pasqualini, Barry Podd, Ann Shields, Wim Stevels, and Steve Vinciguerra.

## References

1. Miller, A. J.; Stubbs, B. M. *US Patent* 5,171,795, assigned to Hercules Incorporated, December 15, **1992**. Bower, B. K. *US Patent* 5,714,552, assigned to Hercules Incorporated, February 3, **1998**.
2. XXXVI Empfehlung der BgVV - Papiere, Kartons und Pappen für den Lebensmittelkontakt - Stand 1.1.2002 - Seite 121 - Fussnote 6a. XXXVI/1 Empfehlung der BgVV - Koch- und Heissfilterpapiere und Filterschichten - Stand 1.1.2002 - Seite 121 - Fussnote 8. XXXVI/2 Empfehlung der BgVV - Papiere, Kartons und Pappen für Backzwecke - Stand 1.1.2002 - Seite 121 - Fussnote 9.
3. Bull, A.; Hardman, D. J.; Stubbs, B. M.; Sallis, P. J. *US Patents* 5,470,742, 5,843,763, and 5,871,616, assigned to Hercules Incorporated, November 28, 1995, December 1, **1998**, and February 16, **1999**. Fauzi, A. M.; Hardman, D. J.; Bull, A. T. *Appl. Microbiol. Biotechnol.* **1996**, *46*, 660.
4. Bigorra Llosas J.; Pinilla, J.-A.; Pi Subirana, R. *European Patent* EP 1 135 427 B1, assigned to Cognis Deutschland GmbH & Co. KG, September 26, **2001**.
5. Amey, R. L. *US Patents* 6,056,855 and 6,057,420, assigned to E. I. du Pont de Nemours and Company, May 2, **2000**. Laurent, H.; Dreyfus, T.; Poulet, C.; Quillet, S. *US Patent* 6,342,580 B1, assigned to Atofina, Puteaux, January 29, **2002**. Constantin, G.; Fouquay, S.; Poulet, C.; Silberzan, I. *World Patent* WO 01/18093 A1, assigned to Ceca S.A., International Publication Date March 15, **2001**.
6. Riehle, R. J.; Allen, A. J.; Hoglen, J. J.; Busink, R.; Cheng, H. N.; Haandrikman, A. J.; Carlin, F. J. Jr.; Hofbauer, M.; Crisp, M. T.; Lapre, J. A.; Jabloner, H. *US Patent* 6,554,961 B1, assigned to Hercules Incorporated, April 29, **2003**. Allen, A. J.; Hoglen, J. J.; Busink, R.; Cheng, H. N.; Haandrikman, A. J.; Carlin, F. J. Jr.; Hofbauer, M.; Crisp, M. T.; Lapre, J. A.; Jabloner, H.; Riehle, R. J. *World Patent* WO 00/77076 A1, assigned to Hercules Incorporated, International Publication Date December 21, **2000**. Riehle, R. J.; Busink, R.; Berri, M.; Stevels, W. Patent Application Publication US 2003/0000667 A1, assigned to Hercules Incorporated, January 2, **2003**.

7. Gorzynski, M.; Pingel, A. *US Patent* 5,516,885, assigned to Akzo Nobel N.V., May 14, **1996**. Gorzynski, M.; Pingel, A. *US Patent* 6,376,578 B1, assigned to Akzo Nobel N.V., April 23, **2002**.
8. Guntelberg, A. V.; Ottesen, M. *Compt. Rend. Trav. Lab. Carlsberg*, **1954**, *29*, 36.
9. Markland, F. S. Jr.; Smith, E. L. In *The Enzymes, 3$^{rd}$ ed.*; Boyer, P. D., Ed.; "Subtilisins: Primary Structure, Chemical and Physical Properties", Chapter 16; Academic Press: New York, NY, **1971**; Vol. III, p 561.

# Condensation Polymers:
# Enzymatic Approaches

# Chapter 22

# Lipase Catalyzed Polyesterifications

Anil Mahapatro[1], Ajay Kumar[2], Bhanu Kalra[2], and Richard A. Gross[1]

[1]NSF I/UCR Center for Biocatalysis and Bioprocessing of Macromolecules, Polytechnic University, 6 Metrotech Center, Brooklyn, NY 11201
[2]Diagnostic Systems Laboratories, Inc., 445 Medical Center Boulevard, Webster, TX 77598

Lipase-catalyzed condensation polymerizations of diols / diacids and linear hydroxy acids was carried out to better understand the influences of monomer building blocks, the reaction medium and reaction time and to investigate the effects of these reaction parameters on product molecular weight averages and chain build up with time. The effects of substrates and solvent on chain formation for lipase-catalyzed polyesterifications were investigated. Diphenyl ether was found to be the preferred solvent for the polyesterification of adipic acid and 1,8-octanediol giving $M_n$ 28 500 (48 h 70$^0$C). The effect of varying the chain length of diols and diacids on the molecular weight distribution and the chain end group structure were assessed. A series of diacids (succinic, glutaric, adipic, sebacic) and diols (1,4-butane, 1,6-hexane, 1,8-octane diol) were evaluated at 70$^0$C in diphenyl ether. It was found that reactions with longer chain lengths of diacids (sebacic and adipic acid) and diols (1,8-octane and 1,6-hexane diol) give higher reactivity than systems with shorter chain length diacids (succinic and glutaric) and 1,4-butanediol. Lipase B (Novozyme-435) was studied for bulk polyesterifications of linear aliphatic hydroxyacids of variable chain length. The products formed were not fractionated by precipitation. The relative reactivity of the hydroxyacids was 16-hydroxyhexadecanoic acid ≈ 12-hydroxydodecanoic acid ≈ 10-hydroxydecanoic acid ($DP_{avg} \cong 120$, $M_w/M_n \leq 1.5$, 48 h, 90°C) > 6-hydroxyhexanoic acid ($DP_{avg} \cong 80$, $M_w/M_n \leq 1.5$, 48 h, 90°C).

The past 15 years have seen great progress in the use of enzymes in organic media to catalyze a wide variety of small molecule transformations.[1] The rationale for using enzymes and its applications as catalysts in polymer synthesis and modification is described elsewhere within comprehensive reviews.[2,3]

Normally, polyester synthesis is performed by ester interchange reactions or by direct esterification of hydroxyacids or diacid/diol combinations.[4] The use of chemical catalysts for these reactions requires harsh reaction conditions (e.g. temperatures > 200°C) as well as metal catalysts that may be problematic for certain product end-uses.[5] These reaction conditions can limit product molecular weight and eliminate the possibility of using building blocks that are not stable at such temperature-catalyst conditions. Lipase-catalyzed polyesterifications have the potential to overcome some or all of these difficulties.

Previous studies on lipase-catalyzed synthesis of polyesters focused primarily between diols and activated diacids,[6] which are expensive and not practical for commercial processes. Furthermore, previous reports of lipase-catalyzed condensation polymerizations of diols / diacids and linear hydroxy acids, may be perceived as discouraging due to long reaction times, excessive catalyst concentrations, and the formation of products of low molecular weight.[7] This paper discloses a small part of a large effort in our laboratory to better understand lipase-catalyzed polyesterifications. The goal of this research was to better understand the influences of monomer building blocks, the reaction medium, temperature and reaction time and to investigate the effects of these reaction parameters on product molecular weight averages and chain dispersity.

## Experimental

### Materials

Chemicals were purchased from Aldrich Chemical Company. Novozyme-435 (specific activity 7000 PLU/g) was provided by Novozymes (Denmark) and consists of *Candida Antarctica* Lipase B (CALB) physically adsorbed within the macroporous resin Lewatit VPOC 1600 (supplied by Bayer). Lewatit consists of poly(methylmethacrylate-co-butylmethacrylate), has a protein content of 0.1% w/w, surface area of 110-150 $m^2g^{-1}$, and average pore diameter of 140-170 A°.[8]

### General Procedure for Polycondensation

Novozyme-435 (1-10% by wt. relative to the total weight of monomer), dried in a vacuum dessicator (0.1 mmHg, 25°C, 24 h), was transferred into a round bottom flask (20-100 mL flask) containing the monomers (20 mmol diacid & diol or hydroxy acid). The reactions were performed in bulk as well as in different solvents. The flasks were capped with a rubber septum and, as applicable, the solvent (2:1 vol/wt of total monomer) was added *via* syringe. The reaction flask was then placed into a constant preset temperature oil bath on a magnetic stirrer (IKA Werke: Rct Basic) at 220 rpm for a predetermined time. Vacuum was applied (10 mm of Hg), to facilitate removal of water. Aliquots of about 3-20 mg were removed at selected time intervals. The reactions were terminated by adding excess cold chloroform, stirring for 15 min, and removing the enzyme by filtration (glass-fritted filter, medium porosity). $1^H$ NMR and GPC were used to characterize the precipitated and non-precipitated samples.

## Results and Discussion

Polyesterifications of various diacids/diols and linear hydroxy acids were studied using lipase-catalysis. The different diols studied include 1,4-butane, 1,6-hexane, and 1,8-octane diol. The diacids studied include succinic, glutaric, adipic, and sebacic acid. The different hydroxy acids studied were 6-hydroxyhexanoic acid, 10-hydroxydecanoic acid, 12-hydroxydodecanoic acid, and 16-hydroxyhexadecanoic. The effect of the chain length of the diol and diacids and hydroxy acids on the build-up of polymer chains by lipase-catalyzed polycondensation reactions was studied. The effect of reaction parameters on product molecular weight averages and chain dispersity were assessed.

### Effect of Solvent

The effects of organic solvents on lipase reactivity are complex. Thus, it is difficult to predict the behavior of different lipases in the same solvent system or of different solvents with the same lipase. Previous work in our laboratory studied solvent properties, such as solvent log $P$ (logarithm of partition coefficient), and its correlation to Novozyme-435 activity for ε-caprolactone ring-opening polymerizations.[9] Of the solvents studied, those with log $P$ values between 1.9 and 4.5 gave more efficient polymerizations (e.g. toluene and isooctane, log $P$ 2.5 and 4.5, respectively).[9] Furthermore, when solvents with log $P$ values between -1.1 and 0.49 were used, the polymerizations were relatively slow.[9] This generally agrees with results by others that have conducted similar experiments with low molar mass substrates and products.[10]

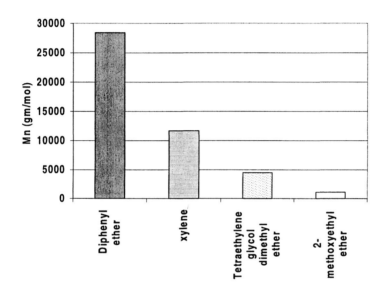

*Figure 1. Effect of solvent on $M_n$ for polymerization of adipic acid and 1,8-octanediol [$70^0C$, 24 h, 1% wt/wt catalyst: monomer, solvent: substrates: 2:1(vol: wt)].*

The boiling point of the solvent is an additional consideration for condensation polymerizations. The solvent should have a sufficiently high boiling point (b.p ≥ $130^0C$) to remain in the reactor during the removal of the by-product (e.g. water). In this paper, four solvents (diphenyl ether, p-xylene, tetraethylene glycol dimethyl ether, 2-methoxyethyl ether) were selected for study using adipic acid/1,8-octanediol polymerization ($70^0C$, 48 h) as a model system. Figure 1 shows that diphenyl ether gave a product having the highest molar mass. The $M_n$ values for polymerizations conducted in diphenyl ether (log $P$: 4.05), xylene (log $P$: 3.09), tetraethylene glycol dimethyl ether (log $P$: -1.03), and 2-methoxyethyl ether (log $P$: -0.48) were 28 500, 11 670, 4 500 and 1 130, respectively.[11] Thus, as was found by Linko et al[12] with a different lipase (from

*M. miehei*) using an activated ester, diphenyl ether was again the preferred solvent.

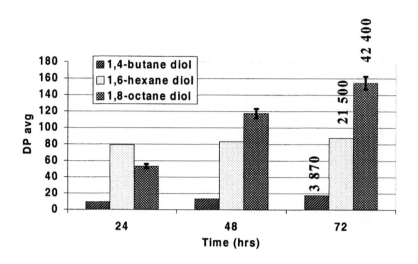

*Figure 2*: Extent of chain growth as a function of time and substrates (diol chain length) [$70^0C$, 1% wt/wt catalyst:monomer] in diphenyl ether (2:1 vol/wt solvent: monomer). The error bars were determined based on triplicate runs.

### Effect of the Monomer Building Block on $DP_{avg}$

The effect of the diol chain length was studied by performing Novozyme-435 catalyzed condensation polymerizations of adipic acid with 1,4-butane, 1,6-hexane, and 1,8-octane diols (Figure 2). In Figure 2, the reactions were conducted in diphenyl ether, at $70^0C$, under vacuum (10 mm of Hg), with 1%-by-wt catalyst (i.e. 0.1% CALB) relative to monomer. To study the progress of the reactions as a function of time, aliquots were withdrawn from reactions after 24, 48, and 72 h. Except for the results at 24 h using 1,6-hexane and 1,8-octane diols, the other data show increased extents of chain growth as the diol length was increased. The average degree of polymerization ($DP_{avg}$) of the product from adipic acid/1,8-octanediol increased with reaction time (Figure 2). However, changes in molecular weight were relatively small between 24 and 72 h when 1,4-butane and 1,6-hexane diols were used.

The highest $M_n$ achieved in this study was that for the 72 h adipic acid/1,8-octanediol polymerization in diphenyl ether ($M_n$= 42 400, Poly dispersity index (PDI)= 1.5). The observation of a continuous increase in $DP_{avg}$ over the 72 h reaction suggests that at least some fraction of the original enzyme activity remains. As in solution, bulk reactions also showed more rapid build-up of chains with increase in the diol length. However, from 48 to 72 h, polymerizations in-bulk showed little or no further increase in $DP_{avg}$. This contrasts markedly to the solution polymerization of 1,8-octanediol where the $DP_{avg}$ increased from 120 to 155 over the same time interval. The most significant difference between 1,6-hexane and 1,8-octanediol polymerizations conducted in solution and bulk is that, without solvent, the increase in the $DP_{avg}$ with time occurs more slowly. This is likely due to greater constraints on chain diffusion for bulk polymerizations. However, it may also be that the enzyme has enhanced activity for substrate polymerization when the reaction is conducted in diphenyl ether rather than in-bulk. Studies are currently looking more closely at the effects of diphenyl ether concentration in reactions as well as process parameters (e.g. agitation rate, shear, viscosity) to improve the kinetics of chain build-up in bulk reactions. Adipic acid/1,8-octane diol polymerizations in diphenyl ether for reaction times ≤ 10 h was also studied.

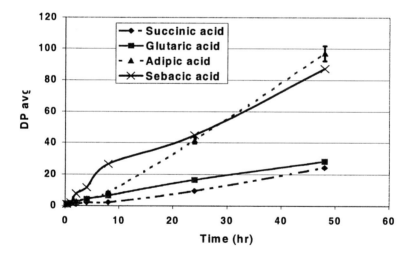

*Figure 3.* Extent of chain growth as a function of time and substrates (diacid chain length) [$70^0C$, 1% wt/wt catalyst:monomer] in diphenyl ether, 2:1 vol/wt solvent:substrates. The error bars were calculated based on triplicate runs

Figure 3 shows the results of CALB-catalyzed polymerizations as a function of diacid chain length in solvent. The more rapid polymerization in diphenyl ether of sebacic ($C_{10}$) acid relative to shorter diacid substrates was evident by 4 and 8 h. However, by 24 and 48 h, the $DP_{avg}$ for sebacic and adipic diacid ($C_6$) polymerizations was statistically equivalent. Similarly, polymerizations of succinic ($C_4$) and glutaric ($C_5$) diacids proceeded at similar rates. Thus, by 24 h, the relative extent of chain-build up for polymerizations with octane-1,8-diol in diphenyl ether was sebacic ≈ adipic > glutaric ≈ succinic acids. It may be that the kinetics of sebacic acid /1,8-octanediol polymerization is more rapid than for the other diacids studied.

The general trends for the solvent-free and solvent-based systems are almost identical. Significant differences for the former are: i) the early stage of sebacic acid copolymerizations do not exhibit the notably more rapid increase in $DP_{avg}$, and ii) by 48 h, the $DP_{avg}$ with glutaric acid is greater than succinic acid. As was found above with different diols, the relative increase in $DP_{avg}$ between 24 and 48 h was larger for solution than bulk polymerizations.

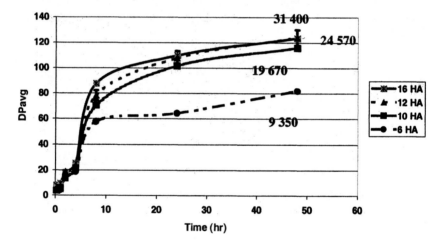

*Figure 4. Lipase-catalyzed polyesterification of linear alphatic ω-hydroxyacids (HA) (10% wt/wt catalyst:monomer substrates, $90^0C$, in bulk, with application of vacuum): extent of chain growth as a function of time The error bars were calculated from triplicate runs*

Hydroxyacids of different chain lengths were used as substrates for Novozyme-435 catalyzed bulk polymerizations (48 h, *in vacuo*). Figure 4 shows that by 4 h, regardless of the substrate chain length, all the products had similar $DP_{avg}$ values (about 20). However, by 8 h and beyond, the relatively slower progress of the 6-hydroxyhexanoic acid polymerization was evident. By 48 h, the $DP_{avg}$ and $M_w/M_n$ of poly(6-hydroxyhexanoate) was 80 and 1.5, respectively. In contrast, by 48 h, the polymerizations of 16-, 12-, and 10-carbon chain-length ω-hydroxyacids gave products with $DP_{avg}$ 115 to 125 and $M_w/M_n$ 1.5. For example, by 48 h, poly(16-hydroxydecahexanoate) was formed with $M_n$ and $M_w/M_n$ of 31 400 and 1.5, respectively. Thus, chain growth catalyzed by Novozyme-435 was similar for ω-hydroxyacids with chain lengths between C-10 and C-16 but was slower for C-6. The difference in chain growth between C-10 and C-6 is consistent with the expected preference by lipases for substrates with chain lengths similar to the fatty acids of triglycerides.

## Summary and Conclusions

The effects of lipase and organic solvents are very complicated and it is difficult to predict the behavior of different lipases in the same solvent system or of different solvents with the same lipase. Diphenyl ether was found to be the preferred solvent that gave the highest molecular weight product ($M_n$ 28 500 in 48 h). Studies on the effect of monomer chain length for polyesterifications with unactivated diacids and diols showed that systems with longer chain length diacids (sebacic and adipic acid) and diols (1,8-octane and 1,6-hexane diol) give higher reactivity than systems with shorter chain length of diacids (succinic and glutaric acid) and 1,4-butanediol. Running bulk lipase-catalyzed condensation reactions is feasible although using diphenyl ether gave higher $M_n$ values. Both in solution and solvent-free, PDI is independent of the diol chain length. However, the PDI values were generally larger for polymerizations in solution. Polymerizations of linear aliphatic hydroxyacids were successfully performed in-bulk, at $90^0C$, using 10% w/w Novozyme-435 relative to monomer. There was little difference in the time profile of $DP_{avg}$ values for ω-hydroxyacids with chain lengths of C-16, C-12, and C-10. These polymerizations gave products with $DP_{avg}$ ~105 and $M_w/M_n$ of 1.5 by 24 h. However, the shorter chain substrate 6-hydroxyhexanoic acid polymerized relatively slower so that by 24 h poly(6-hydroxyhexanoate) was formed with $DP_{avg}$ ~63 and $M_w/M_n$ 1.6

## Acknowledgements

We are grateful to the members of the NSF I/UCRC for Biocatalysis and Bioprocessing of macromolecules at the Polytechnic University for their financial support of this research.

# References

1. Klibanov, A. M., *CHEMTECH* **1986**, 16, 354.
2. Gross, R. A., Kumar, A. and Kalra, B., *Chemical Reviews* **2001**; *101 (7)*; 2097.
3. Kobayashi, S., Uyama, H. and Kimura, S., *Chemical Reviews*; **2001**; *101*(12); 3793.
4. Odian, G., "Principles of Polymerization", John Wiley & Sons, New York, 1981.
5. Ellwood, P.; *Chem. Eng.*, **1967**, 74, 98.
6. a) Magolin. A. L, Creene. J. Y, Klibanov, A. M; *Tetrahedron Lett* **1987**, 28: 1607. b) Wallace, J. S., Morrow, C. J., J Polym Sci Part A: Polym Chem **1989**, 27, 3271-3284. c) Chaudhary, A. K.; Lopez, J.; Beckman, E. J.; Russell, A. J. *Biotechnol. Prog.* **1997**, *13*, 318.
7. a) Linko, Y.-Y.; Wang, Z.-L.; Seppala, J. *J. Biotechnol.* **1995**, *40*, 133. b) Linko, Y.-Y.; Lamsa, M.; Wu, X.; Uosukanum, E.; Seppala, J.; Linko, P. *J. Biotechnol.* **1998**, *66*, 41.c) Binns, F.; Harffey, P.; Roberts, S.M. Taylor, A. *J. Polym. Sci. Part A. Polym. Chem.* **1998**, 36, 2069. d) O'Hagan, D., Zaidi, N. A., *J. Chem. Soc., Perkin Trans.* **1993**, 2389. e) O'Hagan, D., Zaidi, N. A., *Polymer,* **1994**, 35, 3576.f) Shuai. X.,; Jedlinski, Z.; Kowalczuk, M.; Rydz, J.; and Tan, H.; *European Polymer Journal* **1999**, 35, 721.
8. communicated in April 2000 by Morten Christonsen, Novozymes Inc., Denmark.
9. Kumar, A. and Gross, R. A.; *Biomacromolecules*, **2000,** 1, 133.
10. Laane, C., Boeren, S., Vos, K., Veeger, C., *Biotechnology and Bioengineering*, **1997**, 30, 81.
11. Internet reference (http://esc.syrres.com/interkow/kowdemo.htm), demo software for calculating log P.
12. Linko, Y.-Y.; Wang, Z.-L.; Seppala, J. *Enzymol. Microb. Technol.* **1995**, *17*, 506.

Chapter 23

# Versatile Route to Polyol Polyesters by Lipase Catalysis

Ankur S. Kulshrestha, Ajay Kumar, Wei Gao, and Richard A. Gross[*]

NSF Center for Biocatalysis and Bioprocessing of Macromolecules, Department of Chemistry and Chemical Engineering, Polytechnic University, 6 Metrotech Center, Brooklyn, NY 11201
*Corresponding author: http://chem.poly.edu/gross

A simple, environmentally friendly, and practical route is described for the preparation of polyesters from polyols. For the first time, soluble polymers from reduced sugars were synthesized with absolute molecular weights up to 117 000 ($M_w$), without chemical activation of the monomer acid groups or the addition of an organic solvent. Although highly polar reactants were used, they formed a monophasic liquid medium when gently mixed and heated without addition of a polar aprotic solvent. The lipase-catalyst (Lipase B from *Candida antartica* immobilized on Lewatit beads) suspended in the monophasic liquid was selective and highly active. Polyesters rich in hydroxyl functionality, such as poly(sorbityladipate), were prepared without the need to use protection-deprotection chemistry. While the lipase was highly selective for sorbitol polymerizations, it was promiscuous for glycerol copolymerizations. This promiscuity led to poly(1,8-octanoyladipate-co-glyceroladipate) chains that are rich in mono-substituted glycerol terminated branches. Thermal analysis of the sorbitol and glycerol copolymers revealed they are semi-crystalline, low melting, soft materials. The lipase-catalyzed polyol condensation synthetic strategy is anticipated to be broadly applicable to many natural and synthetic polyols.

## Introduction

The incorporation of carbohydrates into non-polysaccharide structures is an important strategy to attain *i*) highly functional polymers *ii*) specific biological functions, and *iii*) complex systems that act as 'smart' materials.[1] Polyesters with carbohydrate or polyol repeat units in the chain can be produced by chemical methods[2a]. However, elaborate protection-deprotection steps[2a-e] are needed to avoid crosslinking between polyol units. Multi-step routes to non-crosslinked polyol polyesters limit the potential of their practical use.

Lipases and proteases are well known to provide regioselectivity during esterification reactions at mild temperatures[3a-b]. These characteristics motivated their study as catalysts for selective polyol polymerizations. The activation of carboxylic acids with electron withdrawing groups was thought to be necessary for enzyme-catalyzed copolymerizations with polyols.[4a-j] For example, the copolymerization of sucrose with bis(2,2,2-trifluoroethyl) diester[4i] (45°C, 30 days) gave in 20%-yield an oligomer with an average degree of polymerization ($dp_{avg}$) of 11. Russell and co-workers[4a] used the lipase-catalyst Novozyme-435 to form polyester with $M_w$ 10 025 from the monomers divinyl adipate and glycerol.

An obstacle to lipase or protease catalyzed polymerizations of polyols is their insolubility in non-polar organic media. Polyols are soluble in polar solvents[4f-j] such as pyridine, dimethylsulfoxide, 2-pyrrolidone, and acetone. However, these solvents cause large reductions in enzyme activity. For example, Patil et.al[4j] copolymerized sucrose with bis (2,2,2-trifluoroethyl) adipate in pyridine that, after nearly a month, gave the corresponding sucrose polyester with an $M_w$ of 2100. Also, Dordick et.al[4f] copolymerized divinyl adipate with sorbitol in acetonitrile for 94h at 45°C to give poly(sorbitol adipate) in 34% yield with $M_w$ 13 660. In addition to the use of activated esters and polar solvents, large quantities of lipases were thought to be necessary. Uyama et al. used 75%-by-wt (relative to total monomers) of the lipase from *Candida antarctica* to copolymerize sorbitol with divinyl sebacate at 60° C in acetonitrile.[4h] This paper describes a simple and versatile strategy to perform selective lipase catalyzed condensation polymerizations between diacids and reduced sugar polyols (Scheme I). Instead of using organic solvents, the monomers adipic acid, glycerol, and sorbitol were solubilized within binary or ternary mixtures. The polymerization reactions were performed without activation of adipic acid. The absolute molecular weight of the polymeric products was analyzed by light scattering and the regioselectivity of monomer esterification reactions was analyzed by inverse-gated $^{13}$C NMR experiments.

The direct condensation of adipic acid and sorbitol was performed in bulk for 48 hrs at 90°C using Novozyme-435 (10%-by-wt. of monomers) (Entry 1, Table 1). The product, poly (sorbityladipate), was water-soluble. Molecular weight averages ($M_n$ and $M_w$) of the non-fractionated poly(sorbityladipate), determined by SEC-MALLS in DMF, were 10 880 and 17 030 g/mol, respectively. The polymer structure was analyzed by $^1$H-NMR and inverse-gated $^{13}$C-NMR spectroscopy in $d$-methanol (See *Figs S-1 and S-2*, respectively). The major signals at 66.6, 67.3, 70.4, 70.5, 72.5 and 73.0 ppm in the spectrum were assigned to the sorbitol C-1, C-6, C-3, C-4, C-2, and C-5

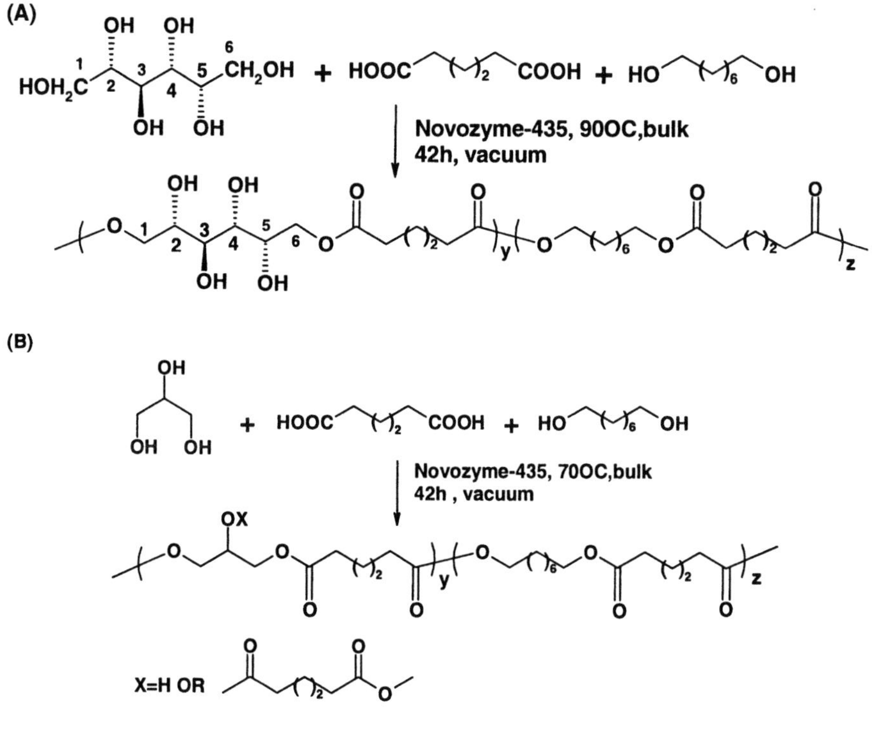

**Scheme 1.** *Novozyme-435 catalyzed polymerization of (A) sorbitol, and (B) glycerol, to form terpolyesters.*

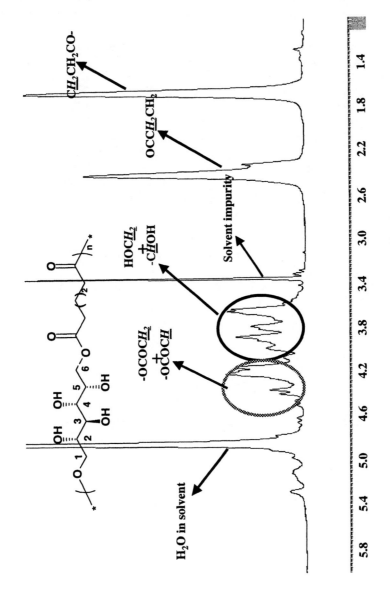

*Figure S-1.* Expanded region (1.0-6.0ppm) from the $^1$H-NMR (300 MHz) spectrum of poly(sorbityladipate) in $d_4$-MeOH

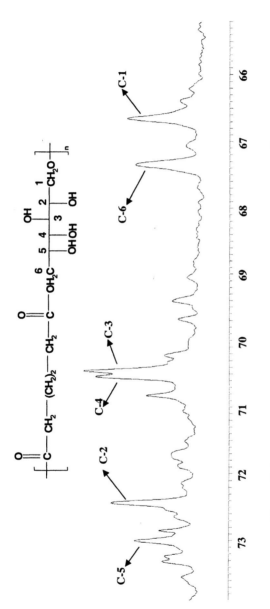

*Figure S-2.* Expanded region (65 to 74 ppm) from the inverse-gated $^{13}C$ NMR spectrum (75 MHz) of poly(sorbityladipate) in $d_4$-MeOH.

carbons, respectively (*see Fig. S-2 and numbering in Scheme 1*).[4h] Analysis of this spectrum revealed that sorbitol was esterified with high regioselectivity (85±5%) at the primary 1- and 6-positions.

Uyama et al.[4h] formed a similar polymer by a polymerization in acetonitrile between divinyl sebacate and sorbitol using 75%-by-wt Novozyme-435. They reported that the polymerization proceeded by exclusive acylation at the sorbitol 1- and 6-positions. This difference in regioselectivity may be due to: i) Uyama et al.[4h] analyzed only a water-insoluble product fraction (64% of total), and/or ii) differences in the reaction conditions used (e.g. solvent vs. bulk).

To obtain water-insoluble sorbitol copolyesters, 1,8-octanediol was used in place of a fraction of sorbitol in the monomer feed. Adipic acid, 1,8-octanediol, and sorbitol were copolymerized in the molar ratio 50:35:15 (Table 1, entry 2). The methanol-insoluble product had an $M_w$ of $1.17\times10^5$. The solubility in water and THF of entry 1 and 2 products, respectively, is direct proof that they have few interchain crosslinks. Furthermore, 0.6~ 1.0% w/v solutions of the products passed through 0.45 μm filters with complete recovery of the products which demonstrates the absence of microgels. This was shown by the fact that 99.9%wt of polymer was recovered when a known concentration of polymer solution was passed through the filter and the solvent was evaporated.

The repeat unit composition and regioselectivity of P(OA-11mol%SA) was analyzed by inverse gated $^{13}$C-NMR (75 MHz). The spectrum in *d*-chloroform (see *Fig.S-3*) showed that for adipate units linked to 1,8-octanediol (A*O), the methylene carbons O(C=O)-$\underline{C}H_2$ and O(C=O)-$CH_2\underline{C}H_2$ have signals at 34.35 and 24.15 ppm, respectively. Similarly, for adipate linked to sorbitol (A*S), the methylene carbons O(C=O)-$\underline{C}H_2$ and O(C=O)-$CH_2\underline{C}H_2$ have signals at 34.30 and 23.88 ppm, respectively. From the relative intensities of the A*O and A*S signals, the content of 1,8-octanediol and sorbitol in the terpolymer was found to be 39 and 11 mol%, respectively. The regioselectivity of esterification at the sorbitol repeat units was also determined from the inverse-gated $^{13}$C-NMR spectrum. The spectrum of P(OA-11mol%SA) in $d_6$-DMSO at 75 MHz (*see fig. S-4*) shows signals at 71.30, 70.40, 69.27, 68.99, 66.30, and 65.52 ppm that were assigned to the sorbitol carbons C-5, C-2, C-4, C-3, C-6, and C-1, respectively[4f]. Furthermore, the signals at 63.40 and 62.35 were assigned to 1,8-octanediol (O=C)O$\underline{C}H_2$ and $\underline{C}H_2$OH carbons, respectively. Weak signals that have a cumulative intensity below ~5±2% of the above-mentioned signals were observed in the 65.5-73 ppm region. These signals are due to sorbitol esters other than those at the primary 1- and 6-hydroxyl positions. Thus, the regioselectivity of esterification at the 1 and 6-carbons of sorbitol was ~95±2%.

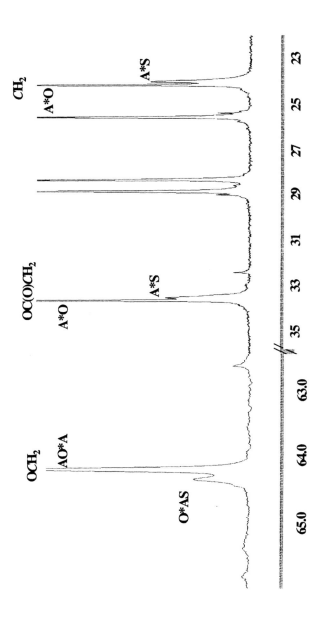

**Figure S-3.** Expanded regions (22 to 36 ppm, 62 to 66 ppm) from the $^{13}C$ NMR spectrum (75 MHz) of fractionated poly(1,8-octanyladipate-co-11mol%sorbityladipate) in d-chloroform.

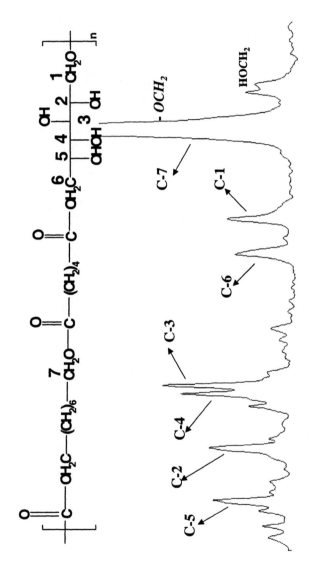

*Figure S-4* Expanded region (61-73 ppm) of the inverse-gated $^{13}C$ NMR spectrum (75 MHz) of poly(1,8-octanyladipate-co-11mol%sorbityladipate) recorded in $d_6$-DMSO.

Table 1.Synthesis of aliphatic polyesters with sorbitol and glycerol repeat units[a].

| entry | X | A:O:X Feed Ratio | Observed A:O:X[b] (mol%) | MeOH insol.[c] (%) | $M_w^d$ $x10^{-3}$ | $M_w/M_n$ |
|---|---|---|---|---|---|---|
| 1 | S | 50:0:50 | 50:0:50 | np[e] | 17 | 1.6 |
| 2 | S | 50:35:15 | 50:39:11 | 80 | 117 | 3.4 |
| 3 | G | 50:0:50 | 50:0:50 | 50 | 3.7 | 1.4 |
| 4 | G | 50:40:10 | 50:41:9 | 90 | 75.6 | 3.1 |

[a]Reaction conditions: bulk, 90°C (entries 1, 2 & 3), 70°C (entry 4), Novozyme-435 (10% wt/wt of monomers), reaction times were 48 hrs for entry 1 and 42 hrs for entries 2, 3 & 4, in-vacuo (40mmHg); [b]A is adipic acid, O is octanediol, X= S (sorbitol) or G (glycerol), the mol % was calculated from $^{13}$C-NMR signals of A*O versus A*S for X=S and from $^1$H-NMR signals of A*O vs. A*G for X=G; [c]percent of product that precipitated in methanol (for entries 2 & 4), [d]determined by size exclusion chromatography (SEC) multi angle laser light scattering (MALLS) measurements in THF (entries 2-4) and DMF (entry 1, 10mM LiBr); [e]np is not-precipitated.

Similar experiments as above were performed where, in place of sorbitol, glycerol was used as the natural polyol. The $M_n$ and $M_w$ values of the resulting poly (glyceryladipate), P(GA), determined by SEC-MALLS in THF, were 2500 and 3700 g/mol, respectively. Thus, substitution of sorbitol by glycerol without 1,8-octanediol resulted in a product of much lower molar mass. The terpolymerization of the monomers adipic acid, 1,8-octanediol and glycerol, in the ratio 50 to 40 to 10 mol%, was performed in-bulk at 70°C using Novozyme-435 as the catalyst (Scheme 1B). Analysis by SEC-MALLS of the resulting glycerol terpolyester P(OA-9mol%GA), See Table 1, entry 4) in THF gave values for $M_w$ and $M_w/M_n$ of 75 600 and 3.1, respectively. The solubility of P(OA-9mol%GA) shows that, without using protection-deprotection chemistry, the product formed has few intermolecular cross-links. The $^1$H-NMR spectrum of P(OA-9mol%GA) (see Fig S-5) has signals due to the glycerol protons -C$H_2$-O(C=O) and -C$H$-O-(C=O) at 4.17 and 4.30 ppm, respectively. The $^1$H-NMR signals due to 1,8-octanediol -C$H_2$-O(C=O) and adipate -C$H_2$-(C=O) protons are at 4.07 and 2.39 ppm, respectively. The mol-% incorporation of glycerol in the copolyester was determined from the additive intensity of the signals at 4.17 and 4.3 relative to that at 4.07 ppm. The inverse-gated $^{13}$C-NMR spectrum (see Fig. S-6) has resolved signals at 71.9, 69.8, 68.8, and 67.5 ppm. These signals were assigned to the glycerol repeat unit methine carbons, A to D, that are 1-substituted, 1,2-disubstituted, 1,2,3-trisubstituted, and 1,3-disubstituted, respectively[4c].The relative intensity of these signals is 17:17:27:39. Thus, the product has a complex substitution pattern at the glycerol units. Unlike the high selectivity at sorbitol primary hydroxyl groups, the selectivity at glycerol

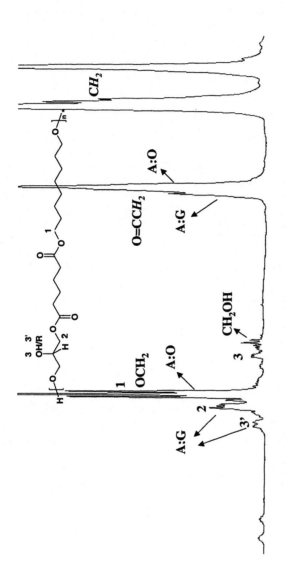

*Figure S-5. Expanded region (1.0-5.4 ppm) from the $^1$H-NMR spectrum (300 MHz) of poly(1,8-octanyladipate-co-9mol%glyceroladipate) recorded in d-chloroform.*

337

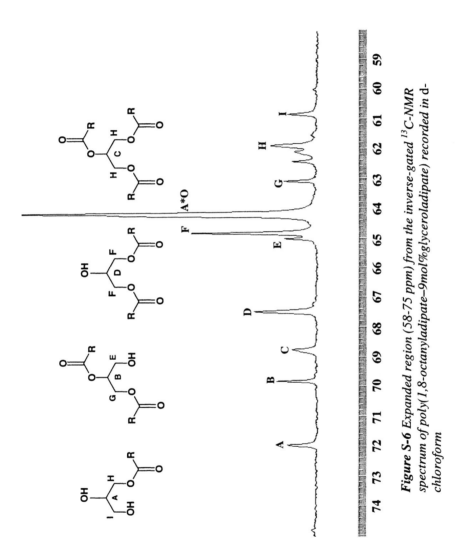

***Figure S-6*** *Expanded region (58-75 ppm) from the inverse-gated $^{13}$C-NMR spectrum of poly(1,8-octanyladipate–9mol%glyceroladipate) recorded in d-chloroform*

primary sites is only 66%.[5] Furthermore, the product has a high level of trisubstituted glycerol repeat units, a high molecular weight, and is organosoluble. Thus, the product is highly branched but has few cross-links. In fact, 27 mol-% of glycerol units are branch sites of which almost 17 mol-% terminate in monosubstituted glycerol units. Work is in progress to determine the average length and substitution of branches.

The role of the enzyme was verified by control reactions. No-enzyme control experiments showed that little (< 2%) acylation of the hydroxyl monomers took place in the absence of Novozyme-435. Furthermore, elsewhere we reported the irreversible inactivation of Novozyme 435 by modification of its active site with diethyl *p*-nitro phenyl phosphate (paraoxon)[6].Novozyme-435 modified in this way was inactive for the catalysis of ε-caprolactone polymerization at 60°C. In contrast, the active enzyme catalyzes ε-caprolactone polymerization to form a high molar mass chain (>10 000 g/mol) in high yield (>80%). When the polymerizations described in entries 2 and 4 were performed using tetrabutyl titanate (1% wt/wt relative to monomers, 60mg) as the catalyst at 180°C for 2 hours in vacuum (10mmHg)[7], the products formed were gels. Hence, by using Novozyme 435 instead of a chemical catalyst, cross-linking was largely avoided and the reaction temperature was dramatically reduced.

The thermal stability, melting transitions and crystallinity ($\chi$) of selected glycerol and sorbitol terpolyesters were studied [8]. Thermo gravimetric analysis of P(OA-11mol%SA) and P(OA-9mol%GA) showed they had onset of decomposition values of 394 and 401°C, respectively. Hence, these copolymers have high thermal stability. Thermograms recorded by differential scanning calorimetry during the first heating scan gave values of the melting enthalpies and peak melting temperatures. These values for P(OA-11mol%SA) are 59 J/g and 58 °C, respectively. The identical studies performed for P(OA-9mol%GA) gave values of 97 J/g and 62 °C. The percentages of crystallinity, measured by wide-angle X-ray diffraction, were 32 and 52%, respectively. Hence, these copolyesters are semi-crystalline, low melting, soft materials. The copolyester from adipic acid and 1,8-octanediol, without natural polyol units, has an enthalpy of melting, peak melting temperature, and %-crystallinity of 136 J/g, 74°C, and 65%, respectively. Thus, incorporation of sorbitol and glycerol into poly(octanyladipate) results in a decrease in the polymer melting point and degree of crystallinity.

The %-activity retained by the enzyme after the above 42-hour reactions to form the 11mol% sorbitol and 9 mol% glycerol copolymers was 82 and 90% respectively. The assay to evaluate the enzyme activity of the recovered catalyst is based on catalysis of propyl laurate formation and is described elsewhere.[9]

## Conclusion

A simple, environmentally friendly, and practical route is described for the preparation of polyesters from polyols. For the first time, soluble polymers from reduced sugars were synthesized with molecular weights up to 117 000 ($M_w$) and without the need to activate monomer acid groups or add an organic solvent. Key to the success of the method is the use of a highly active and selective lipase as the catalyst as well as adjusting the reaction mixture so that it is monophasic. Polyesters rich in hydroxyl functionality such as poly (sorbityladipate) were prepared without the need to use protection-deprotection chemistry. Furthermore, the inability of the lipase to exert high selectivity for glycerol polymerizations led to polymers that are rich in mono-substituted glycerol pendant groups. Thermal analysis of the products showed they have high thermal stability and are low melting. We believe the simplicity of the polyol condensation polymerizations will lead to practical processes and products.

## Experimental Section

### Materials

Adipic acid, 1,8-octanediol, sorbitol and glycerol were all purchased from Aldrich in the highest possible purity and were used without further purification. Novozyme-435 (specified activity 7000PLU/g) was a gift from Novozymes.

### Instrumentation

Proton($^1$H) and inverse gated carbon ($^{13}$C) NMR spectra were recorded on a Bruker DPX300 spectrometer AT 300 AND 75.13MHz , respectively.The $^1$H and $^{13}$C NMR chemical shifts in parts per million (ppm) were referenced relative to tetramethylsilane and chloroform as internal reference.

The absolute molecular weights were measured by SEC-MALLS. The Size Exclusion chromatographic system consisted of a Waters 510 pump, a 717plus autosampler, and a Wyatt Optilab DSP interferometeric refractometer coupled to a Wyatt DAWN DSP multi-angle laser light-scattering photometer (Wyatt Technology, Santa Barbara, CA), a two column set consisting of Polymer Laboratories PL 104A and 500A for THF or Waters Stryragel HR4 and HR2 for DMF in series was used to chromatograph the samples. The Wyatt DSP was calibrated by toluene, and normalized by polystyrene standard with molecular weight 30K in the THF or DMF with 10mmol LiBr. The flow rate is 1.0 mL/min. The software used for data collection and processing was ASTRA (supplied by Wyatt Tech.). The specific refractive-index increments (dn/dc) in corresponding solvent were determined at 632.8nm using Wyatt Optilab DSP interferometeric refractometer by on-line method, in which two assumptions were used, 1) the sample is 100% pure, and 2) the sample is 100% eluted from column.

## 1. General Procedure for lipase-catalyzed synthesis of poly(sorbitol adipate).

Into a 100 mL round bottom flask was transferred sorbitol (2.49g, 0.013mol) and adipic acid (2.0g, 0.013mol). The reactants were heated with stirring to 130°C for not more than 5-10 minutes and the mixture melted. The temperature of the reaction mixture was then lowered to 90-95°C and the reaction components remained as a viscous liquid. Then, Novozyme-435 beads (10% wt/wt relative to monomers, 500mg, dried at 25°C/10mmHg/24 hrs) were added. Within 2 hrs the reaction mixture appeared monophasic with suspended catalyst beads. The flask was sealed with a rubber septum and the reaction was maintained at 90°C with mixing. Furthermore, after the first 6 hrs of the reaction, the contents of the reaction were maintained under reduced pressure (40 mmHg). The polymerization was terminated after 48h by dissolving the reaction mixture in methanol, removing the enzyme by filtration, and stripping off the solvent in vacuo. The product was then dried in a vacuum oven (10mmHg, 30°C, 24 h). A small portion of the product was dialyzed and analyzed by inverse-gated (quantitative) $^{13}$C-NMR spectroscopy.

## 2. General procedure for the lipase-catalyzed synthesis of poly(1,8-octanyladipate-co-sorbityladipate) terpolyesters:

Into a 100 mL round bottom flask was added sorbitol (1.093g, 0.3eq), 1,8-octanediol (2.047g, 0.7eq) and adipic acid (2.92g, 1eq). The reactants were heated with stirring at 115°C for not more than 5 minutes and the mixture melted. The temperature of the reaction mixture was then lowered to 90°C and the reaction components remained as a viscous liquid. Then, Novozyme-435 beads (10% wt/wt relative to monomers, 500mg, dried at 25°C/10mmHg/24 hrs) were added. Within 15 minutes the reaction mixture appeared monophasic with suspended catalyst beads. The flask was sealed with a rubber septum and the reaction was maintained at 90°C with mixing. Furthermore, after the first 2 hrs of the reaction, the contents of the reaction were maintained under reduced pressure (40 mmHg). The polymerization was terminated at 42 h by dissolving the reaction mixture in excess of chloroform, removing the enzyme by filtration and stripping the solvent in vacuo. The resulting product was then dissolved in a minimum volume of chloroform and precipitated by slow addition to methanol. The precipitated product was then dried in a vacuum oven (10mmHg, 30°C, 24h). Inverse-gated (quantitative) $^{13}$CNMR was used to determine the regioselectivity of the ester formation at the sorbitol units.

### 3. General procedure for the lipase-catalyzed synthesis of poly(1,8-octanyladipate-co-glyceroladipate) terpolyesters.

Into a 100 mL round bottom flask was added glycerol (0.368g, 0.2eq), 1,8-octanediol (2.340g, 0.8eq) and adipic acid (2.92g, 1eq). To melt the reactants, the mixture was heated with stirring at 115°C for not more than 5 minutes. The temperature of the reaction mixture was then lowered to 70°C and the reaction components remained as a viscous liquid. Then, Novozyme-435 beads (10% wt/wt relative to monomers, 600mg, dried at 25°C/10mmHg/24 hrs) were added. Within 15 minutes the reaction mixture appeared monophasic with suspended catalyst beads. The flask was sealed with a rubber septum and the reaction was maintained at 70°C with mixing. Furthermore, after the first 2 hrs of the reaction, the contents of the reaction were maintained under reduced pressure (40 mmHg). The polymerization was terminated at 42 h by dissolving the reaction mixture in excess of chloroform, removing the enzyme by filtration and stripping the solvent in vacuo. The resulting product was then dissolved in a minimum volume of chloroform and precipitated by slow addition to methanol. The precipitated product was then dried in a vacuum oven (10mmHg, 30°C, 24h). Inverse gated (quantitative) $^{13}$CNMR was used to determine the regioselectivity of the ester formation at the glycerol units.

## Acknowledgements

We wish to acknowledge members of the NSF Center for Biocatalysis and Bioprocessing of Macromolecules at the Polytechnic University for their financial support.

## References

1. (a) Wang, Q; Dordick, J.S.; Linhardt, J.R.*Chem.Mater.*2002, 14,3232 (b) Blinkovsky, A.M.; Khmelnitsky, Y.L.; Dordick, J.S. *Biotechnol.Technol.* 1994,8,33. (c) Shibatani, S.; Kitagawa, M.; Tokiwa, Y.*Biotechnol.Lett.*1997, 19,511 (d) Kitagawa, M.; Tokiwa, Y.*Carbohydr.Lett* 1997, 2, 343 (e) Martin, B.D.; Ampofo, S.A.; Linhardt. R.J; Dordick, J.S. *Macromolecules*1992, 25,7081

2. (a) Kumar, R; Gao W: Gross R.A *Macromolecules* 2002,35,6835 (b) Shen, Y.; Chen, X.; Gross, R.A. *Macromolecules* 1999,32,2799 (c) Tian, D; Dubois, P; Grandfils, C; Jerome, R. *Macromolecules* 1997,30,406 (d) Haines, A.H. *Adv.Carbohydr. Chem. Biochem.* 1981,39,13 (e) Haines A.H. *Adv.Carbohydr. Chem. Biochem.* 1976,33,11.
3. (a) Therisod, M.; Klibanov, A.M.*J.Am.Chem.Soc.*1986, 108,5638 (b) Patil. D.R.; Dordick, J.S.; Rethwisch, D.G. *Macromolecules* 1991,24,3462
4. (a) Kline, B, J.; Beckman, E.J.; Russell, A.*J.Am.Chem.Soc.*1998, 120,9475(b) Tsujimoto, T.; Uyama, H.; Kobayashi, S., *Biomacromolecules* 2001,2,29 (c) Uyama, H.; Inada, K.; Kobayashi, S., *Macromol.Biosci.*2001, 1,40 (d) Uyama, H.; Inada, K.; Kobayashi, S. *Macromol. Rapid Commun.*1999, 20, 171 (e) Chaudhary, A.K.; Lopez, J; Beckmann, E.J; Russell, A. J.*Biotechnol.Prog.* 1997, 13,318 (f) Kim, Dae-Yun; Dordick, J. S.; *Biotechnol. Bioeng.* 2001,76(3), 200 (g) Park, Oh-jin; Kim, Dae-Yun; Dordick, J. S.; *Biotechnol. Bioeng.*2000, 70, 208 (h) Uyama, H.; Klegraf, E.; Wada, S.; Kobayashi, S.*ChemLett.*2000, 800. (i) Morimoto, T.; Murakami, N.; Nagatsu, A.; Sakakibara, *J., Chem.Pharm.Bull.*1994, 42(3), 751 (j) Patil, D. R.; Rethwisch, D. G.; Dordick J. S. *Biotechnol. Bioeng.* 1991,37,639.
5. The regioselectivity of acylation at the primary hydroxyl sites (1and 3 substitution) is calculated as sum of 1×(1,3-disubstituted) +2/3×(1,2,3-trisubstituted)+1/2× (1,2-disubstituted)
6. Ying Mei; Kumar.A; Gross, R.A. *Macromolecules*, 2003,36,5530.
7. Takiyama et.al US Patent No. 5310782,1994.
8. Fu, H.; Kulshrestha, A.; Kumar, A.; Gross, R.A. *Macromolecules* submitted August 4th, 2003.
9. Mahapatro, A.; Kalra, B.; Kumar, A.; Gross, R.A., *BioMacromolecules,* 2003, 4,544.

## Chapter 24

# Biocompatibility of Sorbitol-Containing Polyesters: Synthesis, Surface Analysis, and Cell Response In Vitro

Ying Mei[1], Ajay Kumar[1], Wei Gao[1], Richard Gross[1,*], Scott B. Kennedy[2], Newell R. Washburn[2], Eric J. Amis[2], and John T. Elliott[3]

[1]Department of Chemistry and Chemical Engineering, NSF I/UCRC Center for Biocatalysis and Bioprocessing of Macromolecules, Polytechnic University, 6 Metrotech Center, Brooklyn, NY 11201
[2]Polymers Division, National Institute of Standards and Technology, 100 Bureau Drive, Stop 8540, Gaithersburg, MD 20899-8540
[3]Biotechnology Division, National Institute of Standards and Technology, 100 Bureau Drive, Stop 3460, Gaithersburg, MD 20899-3460

A series of sorbitol-containing polyesters were synthesized via a one-pot lipase-catalyzed condensation polymerization. Thin films were prepared by spin coating on silicon wafers and surfaces were analyzed by tapping mode atomic force microscopy (AFM) and contact angle measurements. Surface morphologies and surface energies across the series of polyester films, including a poly($\varepsilon$-caprolactone) (PCL) control were nearly indistinguishable. Biocompatibility of the sorbitol-containing polyester series was evaluated against a PCL control by measuring cell spreading of a mouse fibroblast 3T3 cell line *in vitro*. The initial results suggested that the sorbitol-containing polyester surfaces elicited cell behavior similar to the PCL control. These preliminary results indicated the sorbitol-containing polyester series as a promising material for tissue engineering research and development.

© 2005 American Chemical Society

Tissue engineering has received much attention in recent years because it promises a method to repair or regenerate damaged tissue and organs. This is evidenced both in academia and industry by the escalating number of publications and increased resource investment *(1,2)*. It is expected that the next generation of biomaterials will exhibit biodegradability, biocompatibility, and the ability to stimulate specific cellular response at the molecular level *(3,4)*.

Polyesters such as poly(lactic acid) (PLA) and poly($\varepsilon$-caprolactone) (PCL) are used as biomaterials because they demonstrate both biodegradability and biocompatibility. However, the lack of tissue integration and specific bioactivity remains problematic for developing more advanced biomaterials.*(5,6)*. As such, incorporation of various functional groups into polyester matrices has been investigated for the purpose of attaching biological molecules such as oligopeptides or oligosaccharides. While these investigations are informative, the preparation of functional polyesters usually involves tedious protection/deprotection chemistry and the use of organic solvents *(7,8)*. As a result, such systems present significant challenges for transferring technology from the lab to development.

Lipase-catalyzed polymerizations have attracted much attention in recent years because they can provide high efficiency, good enantio- and regio-selectivity, proceed in the absence of solvents, and circumvent potentially toxic catalysts *(9,10)*. Previously, we reported a one-pot synthesis of novel sugar-containing polyesters benefiting from lipase regioselectivity *(11)*. Sugar segments along the backbone provide functional groups for further modification without the use of tedious protection/deprotection chemistry and organic solvents. We begin the process of investigating these materials by establishing the baseline characteristics of unmodified, sugar-containing polyesters. Surface properties are characterized by atomic force microscopy (AFM) and contact angle measurements, and biocompatibility is assessed by measuring cell morphology, spreading and cell proliferation of mouse fibroblast 3T3 *in vitro*.

## Materials and Methods

Certain equipment and instruments or materials are identified in the paper to adequately specify the experimental details. Such identification does not imply recommendation by the National Institute of Standards and Technology, nor does it imply the materials are necessarily the best available for the purpose. The detail of the material and method could be found in the reference *(12)*.

## Results and Discussion

### Synthesis and Characterization of the Polyester

The purpose of this study was to prepare and evaluate a promising new class of biomaterials, sugar-containing polyesters. Sorbitol was chosen as a functional comonomer in this study because it has a low melting point that facilitates the polymerization in bulk. Furthermore, the unreacted pendant hydroxyl groups make these promising candidates for further chemical modification. Figure 1 illustrates the chemical structure of the polyester series reported herein. Table 1 lists the feed monomer ratios along with the corresponding product compositions and molecular weights.

Table 1. Monomer feed ratios (mol to mol), polymer compositions (mol to mol), and molecular weights of sorbitol containing polyesters. S: Sorbitol, O: Octanediol, A: Adipic acid.

| Feed Ratio (S:O:A) | Composition Ratio (S:O:A) | Absolute Mw (x 10-3) | Polydispersity (Mw/Mn) |
|---|---|---|---|
| 1:99:100 | 1:99:100 | 60 | 2.60 |
| 5:95:100 | 5:95:100 | 43 | 2.10 |
| 10:90:100 | 12:88:100 | 44.2 | 2.26 |
| 20:80:100 | 18:82:100 | 47.6 | 2.27 |
| 30:70:100 | 27:63:100 | 117 | 3.40 |

Reproduced with permission from *Tetrahedron Asym.*, **2002**, *13*, 129–135. Copyright 2002 Elsevier.

The compositions and absolute molecular weights reported in Table 1 were determined by $^1$H NMR and GPC-MALS, respectively. Poly(apidic acid-co-octanediol-co-sorbitol) (PAOS) containing (1, 5, 12 and 18) % (mol fraction) sorbitol units have relative molar mass values that are similar (43,000 to 60,000). However, the 27 % (mol fraction) copolymer has a relative molar mass value of 117,000. We believe that the high molecular weight reported for 27 % PAOS is due to branching and an increased propensity for interchain coupling reactions that occur between non-chain end repeat units.

Figure 1. Preparation of PAOS

(Reproduced with permission from *Tetrahedron Asym.*, **2002**, *13*, 129–135. Copyright 2002 Elsevier.)

## Characterization of Polyester Surface

Recent investigations of cell behavior on thin polymer films have shown that surface roughness and surface energy play an influential role in eliciting cell behavior. To study the effect of sorbitol content in the polyester on the surface morphology, the surface morphology of the PAOS were studied by tapping mode AFM. Figure 2 shows the AFM height images of the top surfaces of the polymer films in a two-dimensional form over an area of (20 x 20) $\mu m^2$. Polygonal spherulite structures are seen in all samples to varying degrees. When the sorbitol content reaches 18 % by mol fraction the polygon boundary is no longer fully clear. When the sorbitol content is 27 % by mol fraction, the boundary is difficult to find. This observation indicates that increased sorbitol content inhibits the formation of the spherulite structure because the sorbitol units do not fit within the poly(octanyladipate) crystalline lattice. The disruption of the spherulite structure may be further exacerbated by the rigidity of sorbitol and the hydrogen bonding between sorbitol segments. Indeed, x-ray diffraction analysis by comparing crystalline phase peak area to crystalline and non-crystalline phase area showed that with an increase in the sorbitol content from 0 to 27 % mol fraction the crystallinity decreased from 64 % to 47 % *(13)*.

Differences in surface morphology are quantified by measuring RMS and Ra and the results are reported in Figure 3. The error bars in Figures 3, 4, 9, and 10 represent the standard uncertainty based on one standard deviation of the data measured. RMS is the root mean square average of height deviations taken from the mean data plane, and Ra is the arithmetic average of the absolute values of the height deviations taken from the mean data plane. Notice that the RMS monotonically decreases from 16 nm to 8 nm as the sorbitol content of the PAOS increases from 1 % to 27 % mol fraction. Ra follows the same trend as RMS. These measurements are consistent with the hypothesis that the presence of sorbitol units along the polymer backbone disrupts crystalline ordering.

In addition to altering crystallinity, the presence of sorbitol in the polymer backbone could also affect surface energy. Contact angle measurements are widely used as a simple, sensitive technique for quantifying the hydrophilic / hydrophobic property of a surface. Therefore, aqueous static and dynamic contact angle measurements were carried out to investigate whether the presence of sorbitol would cause the water contact angle to decrease. Results are shown in Figure 4. The static and advancing angles of the different PAOS polymers remained similar, even after soaking in water for 24 h. This indicates that, in air, the hydrophobic segments made from apidic acid and octanediol are oriented towards the surface to minimize the surface energy and stabilize the polymer-air interface. This inability of the surface to rearrange and respond to the water droplet by altering the contact angle occurs irrespective of the fact that these

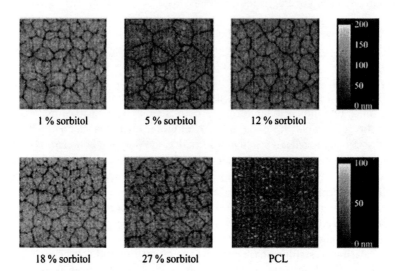

Figure 2. AFM images of the top surfaces of PAOS polymers and PCL. Percentages represent the mole fraction of sorbitol. All images are 20 μm x 20 μm.

(Reproduced with permission from *Tetrahedron Asym.*, **2002**, *13*, 129–135. Copyright 2002 Elsevier.)

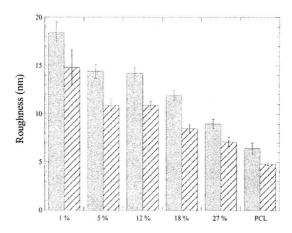

Figure 3. RMS (shaded) and Ra (single hatch) values of PAOS and PCL films. Percentages represent the mole fraction of sorbitol.

(Reproduced with permission from *Tetrahedron Asym.*, **2002**, *13*, 129–135. Copyright 2002 Elsevier.)

copolymers have $T_g$ values well below room temperature ($\approx$-15 °C) *(13)*. It may be that the crystallization process during solvent evaporation at room temperature occurs so that most of the hydrophilic sorbitol segments are isolated within crystalline phases or are restricted by hydrogen bonding between inter/intra polyester chains. In any case, sorbitol units appear to be restricted from reorienting to the surface when it is wetted.

In this study, PCL was chosen as a control because it is an FDA approved biodegradable and biocompatible polymer and PCL has a similar chemical structure with the PAOS. Unsurprisingly, the PCL appears to have a similar surface morphology and hydrophobicity as the PAOS copolymers (Figures 2, 3, and 4).

## Cell Spreading and Proliferation

The NIH 3T3 fibroblast cell line was chosen for this study because it provides a robust and durable platform for investigating common cellular functions: attachment, viability, proliferation, cellular properties, membrane states, etc.*(14)*. Figure 5 illustrates the morphology of mouse fibroblast 3T3 cells cultured on a PCL film and PAOS films. To a first, visual approximation cells grown on the sorbitol-containing polyester films and the PCL control exhibit no morphological differences. This indicates qualitatively comparable biocompatibility between PAOS and PCL.

We chose to quantify cell response by measuring cell density and cell area distributions. Area distributions are considered because cells are complex, living organisms and measured responses may not always follow a normal distribution. As such, reducing the data to a mean and standard deviation may forfeit critical information. Automated fluorescence microscopy was utilized to avoid introducing inadvertent sampling bias *(15)*. [Figure 6]. Also, the automated process for acquiring and analyzing data facilitates larger sampling sizes thereby enhancing the statistical significance of our measurements.

To quantify comparisons we employ a non-parametric Kolmogorov-Smirnov Two-Sample (K-S) Test *(16)* that tests the null hypothesis that two samples are randomly selected from the same *general* (i.e. not normal) population against the alternative that the two samples are drawn from different general populations. Our current algorithm reports whether the null hypothesis can be rejected at one of three discreet confidence levels: 90 %, 95 %, or 99 %. Any level less than 90 % is regarded as insufficient for rejection. Table 2 summarizes the results of the K-S Test where each sorbitol containing polyester is tested against the others as well as the PCL control. A numerical entry indicates the confidence level at which the null hypothesis can be rejected; the letter "N" indicates that the null hypothesis cannot be rejected. Table 2 demonstrates that four of the five sorbitol containing polyesters elicit similar cell

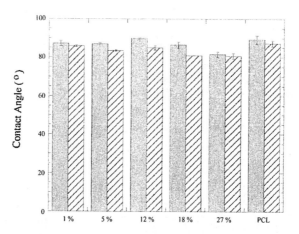

Figure 4. Static (single hatch) and advancing (shaded) water contact angles of PAOS and PCL films. Percentages represent the mole fraction of sorbitol.

(Reproduced with permission from *Tetrahedron Asym.*, **2002**, *13*, 129–135. Copyright 2002 Elsevier.)

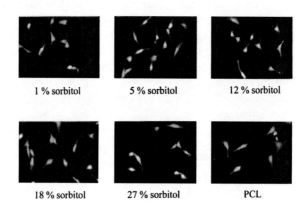

Figure 5. Fluorescence microscopy images of fibroblast on PCL and PAOS films. Percentages represent the mole fraction of sorbitol.

spreading to the PCL control and that only the 27 % sorbitol sample elicits a response statistically different than PCL.

Table 2. K-S Test results comparing each PAOS against the others as well as the PCL control. Percentages represent the mole fraction of sorbitol.

|     | 1 % | 5 % | 12 % | 18 % | 27 % | PCL |
|-----|-----|-----|------|------|------|-----|
| 1 % | --  | N   | N    | N    | N    | N   |
| 5 % | N   | --  | N    | N    | N    | N   |
| 12 %| N   | N   | --   | N    | N    | N   |
| 18 %| N   | N   | N    | --   | N    | N   |
| 27 %| N   | N   | N    | N    | --   | 95  |
| PCL | N   | N   | N    | N    | 95   | --  |

However, the 27 % sorbitol sample does elicit a morphological response similar to all other POAS samples. Furthermore, cell counting indicates that all samples elicit a similar proliferative response. The measured cell densities on all materials after 24 hours in culture are shown in Figure 7. All variations fall within the scope of experimental error introduced by the unavoidable variance in seeding density.

As we seek to establish the PAOS polyesters as viable biomaterials, we interpret the whole of our results and conclude that the PAOS copolymers offer similar biocompatibility to PCL. The outlying morphological response of the 27 % sorbitol sample will be investigated in greater detail as the functional PAOS backbones are used to develop a platform for more advanced biomaterials. These initial results lay a foundation upon which this platform can be built.

## Conclusions

In this study, a series of sorbitol-containing polyesters were successfully prepared by lipase-catalyzed condensation polymerization. The polymer films were spin-coated and surface morphology was assessed by AFM. The AFM measurements found a slight decrease in surface roughness attributed to lower

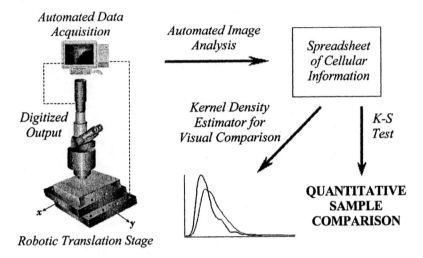

Figure 6. Scheme for automated sample collecting and sample analysis

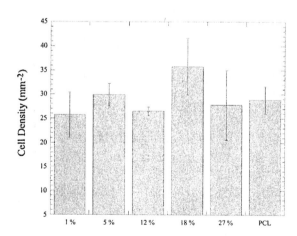

Figure 7. Cell density on PCL and PAOS films after 24 h culture. Percentages represent the mole fraction of sorbitol.

crystallinity from (1 % to 27 %) sorbitol content. The water contact angle measurements indicated that PAOS samples with (1 % to 27 %) mol fraction sorbitol units as well as PCL had a similar hydrophobic surface. For cell culture studies that compared the sorbitol-containing polyester and PCL, the cells were shown to spread and proliferate similarly. This investigation indicates that the sorbitol-containing polyesters and PCL have comparable biocompatibility. These results establish the sorbitol-containing polyester series as a promising material for tissue engineering research and development

## References

1. Griffith, L. G.; Naughton, G. *Science* **2002**, *295*, 1009-1014.
2. Temenoff, J. S.; Mikos, A. G. *Biomaterials* **2000**, *21*, 431-440.
3. Hench, L. L.; Polak, J. M. *Science* **2002**, *295*, 1014-1017.
4. Cook, A. D.; Hrkach, J. S; Gao, N. N. *J. Biomed. Mater. Res.* **1997**, *35*, 513-523.
5. Zhu, H. G.; Ji, J.; Lin, R. Y.; Gao, C. Y.; Feng, L. X.; Shen, J. C. *Biomaterials* **2002**, *23*, 3141-3148.
6. Quirk, R. A..; Chan, W. C.; Davies, M. C.; Tendler, S. B.; Shakesheff, K. M. *Biomaterial* **2001**, *22*, 865-872.
7. Dong, T.; Dubois, P.; Grandfils, C.; Jerome, R. *Macromolecules* **1997**, *30*, 406-409.
8. Chen, X. H.; Gross, R. A. *Macromolecules* **1999**, *32*, 308-314.
9. Gross, R. A.; Kumar, A.; Kalra, B. *Chem. Rev.* **2001**, *101*, 2097-2124.
10. Kumar, R.; Gross, R. A.; *J. Am. Chem. Soc.* **2002**, *124*, 1850-1851.
11. Kumar, A.; Kulshrestha, A.. S.; Gao, W.; Gross, R. A. *Macromolecules*, **2003**, *36*, 8219-21.
12. Mei, Y.; Kumar, A..; Gao, W.; Gross, R. A.; Kennedy, S. B.; Washburn, N. R.; Amis, E. J.; Elliott, J. T. . *Biomaterials*, accepted.
13. Fu, H.; Kulreshrestha, A. S.; Gao, W.; Gross, R. A.; Bairado, M.; Scandola, M. *Macromolecules* **2003**, *36*, 9804.
14. Trentani, L.; Pelillo, F.; Pavesi, F. C.; Ceciliani, L.; Cetta, G.; Forlino, A.. *Biomaterials.* **2002**, *23*, 2863-2869.
15. Elliott, J. T.; Tona, A.; Woodward, J. T.; Jones, P.L.; Plant, A. L. *Langmuir* **2003**, *19*, 1506-14.
16. Hollander, M.; Wolfe, D. A. *Nonparametric Statistical Methods*. John Wiley & Sons, Inc., New York, 1973. p.503.

## Chapter 25

# Enzyme-Catalyzed Synthesis of Hyperbranched Aliphatic Polyesters

### Ingo T. Neuner[1], Mihaela Ursu[2], and Holger Frey[1,*]

[1]Institute of Organic Chemistry, Johannes-Gutenburg-University, D–55099 Mainz, Germany
[2]Department of Macromolecular Chemistry, Technical University "Gh. Asachi", Bulevardul Mangeron 71A, 6600 Iasi, Romania

Immobilized CALB (Novozym 435) is an appropriate catalyst to synthesize polyesters. In this paper, we present the general concept of concurrent ring-opening polymerization and polycondensation to form hyperbranched aliphatic polyesters. The concept is exemplified for the synthesis of $hb$-poly(ε-caprolactone) copolyesters. The route permits variation of the degree of branching by the ratio of ε-CL and the $AB_2$-comonomer. Specific issues of hyperbranched materials such as degree of branching DB, intrinsic viscosity and molecular weight determination are addressed.

Although the structure of polysaccharides such as dextrane and glycogen was identified to be hyperbranched already in the 1930's[1] and Flory introduced his theoretical work on random $AB_m$ polycondensates in the early 1950's[2], only since the beginning of the 1990's a rapidly increasing interest in dendritic and hyperbranched polymers can be noticed.

The perfectly branched dendrimers have to be prepared by tedious stepwise organic synthesis that involves protecting group chemistry and purification procedures in each step. Therefore, perfectly branched structures of higher, monodisperse molecular weight involve demanding, time consuming and cost-intensive procedures. In contrast to dendrimers, hyperbranched polymers are not perfectly, but randomly branched, which usually reduces the synthesis protocol to a one-pot set-up[3]. As a result, highly branched materials can be produced at low cost. In general, dendrimers and hyperbranched polymers have some properties in common that renders them different in comparison to their linear analogues, such as a rather globular or bulky shape, resulting in the absence of entanglements[3,4]. These characteristics lead to low viscosity in bulk and solution, strongly hindered crystallization, high functionality, eventually at the outer sphere or throughout the whole molecule.

Hyperbranched structures are usually obtained by polycondensation of either $AB_m$ monomers or by polymerization of inimer-type structures, i.e., monomers bearing an initiating site.

Since biocompatibility is a precondition for medical and pharmaceutical application and moreover controlled degradability, by UV radiation or in vivo, it is highly desirable for use of polymers as drug carrier in subdermal implants or film former in ointment formulations the choice of suitable co-monomers is limited. Another prerequisite for this segment of application is the complete absence of heavy-metal catalysts and other, potentially toxic, organic residues such as solvents or residual monomer in the final product that has to be addressed by the synthetic procedure.

In 1999 Trollsas as well as Fréchet reported the synthesis of hyperbranched copolyesters using hydroxy-functional ε-CL as cyclic inimer[5,6]. In 2002, our group introduced a concurrent Ring-Opening Polymerization / Polycondensation concept where an $AB_2$ type co-monomer, BHB, and an AB type monomer, in this case ε-caprolactone serving as latent AB monomer are polymerized[7]. By the use of Novozym 435 they were able to synthesize hyperbranched copolyesters of higher molecular weights (more than $5 \times 10^3$ gmol$^{-1}$) and avoided heavy-metal catalysts, e.g. Sn(Oct)$_2$. Post reaction enzyme removal was facilitated by its adsorption on a macroporous resin that allows easy filtration and regeneration of the enzyme coated beads for repeated use.

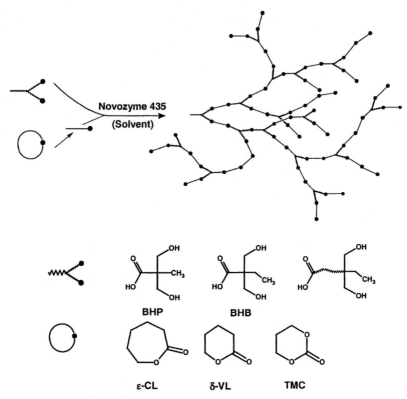

*Figure 1: General Scheme of Combined Ring-opening Polymerization/Polycondensation and Possible Monomers Representing Branched and linear units.*

It should be pointed out that this principle is general, opening access to a large variety of hyperbranched analogues of aliphatic polyesters, as is obvious from screening experiments and subsequent studies with other lactones and trimethylene carbonate[8]. Furthermore, enzymatic copolymerization avoids complex multistep monomer synthesis required for cyclic inimers used in previously reported self-condensing ring-opening-type polymerizations[5,6].

Another concept that leads to dendritic structures containing a defined number of linear units was introduced in 1998[9]. It was right now investigated with respect to the influence of the length of the linear backbone segments[10].

## *hb*-Caprolactone Copolyesters

The results presented in this paper deal with hyperbranched poly (ε-caprolactone) copolyesters (*hb*-PCL), based on the combination of ring-opening polymerization of ε-caprolactone (ε-CL) as AB monomer and polycondensation of 2,2´-bis(hydroxymethyl) butyric acid (BHB) as branching AB$_2$ comonomer unit. By systematic variation of the fraction of the AB$_2$ comonomer BHB, series of hyperbranched copolyesters with different degree of branching have been prepared. Immobilized Lipase B from Candida Antarctica (commercialized as Novozym 435) is known to catalyze both ring-opening polymerization of ε-CL and the concurrent polycondensation of BHB, leading to the desired hyperbranched structure (Figure 1). The copolyester synthesis was performed either in toluene or in bulk.

The choice of toluene as solvent and the reaction temperature of 90 °C is based on the results published by Gross and Kumar[11]. However, a mixture of toluene-dioxane had to be used, when the fraction of the polar AB$_2$ comonomer, BHB, exceeded 20 mol% in order to solubilize BHB. As the activity of the enzyme is considerably lower in dioxane than in toluene[12], the amount of dioxane used was kept to a minimum.

In the case of bulk polymerization it should be mentioned that BHB and ε-CL form a homogenous solution above 76 °C. Both techniques, solution and bulk polymerization lead to similar results with respect to polymer properties. In terms of sustainable chemistry bulk techniques are more ecologically efficient and therefore favorizable.

Table 1: Experimental data. a) determined from $^1$H-NMR; b) in CHCl$_3$; c) CHCl$_3$; T=20 °C; d) commercial sample of Solvay Caprolactones.

| Sample | BHB (feed) [mol %] | BHB (polym)$^a$ [mol %] | Yield [%] | GPC$^b$ $\overline{M_n}$ [g/mol] | GPC$^b$ $\overline{M_w}$ [g/mol] | GPC$^b$ $\overline{M_w}/\overline{M_n}$ | VPO$^c$ $\overline{M_n}$ [g/mol] |
|---|---|---|---|---|---|---|---|
| L-PCL$^d$ | - | - | - | 62000 | 100300 | 1.6 | - |
| hb-PCL 99 | 1 | 1.0 | 90 | 9000 | 20200 | 2.1 | 2000 |
| hb-PCL 98 | 2 | 1.7 | 95 | 7200 | 14500 | 2.0 | 1750 |
| hb-PCL 94 | 6 | 4.5 | 97 | 3500 | 6300 | 1.8 | 1700 |
| hb-PCL 90 | 10 | 8.4 | 85 | 2500 | 4100 | 1.6 | 1400 |
| hb-PCL 85 | 15 | 14 | 80 | 2500 | 4300 | 1.7 | 1350 |
| hb-PCL 75 | 25 | 25 | 73 | 1100 | 1200 | 1.1 | 1200 |

The data compiled in Table 1 illustrate that the yield of the hyperbranched copolyesters decreases as the BHB fraction in the feed increases. This issue was already observed by Skaria et al.[7] and ascribed to the antagonistic solubility properties of BHB and CALB. This is best described by the solubility of BHB being best in polar solvents, the difficulties in BHB homopolymerization[7] and the decrease in CALB's activity the more polar its close surroundings [11].

Calculations of DB were made based on expressions described previously[13] for copolymerization of $AB_2$ - and AB monomers while values of characteristic structural information were extracted from $^1$H-NMR analysis (Figure 2a-c). In the equations below, D represents the dendritic units and $L_{co}$ the total number of linear units in the $AB/AB_2$ copolymers:

$$DB_{th\,AB/AB_2} = \frac{2D}{2D + L_{co}}$$

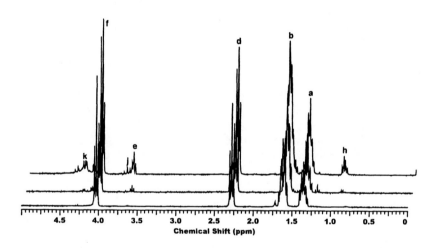

Figure 2: $^1$H-NMR of (a) Linear PCL, (b) *hb*-PCL, DB = 10%, (c) *hb*-PCL, DB = 30%.

Molecular weights characterization of hyperbranched polymers which usually possess a large number of hydroxyl or other end groups is problematic and can lead to erroneous results, if common linear standards are employed[14].

Since the hydrodynamic volume of hyperbranched polymers is a specific property influenced by structural features as the degree of branching, the shape or the rigidity of the backbone and by chemical characteristics such as polarity and type of functionalization, appropriate GPC-calibration for molecular weight measurement is difficult. Although the hydrodynamic volume of compact hyperbranched macromolecules can be smaller than for linear chain analogues, interaction of the highly polar end groups with solvent and/or column often leads to strong overestimation of molecular weights.

Number-average molecular weights ($\overline{M_n}$) determined by means of GPC (polystyrene standards) are in the range of 1100 to 9000 g/mol, weight-average molecular weights ($\overline{M_w}$) in the range of 1200 to 20200 g/mol. All the samples showed monomodal molecular weight distributions with apparent polydispersities in the range of 1.6 to 2.1. Due to the relatively low polydispersities as well as the monomodal molecular weight distribution, formation of aggregates does not seem very likely. As determined with respect to yield, molecular weight of the synthesized *hb*-PCLs also exhibits a decreasing tendency the higher the fraction BHB in feed (table 1). But we are skeptic that this reflects reality but a limitation of GPC as analytical method for molecular weight determination of *hb*-polymers.

As vapor pressure osmometry (VPO) is independent of the structure of the samples, it is more appropriate for molecular weight determination of *hb*-polymers but limited by a maximum number average molecular weight of approximately 25-100 kDa depending on the solvent. Number average molecular weights ($\overline{M_n}$) of *hb*-PCL are in the range of 1200 to 2000 g/mol, depending of the amount of BHB incorporated. As observed in GPC molecular weights decrease the higher the amount of BHB although this effect is considerably less pronounced. Since all samples show excellent linear correlation in VPO measurements (Figure 3), aggregation due to polar hydroxyl groups can be excluded. Concluding, absolute molecular weights determined by VPO are more reliable compared to GPC analysis but limited by an upper maximum molecular weight.

As important as elucidation of structure and actual molecular weights is determination of the materials' properties in solution by intrinsic viscosity. On the one hand information on the shape of the polymer chains in the respective solvent is obtained, but on the other hand knowledge for processing the materials is achieved as well.

The dependence of the intrinsic viscosity [η] on molecular weights as well as the DB has been investigated for all copolymers. Figure 4 depicts the results of viscosity measurements in chloroform at 20 °C.

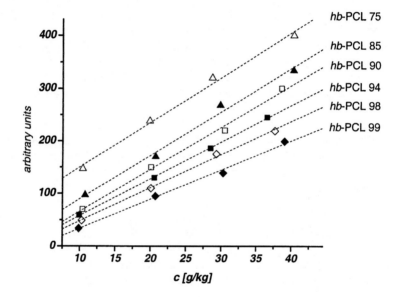

*Figure 3. VPO measurements for HB-PCLs*

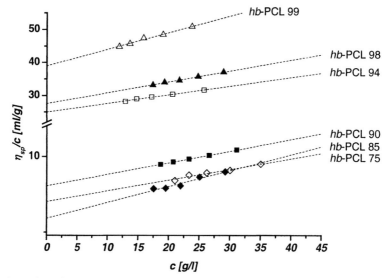

*Figure 4. Reduced viscosity ($\eta_{sp}/c$) of hyperbranched copolyesters as a function of concentration in chloroform at 20°C*

The linear relationship between $\eta_{sp}/c$ and concentration demonstrates that no aggregation occurs in the investigated range which corresponds to observations discussed in the GPC section. The slope of the lines flattens out the higher the DB of the sample which means that higher branched samples have a very compact structure in solution.

**Table 2. Intrinsic viscosities for hyperbranched polymers of different branching degrees in CHCl$_3$ and THF solutions, at 20 and 50 °C resp.**

| Sample | Intrinsic viscosity, [η] [cm$^3$/g] | | | |
| --- | --- | --- | --- | --- |
| | Chloroform | | Tetrahydrofuran | |
| | T=20°C | T=50°C | T=20°C | T=50°C |
| hb-PCL 99 | 40.1 | 35.5 | 17.5 | 15.8 |
| hb-PCL 98 | 24.8 | 15.3 | 15.8 | 10.4 |
| hb-PCL 94 | 16.7 | 10.2 | 11.3 | 7.8 |
| hb-PCL 90 | 8.0 | 7.2 | 9.5 | 6.6 |
| hb-PCL 85 | 6.3 | 7.0 | 8.1 | 4.5 |
| hb-PCL 75 | 4.0 | 3.3 | 3.8 | 3 |

The interaction that exists in the hyperbranched PCL - solvent system is due to the physical interactions between the carbonyl group of hyperbranched PCL and the characteristic functional groups of the solvents. As illustrated in table 2, the intrinsic viscosity of hyperbranched PCL in $CHCl_3$ is larger than the intrinsic viscosity of hyperbranched PCL in THF at 20 and 50°C respectively which leads to the conclusion that $CHCl_3$ is a better solvent for the hyperbranched PCL than THF. Polymer-solvent interactions in $CHCl_3$ are larger than in THF, hence the coil swells in solution and the hydrodynamic volume is increased, which is translates into higher intrinsic viscosity. Raising the temperature from 20°C to 50°C enlarges the intensity of polymer-solvent interactions, which leads to a decrease of the intrinsic viscosity. .

Figure 5 shows a plot of intrinsic viscosities versus degree of branching. It is obvious that the intrinsic viscosity of the hyperbranched samples decreases the higher the degree of branching, in analogy to the plot of $M_n$ versus DB.

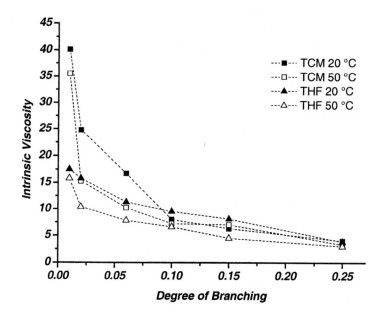

*Figure 5. Plots of intrinsic viscosities vs. degree of branching of hyperbranched polycaprolactones in $CHCl_3$ and THF at 20 and 50 °C resp.*

As before the values of intrinsic viscosity of the *hb*-PCLs with different DB are lower than those of linear PCL. This is due to the higher molecular weight of linear PCL and also related to the different structure and architecture of the linear and hyperbranched samples. Intrinsic viscosity measures the ratio of hydrodynamic volume to molecular weight. The structures of hyperbranched polymers are more densely packed, resulting in smaller hydrodynamic volumes compared to those of linear polymers with comparable molecular weights, leading to smaller viscosities.

Thermal properties of the hyperbranched copolyesters have some importance for their applications and have been investigated using differential scanning calorimetry (DSC). A typical DSC diagram obtained after repeated heating/cooling cycles is shown in Figure 6. All the DSC curves showed a characteristic double peak up to DB = 0.35, from which the more intensive upper peak was confirmed to be the melting peak by microscopy. The shape of the double peak results from an annealing or reorganization process in the DSC, presumably through a melting – recrystallization – final melting mechanism that is due to a recrystallization point at 30 °C[15].

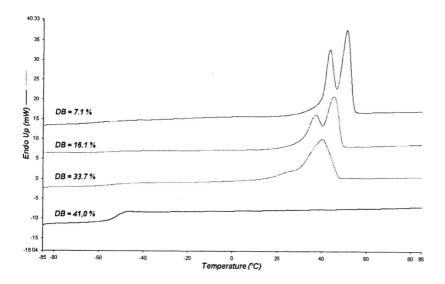

*Figure 6: DSC curves for hb-PCl of different DB*

Further more the dependence of the melting points on the branching degree is easily observed. As expected, raising the incorporation of the $AB_2$ monomer from 1 to 33 % mol causes a decrease of the melting point from 61 to 27 °C and a decrease of the melting enthalpy from 83 to 60 J/g. We attribute these trends to the influence of hydroxyl end groups on the crystallinity of copolyesters. Also the values of $T_g$ showed slightly increasing tendency the higher the degree of branching.

## Conclusions

The universal concept of concurrent ring-opening polymerization / polycondensation was introduced for branched polymers[16] and two routes to green hyperbranched poly(ε-caprolactone) as ecologically friendly and potential medical polymer were discussed. The sustainable approach of this chemistry lies in the use of immobilized CALB (Novozym 435) and the eventual sacrifice of solvents during synthesis and work-up.

In the discussion of characterization techniques that were used to describe $hb$-poly(ε-caprolactone) samples, systematic phenomena such as yield and molecular weight dependence on the branching co-monomer's feed (BHB) as molecular weight determination in general were addressed.

## Acknowledgements

We are grateful to the SFB 428 of the DFG for financial support. Immobilized CALB was kindly provided by Novozymes A/S (Novozym 435™).

## References

1. Geddes, R. In *The Polysaccharides*; Aspinall, G. O., Ed.; Academic Press: London, New York, 1985; Vol. 3, p 209.
2. Flory, P. J. *J. Am. Chem. Soc.* **1952**, *74*, 2718-2723.
3. Newkome, G. R. *Advances in Dendritic Molecules*; JAI Press: Greenwich, CT, 1994; Vol. 1.
4. Wooley, K. L.; Frechet, J. M. J.; Hawker, C. J. *Polymer* **1994**, *35*, 4489-4495.
5. Liu, M.; Vladimirov, N.; Frechet, J. M. J. *Macromolecules* **1999**, *32*, 6881-6884.
6. Trollsas, M.; Loewenhielm, P.; Lee, V. Y.; Moeller, M.; Miller, R. D.; Hedrick, J. L. *Macromolecules* **1999**, *32*, 9062-9066.
7. Skaria, S.; Smet, M.; Frey, H. *Macromol. Rapid Commun.* **2002**, *23*, 292-296.
8. Neuner, I.; Ursu, M.; Frey, H. *Polym. Mat. Sci. Eng.* **2003**, *88*, 342.
9. Trollsas, M.; Hedrick, J. L. *Macromolecules* **1998**, *31*, 4390 - 4395.
10. Choi, J.; Kwak, S.-Y. *Macromolecules, submitted* **2003**.

11. Gross, R. A.; Kumar, A.; Kalra, B. *Chem. Rev.* **2001**, *101*, 2097 - 2124.
12. Kumar, A. G., R.A. *Biomacromolecules* **2000**, *1*, 133 - 138.
13. Frey, H.; Holter, D. *Acta Polym* **1999**, *50*, 67-76.
14. Burgath, A.; Hanselmann, R.; Holter, D.; Frey, H. *Abstr Pap Am Chem S* **1997**, *214*, 150-Pmse.
15. Runt, J.; Harrison, I. R. *Methods Exp. Phys.* **1980**, *16B*.
16. Note: Combined ring-opening and polycondensation has been previously employed for the synthesis of linear polymers: Namekawa, S.; Uyama, H.; Kobayashi, S. *Biomacromolecules* **2000**, *1*, 335.

Chapter 26

# Enantioenriched Substituted Polycaprolactones by Enzyme Catalysis

Kirpal S. Bisht[1,*], Leelakrishna Kondaveti[1,3] and Jon D. Stewart[2]

[1]Department of Chemistry, University of South Florida, 4202 East Fowler Avenue, Tampa, FL 33620
[2]127 Chemistry Research Building, University of Florida, Gainesville, FL 32611
[3]Current address: Mclean Hospital/Harvard Medical School, 115 Mill Street, Belmont, MA 02478
*Corresponding author: telephone: (183) 974-0350; email: kbisht@cas.usf.edu

The kinetic resolution of racemic seven-membered lactones substituted either at the 4- or the 6-position is achieved in bulk (solvent free) conditions, by lipase-catalysed butanolysis. Novozym-435 lipase (from *Candida antarctica*) induced (S)-selective butanololysis, and (R)-lactones were recovered unreacted with >97% enantiomeric excess (e.e.). The (S)-selective Novozyme-435 catalysis is also employed for the ring opening polymerization of 4-methyl-ε-caprolactone (4-MeCL) and 4-ethyl-ε-caprolactone (4-EtCL) to obtain highly (S)-enriched substituted poly ε-caprolactones. The polymerizations were performed in bulk, thus eliminating the need for solvents in the polymerization process. Poly (4-EtCL) and poly (4-MeCL) having >95% enantiomeric purity ($ee_p$) have been prepared. The effect of reaction temperature on enzyme enantioselectivity, polymer molecular weight, and monomer conversion was also investigated at 45 and 60 $^0$C. The optically enriched (R)- 4-MeCL and 4-EtCL obtained were mixed with *rac*-4-MeCL and 4-EtCL to prepare mixtures of variable enantiopurities and the mixtures were subjected to chemical polymerization initiated by Sn(Oct)$_2$ to obtain poly (4-EtCL) and poly (4-MeCL) of variable enantiopurities. The effects of optical purity of the polymers glassy state were evaluated by DSC.

Optically active polymers are important functional materials for several industrial and bio-medical applications and are extensively used as chiral catalysts for asymmetric synthesis, packing materials of chromatographic columns and chiral materials for the preparation of liquid crystal polymers (*1*). Polymers such as poly hydroxy alkanoates (PHAs), naturally occurring microbial optically active polyesters, are important materials in biomedical applications owing to their biodegradability (*2*). In synthetic polymer chemistry, synthesis of optically active polymers has been one of the most challenging tasks. Most synthetic chiral polymers are prepared from optically pure starting materials which are, except when isolated from nature, in limited supply and difficult to prepare (*1, 3*).

Optically active lactones are valuable building blocks in organic synthesis (*4*) and in the preparation of optically active biodegradable polymers (*1,5*). Several chemical methods for producing these compounds and their corresponding polymers have been explored (*6*) but unfortunately all of these methods are either experimentally cumbersome or afford the lactones with only modest enantioselectivities. Examples of chemically prepared optically active polyesters include poly(α-phenyl-β-propiolactone) (*7*), poly(α-ethyl-α-phenyl-β -propiolactone) (*8, 9*), poly(α-methyl-α-ethyl-β-propiolactone) (*10*) and poly(lactic acid) (*11, 12*). Use of enantioselective polymerization catalysts to carry out stereoelective polymerizations of racemic lactones has produced mixed results. For example, stereoelective polymerization of [*R*,*S*]- β-methyl-β-propiolactone with a catalyst from Zn (C$_2$H$_5$)$_2$ and [*R*]-(-)-3,3-dimethyl-1,2-butanediol showed only a small enantiomeric enrichment in the final polymer (*13*). Stereoselective copolymerizations of racemic (LL/DD monomers) and meso (LD monomer) lactides using chiral catalyst that gives heterotactic and syndiotactic PLA, respectively have also been studied (*11*).

The use of isolated enzymes in organic solvents has already found its niche in organic synthetic laboratories, both in academia and industry (*14, 15*). Enzymes have emerged as very useful tools for the synthesis of optically active lactones by enantioselective lactonization of racemic γ-hydroxy esters, ω-hydroxy esters, and δ-hydroxy esters into enantioenriched γ-butyrolactones, ω-lactones, and δ-lactones, respectively (*16,17,18*). Isolated enzymes, such as lipases (*20*) and esterases (*21*) have also been exploited for preparing optically active four-, five- and six-membered lactones by enantioselective ring-opening of racemic lactones. In addition,enantioselective Baeyer–Villiger oxidations catalyzed by enzymes offer another route to homochiral lactones (*19*) Enzymatic polymerization has received increased attention as a new methodology for metal free polymer synthesis in recent years (*22*). Isolated enzymes are fast gaining acceptance as attractive alternative to conventional polymerization catalysts because of their selectivity, ability to operate under mild reaction conditions, recyclability, nontoxicity, and biocompatibility (*23*). Lipases are reported to efficiently catalyse the ring-opening polymerization of achiral lactones and cyclic carbonates to obtain polyesters (*24*) and polycarbonates (*25*). Lipase catalysis has also been useful for enantioselective polycondensation of racemic diester with a diol (*26*), enantioelective ring-opening of (R,S)-α-methyl-β-propiolactone (*27*), enantioselective polymerization of β-butyrolactone (*28*) and enantioselective copolymerization of (R,S)-β-butyrolactone and δ-caprolactone with other lactones (*29*). Interestingly, only a few literature reports describe the enzymatic resolution of substituted seven-membered lactones (*21c, 30*).

The present article describes our efforts (*31, 32, 33*) towards the synthesis of enantioenriched polycaprolactones by lipase Novozym-435 (from *Candida antartica*) catalysis in solvent free conditions. The use of organic solvents (*15*) for enzymatic resolution addresses solubility issues of organic substrates in aqueous (natural) environment, but their use renders an otherwise green technology non-biocompatible. Addressing some of these concerns, we recently carried out lipase-catalyzed kinetic resolution and polymerization of ε-caprolactones substituted at the 4- or the 6-position in a solvent-free environment (*31, 32*). The reactions were highly enantioselective despite the large distance (three/five bonds) between the stereo center at the 4/6-position and the lactone carbonyl group. Solvent free kinetic resolution of substituted racemic ε-caprolactones to obtain enanatioenriched (*R*) caprolactones, enantioelective polymerization of substituted caprolactones in bulk to obtain (*S*) enriched poly ε-caprolactones, and chemo-enzymatic synthesis of enantioenriched substituted polycaprolactones is discussed in detail.

## Results and Discussion

### Lipase-catalyzed solvent-free kinetic resolution of substituted racemic ε-caprolactones

The substrates, (±)-ε-caprolactones **1, 2** and **3** (Figure 1) were synthesized by Baeyer–Villiger oxidation of the corresponding cyclohexanones and kinetic resolutions were examined using Novozym-435 (immobilized lipase-B from *Candida antarctica* catalyzed ring-opening of the lactones **1, 2,** and **3** with 1.5 molar equivalents of *n*-butanol (Scheme-1). Authentic samples of the racemic hydroxy butyl esters (±)-**4**, (±)-**5** and (±)-**6** were prepared by TFA-catalyzed ring-opening of the lactones **1**, **2** and **3**, respectively with *n*-butanol.

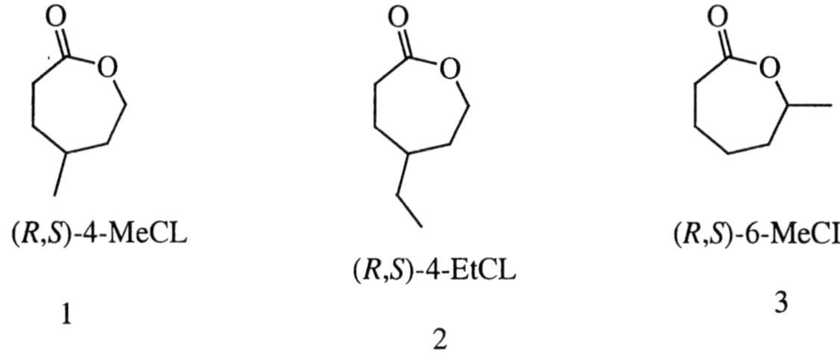

(*R,S*)-4-MeCL
**1**

(*R,S*)-4-EtCL
**2**

(*R,S*)-6-MeCL
**3**

**Figure 1:** (±)-ε-caprolactones

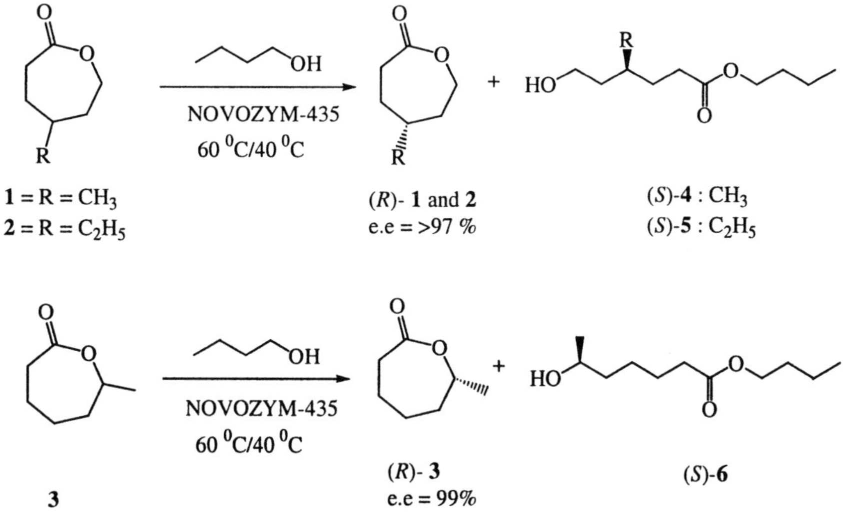

**1** = R = $CH_3$
**2** = R = $C_2H_5$

NOVOZYM-435
60 °C/40 °C

(*R*)- **1** and **2**
e.e = >97 %

(*S*)-**4** : $CH_3$
(*S*)-**5** : $C_2H_5$

**3**

NOVOZYM-435
60 °C/40 °C

(*R*)- **3**
e.e = 99%

(*S*)-**6**

**Scheme 1.** Lipase, Novozym-435 catalysed enantioselective ring opening of lactones 1, 2 and 3 with n-butanol

All reactions were carried out in bulk and monitored by collecting samples periodically, which were then analyzed by gas chromatography (GC) (Figure 2). For all three lactones, the (*S*) enantiomer reacted faster, Novozym-435 lipase induced the (*S*) selective butanolysis and (*R*) lactones were recovered unreacted with > 97 % enantiomeric excess (e.e).

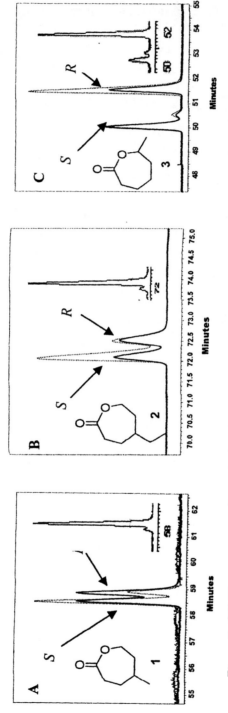

**Figure 2.** GC Chromatograms of racemic (solid line) lactones and their coinjections with authentic enriched (dashed line) lactones. Inserts: GC chromatograms of the recovered lactones (*R*)-**1** (e.e. >97%), (*R*)-**2** (e.e. >95%), and (*R*)-**3** (e.e. >70%) at ~50% conversion.

(Reproduced with permission from reference 31. Copyright 2002 Elsevier.)

**Table 1.** Novozym-435 catalyzed solvent-less kinetic resolution of **1**, **2**, and **3** with *n*-butanol.[a]

| | REACTION AT 60 °C | | | | REACTION AT 40 °C | | | | REACTION AT 23 °C | | | |
|---|---|---|---|---|---|---|---|---|---|---|---|---|
| | Conv % (Time) | % ee[b] | | $E^c$ | Conv % (Time) | % ee[b] | | $E^c$ | Conv % (Time) | % ee[b] | | $E^c$ |
| | | Lactone (R) ($[\alpha]_D^{23}$ c 0.5) | Ester (S) | | | Lactone (R) | Ester (S) | | | Lactone (R) | Ester (S) | |
| **1** | 25 (1h) | 70 | >99 | | 21 (1h) | 27 | >99 | | 36 (4h) | 40 | 71 | |
| | 45 (3h) | 85 | >99 | | 35 (3h) | 41 | 76 | | 39 (6h) | 56 | 88 | |
| | 51 (6h) | 97 (+ 40°) | 93 | >125 | 45 (6 h) | 69 | 84 | 24 | 44 (8 h) | 68 | 86 | 28 |
| **2** | 12 (1h) | 17 | >99 | | 10 (1h) | 23 | >99 | | 38 (4h) | 40 | 65 | |
| | 44 (3h) | 73 | 93 | | 26 (3h) | 30 | 85 | | 45 (6h) | 56 | 68 | |
| | 54 (4 h) | 99 (+42°) | 84 | 60 | 45 (6 h) | 60 | 73 | 12 | 47 (8 h) | 64 | 72 | 12 |
| **3** | 43 (2h) | 35 | 46 | | 17 (2h) | 12 | 59 | | 12 (4h) | 8 | 59 | |
| | 59 (5h) | 80 | 56 | | 33 (5h) | 26 | 53 | | 26 (12h) | 22 | 63 | |
| | 80 (7 h) | 99 (+18°) | 25 | 7 | 47 (7 h) | 35 | 40 | 3 | 46 (24 h) | 30 | 35 | 3 |

[a] *No reaction was observed at 0 °C even after 24 h. Enzyme to substrate ratio (w/w) = 0.5 and n-butanol to lactone molar ratio = 1.5.* [b] *Calculated from chiral gas chromatographic analysis (see experimental section).* [c] *Calculated from conversion and e.e values* (Reproduced with permission from reference 31. Copyright 2002 Elsevier.)

The Novozym-catalyzed ring-openings were investigated at four different reaction temperatures viz. 0, 23, 40 and 60 °C. In Figure 3, the effect of reaction temperature on lactone conversion is shown at 5 h. The results of kinetic resolutions of the lactones **1**, **2** and **3** are summarized in Table 1. Generally, the rates of monomer conversion increased with increasing reaction temperatures. For example, in the ring-opening of **1** at 0 °C, there was no reaction even after 12 h, but at room temperature (23 °C), ~45% of the lactone was converted to the (*S*)-enriched hydroxy ester in 8 h. A further increase of 17 °C in the reaction temperature to 40 °C only affected the rate of the reaction slightly. The reactions at 60 °C, however, were more efficient, i.e., more than 8 h was needed for 50% conversion of **1** at 40 °C compared to less than 4 h at 60 °C. The higher rate of monomer conversion at a higher temperature, i.e. 60 °C, coupled with the observation that the lipase was even more selective at 60 °C (Table 1), suggests that the reaction is a diffusion-controlled process. Interestingly, the ring-opening of **1** in solvent-free reactions was still faster than its hydrolysis in water-saturated diisopropyl ether (*30*). Novozym-435 lipase was highly enantioselective for the resolution of racemic lactones **1** and **2**, in which the (*S*)-enantiomers were preferentially esterified and the unreacted (*R*)-lactones were recovered with e.e. of >97%. As expected, to about 50% monomer conversion, the enantiomeric excess of the unreacted lactones **1** and **2** increased with increasing lactone conversion. The corresponding hydroxy esters also had rather high e.e. values (90%) at ~50% lactone conversion. The enantioselectivity in esterification of the racemic lactone **3**, however, was only moderate. At 60 °C, 80% of the lactone had to be converted to the hydroxy ester to recover (*R*)-**3** with 99% e.e. The effect of reaction temperature on enantioselectivity, $E$ was more intriguing as it varied greatly with temperature (Figure 4).

When the reaction yield was lower than 50%, a temperature drop from 60 to 40 °C diminished the $E$ and enantiomeric excess, but a similar trend was not observed in the reactions at 23 and 40 °C. The enantioselectivity of the lipase in reactions performed at 60 °C was higher (for **1**, $E > 125$) than that at 40 °C (for **1**, $E \sim 25$). Usually an $E$ value of 100 is considered a minimum satisfactory value for the industrial application of an enzymatic process. Thus, in the resolution of **1**, a 20 °C temperature increment from 40 to 60 °C made an otherwise inadequate reaction process a very efficient one. This effect was also observed for **2**, where the $E$ value increased from 8 to 60 with an increase in temperature from 40 to 60 °C.

The enantioselectivity ($E$) of the lipase was lowest ($E=4$ at 60 °C) for the 6-methyl-ε-caprolactone **3** compared to that for the 4-ethyl- **2** ($E=60$) and 4-methyl- **1** ($E > 125$) analogues. This is in contradiction to previous findings in

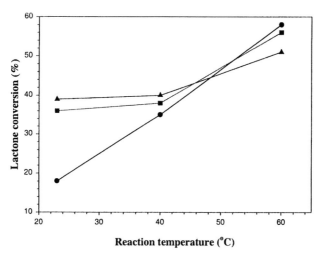

**Figure 3.** Plots of reaction temperature versus lactone conversion in Novozym-435-catalyzed resolution of lactones **1**(▲), **2** (■) and **3**(●) at 23, 40 and 60°C after 5 h. No reaction was observed at 0°C even after 24 h.

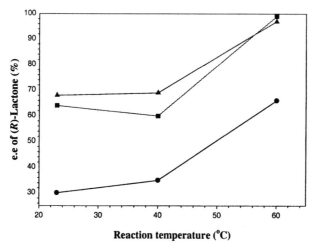

**Figure 4.** Plots of reaction temperature versus % e.e. of recovered (R)-lactone in Novozym-435 catalyzed resolution of lactones **1** (▲), **2** (■) and **3** (●)at 23, 40 and 60 °C a

water-saturated ether (30) in which the enantioselectivity of the enzyme increased with decreasing distance between the lactone carbonyl group and the substituent on the lactone ring, i.e. E for substitution at position 6~2>5~3>4 which may be a result of the conformational differences in the enzyme in two very different reaction environments.

## Solventless Enantioelective Ring-Opening Polymerization of Substituted ε-Caprolactones by Novozym-435 (from *Candida antartica*) Catalysis

**(S)-Poly(4-ethyl-ε-caprolactone):** Optically active poly(4-ethyl-ε-caprolactone), ee > 97%, was synthesized by lipase Novozym-435 catalyzed ring-opening polymerization of (R,S)-4-ethyl-ε-caprolactone at 60 °C in bulk (see Scheme 2).

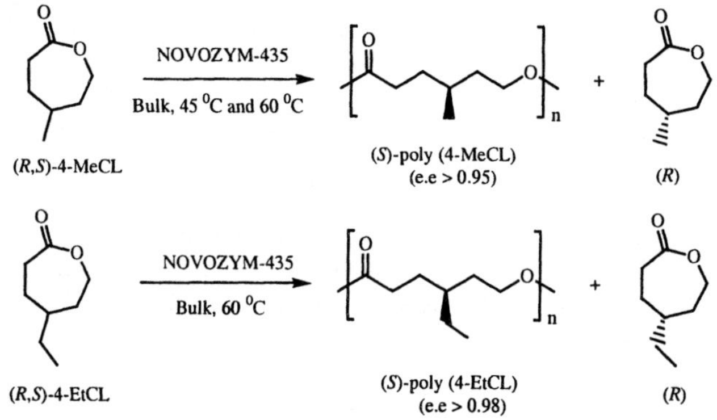

**Scheme-2.** Novozym-435 catalysed Enantioselective ring opening of 4-methyl-ε-caprolactone and 4-ethyl-ε-caprolactone.

Table 2 lists the results of the polymerization reactions. The molecular weights were determined by GPC on the basis of polystyrene standards and were in good agreement with those determined from the $^1$H NMR spectral analysis.

In Figure 5, a plot of percent monomer conversion as a function of time is presented. The monomer conversion increased steadily with time. For example, after 1.5 h the conversion was 19% and reached 74% after 9 h. The monomer conversion kept increasing beyond 50% because the enzyme polymerized both enantiomers in the racemic monomer mixture although the (S)-enantiomer was polymerized in preference to the (R)-enantiomer.

Number-average molecular weight ($M_n$) increased with increasing percent monomer conversion (Figure 6). Up to 2.5 h, when the monomer conversion was ~29%, the molecular weight ($M_n$ 500) did not increase noticeably. However,

**Table 2.** Lipase Novozym-435-Catalyzed Enantioelective Ring-Opening Polymerization of 4-Ethyl-ε-caprolactone at 60 °C[a]

| # | time (h) | Conv (%)[b] | (R)-4-Et CL $ee_m$[c] | $ee_p$[d] | $(S_p/R_p + S_p)$[e] | $[\alpha]_D^{23}$ | $M_n$[f] (g/mol) | PDI[f] |
|---|---|---|---|---|---|---|---|---|
| 1 | 1.5 | 19 | 0.20 | 0.85 | 0.93 |  | 900 | 1.62 |
| 2 | 2 | 29 | 0.41 | 0.98 | 0.99 |  | 1200 | 1.57 |
| 3 | 4 | 38 | 0.58 | 0.95 | 0.98 | -3.80 | 4400 | 1.84 |
| 4 | 5 | 52 | 0.85 | 0.79 | 0.90 | -1.80 | 5000 | 2.09 |
| 5 | 6 | 63 | 0.77 | 0.45 | 0.73 | -0.73 | 6200 | 2.13 |
| 6 | 7 | 67 | 0.73 | 0.36 | 0.68 | -0.55 | 6600 | 1.80 |
| 7 | 8 | 69 | 0.78 | 0.35 | 0.67 | -0.49 | 7500 | 1.69 |
| 8 | 9 | 74 | 0.67 | 0.23 | 0.62 | -0.20 | 8500 | 1.65 |

[a] Reactions were carried out in bulk; monomer/enzyme (w/w) = 2. [b] Monomer conversion was determined from $^1$H NMR data. [c] Enantiomeric excess of recovered monomer was calculated from chiral GC. [d] Calculated from eq 1 (See experimental section). [e] The fraction of (S) enantiomer in the poly(4-EtCL). [f] Determined by GPC (PDI = $M_w/M_n$).

beyond 29% conversion the molecular weight rose sharply and was 4400 g/mol at 38% monomer conversion and reached 8500 at 74% (9 h) conversion. The molecular weight profile was in accordance with the chain polymerization mechanism proposed for the lipase-catalyzed ring-opening polymerizations (24 c).

The initial stagnation in molecular weight may very well be indicative of the initiation stage of the polymerization when mostly new chains were being formed. However, as soon as the initiator, i.e., water was consumed (at about 29% conversion), the chain propagation dominates and was reflected in sharp molecular weight increase. The polydispersity index ($M_w/M_n$) registered only a slight increase when monomer conversion increased from 19 to 63% and then decreased with further increase in the monomer conversion to 74% (Figure 6). Importantly, the variation in the polydispersity index coincided with the molecular weight profile. In agreement with the previous reports (22 g), we found that water in these reactions does have a significant effect on the polymer molecular weight. Figure 7 shows a plot of enantiomeric excess of the monomer ($ee_m$) against monomer conversion.

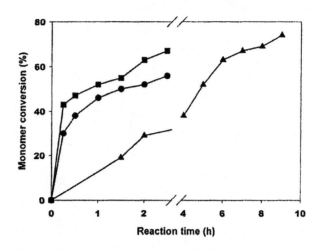

**Figure 5.** Percent monomer conversion as a function of time for novozym-435 catalyzed ring-opening bulk polymerization of 4-ethyl-ε-caprolactone at 60 °C (▲), 4-methyl-ε-caprolactone at 60 °C (■) and 4-methyl-ε-caprolactone at 45 °C (●).

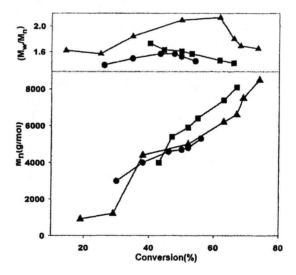

**Figure 6.** Number average molecular weight ($M_n$) as a function of percent monomer conversion during Novozym-435-catalyzed ring-opening bulk polymerization of 4-ethyl-ε-caprolactone at 60 °C (▲), 4-methyl-ε-caprolactone at 60 °C (■) and 4-methyl-ε-caprolactone at 45 °C (●).

The $ee_m$ increased with increasing percent monomer conversion and reached a maximum ~50% conversion. As can be seen from the plot, before 50% conversion, the enzyme can choose freely the "(S)"-enantiomer from the racemic mixture and the $ee_m$ of the recovered "(R)"-enantiomer increases. During the course of the reaction, the "(S)"-enantiomer is gradually depleted, leaving behind the "(R)"-enantiomer. Close to 50% conversion, the enhanced relative concentration of the "(R)"-enantiomer leads to its increased transformation by the lipase. As a consequence, beyond 50% the $ee_m$ rapidly decreases.

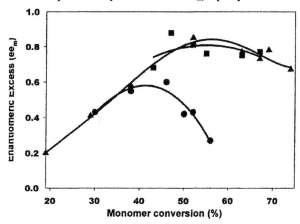

**Figure 7.** Plots of enantiomeric excess of the recovered monomer ($ee_m$) as a function of percent monomer conversion during lipase Novozym-435-catalyzed ring-opening bulk polymerization of 4-ethyl-ε-caprolactone at 60 $^0$C (▲), 4-methyl-ε-caprolactone at 60 $^0$C (■) and 4-methyl-ε-caprolactone at 45 $^0$C (●).

**Enzymatic Synthesis of (S)-Poly(4-methyl-ε-caprolactone):** (S)-Enriched poly(4-methyl-ε-caprolactone) was synthesized by lipase Novozym-435 catalyzed ring-opening polymerization of 4-methyl-ε-caprolactone at 45 and 60 $^0$C, in bulk (Scheme 2). Tables 3 and 4 list the results of the polymerizations at 60 and 45 $^0$C, respectively. The 4-methyl-ε-caprolactone was a very good substrate for the lipase Novozym-435 (Figure 5). The polymerization reactions were very fast, and in only 15 min, 43% monomer conversion had resulted at 60 $^0$C (Figure 5). The monomer conversion was affected by change in the reaction temperature and the rate of monomer conversion increased at higher reaction temperature. For example, the conversions were 30% and 43% at 45 and 60 $^0$C, respectively, after 15 min. The rate of 4-MeCL conversions was considerably reduced once the conversions reached 56 and 67% at 45 and 60 $^0$C, respectively. The decrease in the rate of the monomer conversion could be attributed to the high enantioelectivity of the lipase, i.e., as the faster reacting (S)-enantiomer is depleted the rate of polymerization decreases. It is also possible that it was an artifact of the increasing diffusion constraints in the reaction mixture, resulting from higher viscosity of the polymer, which limited availability of the monomer to the active site of the lipase. However, increased monomer conversion despite higher enzyme selectivity in these polymerizations at the higher reaction temperature suggests the latter being a more dominant factor that influences the rate of the monomer conversion

**Table 3.** Lipase Novozym-435-Catalyzed Enantioelective Ring-Opening Polymerization of 4-Methyl-ε-caprolactone at 60 $^0$C$^a$

| # | time (h) | Conv (%)$^a$ | (R)-4-MeCL ee$_m$$^b$ | ee$_p$$^c$ | $(S_p/R_p + S_p)$$^d$ | $[\alpha]_D^{23}$ | $M_n$$^e$ (g/mol) | PDI$^e$ |
|---|---|---|---|---|---|---|---|---|
| | | | | (S)-poly(4-EtCL) | | | | |
| 1 | 0.25 | 43 | 0.68 | 0.90 | 0.95 | -7.21 | 4000 | 1.73 |
| 2 | 0.5 | 47 | 0.88 | 0.95 | 0.98 | -7.90 | 5400 | 1.63 |
| 3 | 1.0 | 52 | 0.81 | 0.75 | 0.87 | -5.20 | 5900 | 1.61 |
| 4 | 1.5 | 55 | 0.76 | 0.62 | 0.81 | -4.19 | 6400 | 1.57 |
| 5 | 2.0 | 63 | 0.75 | 0.44 | 0.72 | -3.70 | 7400 | 1.47 |
| 6 | 2.5 | 67 | 0.77 | 0.38 | 0.69 | -2.00 | 8100 | 1.43 |

$^a$ Reactions were carried out in bulk; monomer/enzyme (w/w) = 2. $^b$ Monomer conversion was determined from $^1$H NMR data. $^c$ Enantiomeric excess of recovered monomer was calculated from chiral GC. $^d$ Calculated from eq 1 (see experimental section). $^e$ The fraction of (S)-enantiomer in the poly(4-EtCL). $^f$ Determined by GPC (PDI = $M_w/M_n$).

The molecular weight ($M_n$) of the poly(4-MeCL) was also affected by changing the reaction temperature from 45 to 60 $^0$C. Figure 6 shows variation in polymer number-average molecular weight ($M_n$) as a function of percent monomer conversion during the polymerization of 4-methyl-ε-caprolactone at 45 and 60 $^0$C. Although polymerization of MeCL catalyzed by Novozym-435 was highly efficient at both reaction temperatures, higher reaction temperature (60 $^0$C) generally led to higher molecular weight polymers. The polydispersity index ($M_w/M_n$) registered a decrease with increasing molecular weight of the polymer chains. The polydispersity index decreased from 1.73 ($M_n$ = 4000) to 1.43 ($M_n$ = 8100) and from 1.57 ($M_n$ = 4600) to 1.46 ($M_n$ = 5300) when the polymerizations were conducted at 60 and 45 $^0$C, respectively.

In polymerizations conducted at 45 and 60 $^0$C, the enantioselectivity of the enzyme was also affected considerably by the change in reaction temperature. Interestingly, the lipase enantioselctivity, $E$, increased with increase in temperature from 45 to 60 $^0$C. This finding is in contrast to several papers reporting an inverse correlation between the temperature and the lipase enantioselectivity (34). Unfortunately, the effect of temperature on lipase selectivity is not very well understood (35), partly because enzymes are temperature labile and variation in temperature is a rather less obvious choice to improve lipase enantioselectivity. In Figure 7, a plot of enantiomeric excess of the recovered lactone as a function of percent conversion of 4-MeCL at 45 and 60 °C is shown. At 45 °C, the enantiomeric excess of the recovered (R)-lactone increased with conversion and peaked at about 47% to 0.58. The polymer $ee_p$, as expected, was high in the initial stages of the polymerization at 45 °C with enantiomeric excess of the polymer, $ee_p$ being 0.98 at 30% monomer conversion. The polymer $ee_p$ subsequently decreased rapidly beyond the 50% monomer conversion. The ring-opening polymerization of 4-MeCL at 60 °C was highly efficient; not only the monomer conversion was 43% but also the enantiomeric excess of the recovered monomer ($ee_m$) was 0.68 after 15 min, which increased to 0.88 at 47% monomer conversion. The enantiomeric excess of the polymer ($ee_p$) also reached a maximum of 0.95 ($E = 122$) at 47% monomer conversion when the reaction temperature was 60 °C.

**Table 4.** Lipase Novozym-435-Catalyzed Enantioelective Ring-Opening Polymerization of 4-Methyl-ε-caprolactone at 45 °C[a]

| # | Time (h) | Conv (%)[b] | (R)-4-MeCLee$_m$[c] | (S)-poly(4-MeCL) | | | | |
|---|---|---|---|---|---|---|---|---|
| | | | | $ee_p$[d] | $(S_p/R_p + S_p)$[e] | $[\alpha]_D^{23}$ | $M_n^f$ (g/mol) | PDI[f] |
| 1 | 0.25 | 30 | 0.43 | 0.98 | 0.99 | -8.10 | 3000 | 1.40 |
| 2 | 0.5 | 38 | 0.55 | 0.85 | 0.92 | -6.22 | 4000 | 1.50 |
| 3 | 1.0 | 46 | 0.60 | 0.70 | 0.85 | -4.82 | 4600 | 1.57 |
| 4 | 1.5 | 50 | 0.42 | 0.42 | 0.71 | -3.20 | 4700 | 1.57 |
| 5 | 2.0 | 52 | 0.43 | 0.40 | 0.70 | -2.96 | 4800 | 1.53 |
| 6 | 2.5 | 56 | 0.27 | 0.13 | 0.57 | -0.53 | 5300 | 1.46 |

[a] Reactions were carried out in bulk; monomer/enzyme (w/w) = 2. [b] Monomer conversion was determined from $^1$H NMR data. [c] Enantiomeric excess of recovered monomer was calculated from chiral GC. [d] Calculated from eq 1 (see experimental section). [e] The fraction of (S) enantiomer in the poly(4-EtCL). [f] Determined by GPC (PDI = $M_w/M_n$).

Considering the fact that the polymerizations were conducted without added solvents, the enantioselectivity in these reactions was remarkable. The lower reaction rate and the lower ee values in reactions conducted at 45 °C compared to the polymerizations at 60 °C suggested that the increasing viscosity of the reaction mixture, due to the formation of high molecular weight polymers, was a limiting factor in these bulk polymerizations. The higher viscosity in the reaction mixture may limit the accessibility of the unreacted monomer to the lipase active site. The higher reaction temperature, on the other hand, translates into higher fluidity in the reaction mixture and hence explains higher monomer conversion at 60 °C. Interestingly, lipase enantioselectivity in these bulk polymerizations was higher than reported in any lipase-catalyzed polymerizations conducted in organic solvents. Additionally, the high enantioselectivity demonstrated in the polymerization of 4-methyl- and 4-ethyl-ε-caprolactone, monomers that contain a stereocenter as remote as four bonds from the reaction site (lactone carbonyl), are extremely rare.

**NMR Characterization:** The structures of the polymers were confirmed by $^1$H and $^{13}$C NMR spectral data. Stereospecific dyads in the NMR data were not observed because of a four-bond separation between the stereocenter and the carbonyl group. $^1$H and $^{13}$C NMR spectra for both polymers along with the peak assignments are shown in Figures 8 and 9. The peak assignments were based on two-dimensional $^1$H-$^1$H COSY and $^1$H-$^{13}$C HETCOR NMR experiments (Supporting Information). In 4-Me-CL, the diastereotopic protons ($H_a$ and $H_b$) in the two methylenes on either side of the stereocenter were split into two signals. The $^1$H-$^1$H COSY experiment showed connectivities between the protons *a* and *b* in each set of methylene protons

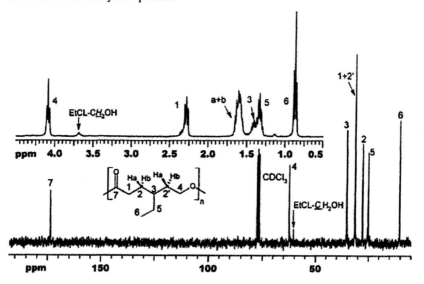

**Figure 8** $^1$H NMR (250 MHz, CDCl$_3$) and $^1$H NMR (250 MHz, CDCl$_3$, inset spectra of poly(4-ethyl-ε-caprolactone) (entry 4, Table 2) obtained by lipase Novozym-435-catalyzed ring-opening polymerization in bulk at 60 °C.

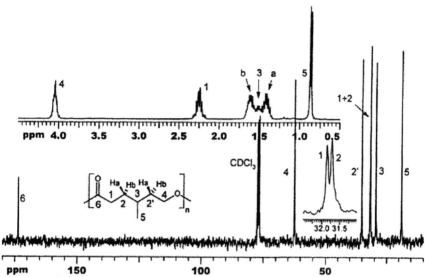

**Figure 9.** ¹³C NMR (62.9 MHz, CDCl₃) and ¹H NMR (250 MHz, CDCl₃, inset) spectra of poly(4-methyl-ε-caprolactone) (entry 6, Table 3) obtained by lipase Novozym-435-catalyzed ring-opening polymerization in bulk at 60 °C.

The $^1$H-$^{13}$C HETCOR experiments also had correlations between the methylene carbon signals and the protons *a* and *b* signals in each set. Interestingly, the *a* and *b* protons appeared as two signals (Δδ~ 0.19 ppm) in poly(4-MeCL), but in the poly(4-EtCL) a broad multiplet instead of two separate signals was observed (Figure 8). Besides prominent resonances due to the repeating units of the polymer, small but characteristic resonances ascribed to the α-methylene of the terminal hydroxy and the carboxylic acid groups were also observed in the NMR spectra. The polymers were thus identified to have a hydroxyl and a carboxylic acid group at each chain terminus.

## Enantioselectivity of the lipase

For all three lactones, the (*S*)-enantiomer reacted faster than the (*R*)-enantiomer. *C. antarctica* lipase (CALB or Novozym-435), like other lipases, has the same mechanism of action as a serine protease and the $Ser_{105}$–$His_{224}$–$Asp_{187}$ triad is responsible for the catalytic action (*36*). Interestingly, *C. antarctica* lipase demonstrates (*R*)-stereoselectivity towards a number of secondary alcohols (*37*). However, substrates with a stereogenic center on the acyl side have not been studied with equal vigor. Contrary to our observations, it has been suggested (*38*) that the acyl side of the active site is more spacious and hence possesses much lower stereo selectivity. Based upon the acyl binding models (*39*) of the lipase and the *S* selectivity of the lipase CALB for lactones, as observed by others (*30*), a cartoon representation of the active site of the enzyme is shown in Figure 10 . We have concluded that there is a certain steric restriction in the small subsite (stereospecificity pocket) to accept the alkyl group instead of the hydrogen (Figure 10). We assumed (in Figure 10) that in the fast reacting (*S*)-lactone, the position of the carbonyl carbon and the oxyanion as well as the oxygen on the lactone must be relatively fixed in order to allow the formation of a hydrogen bond between the catalytic histidine ($His_{224}$) residue and the oxygen. These constraints and the rigidity of the lactone ring then determine how the remainder of the lactone will fit in the active site. In the (*S*)-enantiomer of lactones **1** and **3** (Figure 1), the methyl group effectively fits into the small `stereospecificity' pocket. In the (*R*)-enantiomer, the methyl group is pointing away from the `stereospecificity' pocket and may have many close van der Waals contacts. These steric clashes make this interaction highly unfavorable. The lower lipase enantioselectivity for the 4-ethy-caprolactone compared to the 4-methyl -caprolactone suggested that the `stereosepecificity' pocket is only big enough to accommodate a methyl group effectively. Also, the higher rate of conversion in reactions of **1** over **2** support that a methyl group rather than an ethyl group at the 4-position results in a better fit in the active site of Novozym-435.

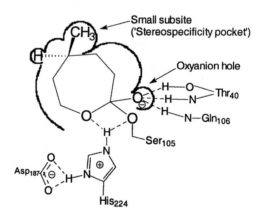

**Figure 10**. A cartoon representation of the lipase active site and its binding to the lactone

## Chemo-enzymatic synthesis of enantio enriched substituted polycaprolactones

To determine the effects of optical purity of polyesters on their physical characteristics we carried out (*33*) synthesis of poly 4-ethyl-ε-caprolactone (4-EtCL) and poly 4-methyl-ε-caprolactone (4-MeCL) of variable enantiopurities. From the enantioelective ring opening polymerization of the racemic 4-ethyl and

4-methyl-ε-caprolactones catalyzed by lipase Novozym-435 at 60 °C in bulk for 6 h (Scheme 2), (R)-4-ethyl-ε-caprolactone and (R)-4-methyl-ε-caprolactone with enantiomeric excess (ee) values of 70% and 99% were obtained. The mixtures of (R)-and rac-4-substituted -ε-caprolactones were used to prepare mixtures of variable enantiopurities and the optical purities of the lactone mixtures were checked by chiral GC to ascertain their exact composition. Several polycaprolactone samples with different (S-contents were prepared by the co-polymerization of the racemic lactones with (S)-enriched lactones (Tables 5 and 6) by ring opening polymerization initiated by Sn(Oct)$_2$ at 120 °C (Scheme 3).

The optical rotations of both polymers decreased with decreasing enantiomeric excess, as expected. The number average molecular weight ($M_n$) ranged from 11,500 to 7,100 g/mol, with polydispersity indexes (PDI) ranging from 1.32 to 1.40 for polymers. The stereo-copolymers were characterized by DSC. From tables 5 and 6, it was obvious that $T_g$ increase with increasing the R-content of the 4-methyl and 4-ethyl polycaprolactones. A similar trend could be expected with increasing (S) content of the ethyl and methyl caprolactones as evident from the $T_g$ of the pure (S)-methyl and ethyl caprolactones (Tables 5 and 6). The change in glass transition temperature ($T_g$) values of racemic and optically enriched polymers is relatively small because the large distance between two chiral centers (7 bonds) in the polymer. A change in the $T_g$, although small, does assure influence of optical purity of these polymers on polymer crystal packing. The change in $T_g$ of poly(4-ethyl caprolactone) is larger than that in poly(4-methyl caprolactone) because larger size of the ethyl group influences crystal packing to a greater extent than a smaller methyl group.

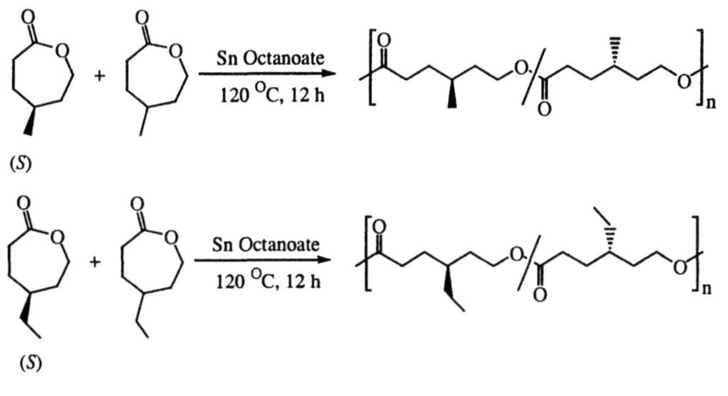

**Scheme 3**. Synthesis of optically enriched substituted ε-caprolactones by ring opening polymerization initiated by Sn(Oct)$_2$.

**Table 5.** Ring-opening of optically active 4-EtCL catalyzed by $Sn(Oct)_2$ at 120 °C for 12 h.

| Entry | Ratio, R: S [a] | Yield (%) [b] | $M_w$ [c] | $M_n$ [c] | PDI [c] | $[\alpha]_D^{23}$ | $T_g$ [d] |
|---|---|---|---|---|---|---|---|
| 1 | 85 : 15 | 75 | 16,000 | 11,500 | 1.39 | + 3.30 | -34.46 |
| 2 | 70 : 30 | 90 | 12,700 | 9,200 | 1.38 | + 2.96 | -35.88 |
| 3 | 60 : 40 | 88 | 13,000 | 9,400 | 1.38 | + 0.40 | -36.82 |
| 4 | 55 : 45 | 85 | 9,600 | 7,100 | 1.36 | + 0.02 | -37.03 |
| 5 | 50 : 50 | 92 | 13,500 | 9,700 | 1.40 | 0.00 | -39.04 |
| 6 | 0 : 100 | 90 | 15,000 | 11,400 | 1.32 | -6.70 | -32.42 |

[a] Monomer composition determined by chiral phase GC. [b] Precipitated polymer yield. [c] Determined by GPC. [d] From DSC.

**Table 6.** Ring-opening of optically active 4-MeCL catalyzed by $Sn(Oct)_2$ at 120 °C for 12 h.

| Entry | Ratio, R: S [a] | Yield (%) [b] | $M_w$ [c] | $M_n$ [c] | PDI [c] | $[\alpha]_D^{23}$ | $T_g$ [d] |
|---|---|---|---|---|---|---|---|
| 1 | 75 : 25 | 85 | 15,000 | 11,400 | 1.32 | + 7.80 | -38.57 |
| 2 | 70 : 30 | 85 | 14,800 | 10,000 | 1.48 | + 3.37 | -39.43 |
| 3 | 62 : 28 | 85 | 15,000 | 10,400 | 1.44 | + 1.77 | -40.57 |
| 4 | 50 : 50 | 92 | 12,600 | 10,800 | 1.16 | 0.00 | -39.99 |
| 5 | 0 : 100 | 90 | 13,800 | 11,200 | 1.23 | -8.70 | -37.52 |

[a] Monomer composition determined by chiral phase GC. [b] Precipitated polymer yield. [c] Determined by GPC. [d] From DSC.

## Conclusions

An efficient Kinetic resolution of substituted caprolactones was achieved in bulk (solvent free) conditions by lipase Novozym-435 catalyzed ring opening with *n*-butanol. The Resolutions were investigated at four different temperatures, i.e. 0, 23, 40 and 60 °C. For all the compounds studied Novozym-435 lipase showed (S)-selectivity, i.e the (S)-enantiomer reacted more rapidly with n-butanol and the (R)-lactones were recovered unreacted with > 97% enantiomeric excess (ee). Contrary to general perception and literature reports that suggest an inverse relationship between the reaction temperature and enantiomeric excess, in the resolution carried out by us, the stereoselectivity increased with increasing reaction temperature. It was demonstrated that 60 °C is the desired temperature for these resolutions. The (S)-selectivity of lipase Novozym-435 for substituted ε-caprolactones is also utilized for the enantioelective ring opening polymerizations of 4-methyl-ε-caprolactone (4-MeCL) and 4-ethyl-ε-caprolactone (4-EtCL) to synthesize highly (S)-enriched poly(4-MeCL) ($ee_p > 0.95$, $M_n = 5400$) and poly(4-EtCL) ($ee_p > 0.98$, $M_n = 4000$). The reaction was also investigated at two different temperatures, viz. 45 and 60 °C. The higher reaction temperature led to higher monomer conversion, polymer molecular weight and enzyme enantioselectivity. Stereo-copolymers from optically enriched 4-ethyl-ε-caprolactone and 4-methyl-ε-caprolactone were synthesized through a chemoenzymatic approach by the co-polymerization of the racemic lactones with R-enriched lactones obtained by enantioelective ring opening by enzyme catalysis. The $T_g$ of the enriched polymers increased with increasing optical purity of the polymers. The change in $T_g$ however was relatively small because of low content of chiral centers and a relatively large distance 97 bonds) between two chiral centers in these polyesters.

## Experimental Procedures

Novozym-435 lipase was kindly provided by Novo Nordisk Bioindustrial Inc. 4-Methylcyclohexanone, 4-Ethylcyclohexanone, and m-chloroperoxybenzoic acid (m-CPBA) were purchased from Acros Chemical Co. and used as received.

### i) Instrumental Methods

*NMR Spectroscopy.* $^1H$ and $^{13}C$ NMR spectra were recorded on a Bruker ARX-360 spectrometer at 360 and 90 MHz and a Bruker DPX-250 spectrometer at 250 and 62.9 MHz, respectively. $^1H$ NMR chemical shifts (ppm) are reported downfield from 0.00 ppm using tetramethylsilane (TMS) as an internal standard. The concentrations used were ~4% w/v in chloroform-d ($CDCl_3$). $^{13}C$ NMR spectral chemical shifts in (ppm) are referenced relative to the internal standard chloroform-d at 77.00 ppm. 4-Methyl-ε-caprolactone monomer conversions were determined from the relative peak areas of $^1H$ NMR signals corresponding to methyl ($-CH_3$) protons in the polymer and the monomer at 0.93 and 0.99 ppm, respectively. 4-Ethyl-ε-caprolactone monomer conversions were determined from the relative peak areas of $^1H$ NMR signals corresponding to methylene (-

(CO)-C*H₂*-) protons in the polymer and the monomer at 2.27 and 2.68 ppm, respectively.

*Gas Chromatography (GC).* Gas chromatographic analyses were conducted on a Shimadzu GC-17A chromatograph equipped with a flame ionization detector (FID) and a cyclodex-β-chiral capillary column (J&W Scientific, Film: 0.25 μm × 30 m × i.d. of 0.25 mm). Helium was used as a carrier gas. Chiral separations used the following conditions: 64 $^0$C (2 min) to 130 $^0$C (5 min) at 1 $^0$C/min followed by a 10 $^0$C/min gradient to 180 $^0$C (15 min). Carrier gas flow was 1.8 mL/min for analysis of 4-ethyl-ε-caprolactone and 2.0 mL/min for 4-methyl-ε-caprolactone. The injector and detector temperatures were maintained at 250 and 300 $^0$C, respectively. Optical rotations were measured on an Autopol IV (Rudolph Instruments) automatic polarimeter at 23 $^0$C in CHCl₃ at a concentration of 0.5.

*Molecular Weight Measurements.* Molecular weights were measured by gel permeation chromatography (GPC) using a Shimadzu HPLC system equipped with a model LC-10AD*vp* pump, model SIL-10A autoinjector, model RID-10A refractive index detector (RI), model SPD-10AV UV-vis detector, and Waters HR 4E styragel column. THF (HPLC grade) was used as an eluent at a flow rate of 1.0 mL/min. The sample concentration and injection volumes were 0.5% (w/v) and 100 μL, respectively. EzChrome Elite (Scientific Software Inc.) was used to calculate molecular weights on the basis of a calibration curve generated by narrow molecular weight distribution polystyrene standards (5.00 × 10², 8.00 × 10², 2.10 × 10³, 4.00 × 10³, 9.00 × 10³, 1.90 × 10⁴, 5.00 × 10⁴, 9.26 × 10⁴, 2.33 × 10⁵, and 3.00 × 10⁵ g/mol, Perkin-Elmer). The degree of polymerization (DP) calculated from the ¹H NMR spectral analysis was in good agreement with the molecular weight obtained using GPC.

## ii) General Procedure for Enzymatic Resolution

*Reaction of 4-ethyl-ε-caprolactone 2 with n-butanol* : All reactions were carried out in bulk. The lipase was dried in a drying pistol over P₂O₅ at 50 °C/0.1 mmHg, 15 h. In a glove bag, maintained under a nitrogen atmosphere, 4-ethyl-ε-caprolactone (0.2 g, 1.40 mmol) and *n*-butanol (0.19 mL, 2.1 mmol) were transferred to a 12 mL reaction vial, and then pre-weighed dried enzyme (0.1 g) was added. The reaction vial was sealed with a rubber septum and placed in a thermostatically controlled oil bath maintained at 40 or 60°C. At predetermined times, samples from the reaction vial were collected through the septum and the progress of the reaction was monitored by chiral phase GC analyses. Reactions were terminated by dissolution of the contents of the reaction vial in chloroform and removal of the enzyme (insoluble) by filtration (glass-fritted filter, medium pore porosity). The filtrates were combined and solvent removed in vacuo. The absolute configuration of the recovered lactone was determined by coinjection with the authentic (S)-**2** (Fig. 2) in chiral phase GC analysis and also by the specific rotation of the enantioenriched unreacted **2** isolated by column chromatography (e.e. >98%, $[\alpha]_D^{23}$=+42 (c=0.5, CHCl$_3$)) with the reported data in the literature (40).

### iii) General Procedure for Enzymatic Polymerization

All reactions were carried out in bulk. The lipase was dried as described above. In a glovebag, maintained under nitrogen atmosphere, the monomer was transferred to a 6 mL reaction vial, and the preweighed enzyme (62 mg/mmol of lactone) was added. The reaction vial was capped with a rubber septum and placed in a constant temperature oil bath maintained at 45 and 60 $^0$C for predetermined times. Progress of the polymerization was monitored using GC analysis. Reactions were terminated by dissolution of the contents of the reaction vial in chloroform and removal of the enzyme (insoluble) by filtration (glass fritted filter, medium pore porosity). The filtrates were combined, solvents were removed in vacuo, and the crude products were analyzed by proton ($^1$H) NMR and gel permeation chromatography (GPC). The polymer was purified by precipitation in methanol, and the recovered monomer was analyzed by gas chromatography (GC).

### iv) General Procedure for the Synthesis of (S) enriched 4-Substituted ε-Caprolactone

*(S)-4-Methyl-ε-caprolactone using an enzymatic Baeyer-Villiger oxidation.* A single colony of *Saccharomyces cerevisiae* 15C(pKR001) was used to inoculate 25 mL of YPD medium (10 g/L Bacto-Yeast Extract, 20 g/L Bacto-Peptone, 2% glucose added after sterilization). The culture was shaken at 30°C until it reached O.D.$_{600}$=4.3, then a 20 mL portion was added to 1 L of YP-Gal medium (10 g/L Bacto-Yeast Extract, 20 g/L Bacto-Peptone, 2% galactose added after sterilization) in a 2 L New Brunswick MultiGen fermenter. The culture was grown at 30°C with vigorous stirring until it reached O.D.$_{600}$=2.9, then neat 4-methylcyclohexanone (10.2 mmoles, 1.25 mL) was added. The culture was stirred vigorously at room temperature. Samples were removed periodically analyzed by GC. When all the ketone had been consumed, the cells were removed by centrifugation (4000 × *g* for 20 min at 4°C). The supernatant was divided into two equal portions and each was extracted with ca. 200 mL of EtOAc in a continuous extractor. The cell pellet was re-suspended in 25 mL of water, then this slurry was extracted with EtOAc (3 × 50 mL). All organic extracts were combined, washed with brine and concentrated by rotary evaporator to yield 1.62 g of crude lactone as an orange oil. The crude product was purified by flash column chromatography over silica get using a hexane/ethylacete gradient.

*(S)-4-Ethyl-ε-caprolactone using an enzymatic Baeyer-Villiger oxidation.* The general procedure described above was also used to oxidize 4-ethylcyclohexanone (10 mmoles, 1.41 mL). In this case, an equimolar quantity of β-cyclodextrin (10 mmoles, 11.35 g) was included in the YP-Gal medium to solubilize the hydrophobic ketone. Extractive workup afforded 1.76 g of crude lactone as an orange oil.

### v) Chemoenzymatic Synthesis of Enantioenriched 4-substituted ε-Caprolactones

The lipase Novozym-435 was dried and the enzyme and caprolactones (Enzyme/ caprolactone; 1 : 2 by weight), were transferred to the reaction vial which was capped with a rubber septum and placed in a constant temperature bath at 60 °C for 6 h. Reactions were terminated upon removal of the enzyme (insoluble) by filtration. The filtrates were concentrated to a viscous solution, which was precipitated in 200 ml of methanol with vigorous stirring in a beaker. Methanol soluble portion was concentrated in vacuum and purified by column chromatography on a silical gel column (100-200 mesh, 10 % EtOAc in petroleum ether) to obtain (R)-4-methyl-ε-caprolactone (99 % ee, $[\alpha]_D^{25}$ = +40 °C; c=0.5/CHCl$_3$) and (R)-4-ethyl-ε-caprolactone (70 % ee, $[\alpha]_D^{25}$ = +28 °C; c=0.5/CHCl$_3$). The mixtures of (R)-enriched and rac-substituted -ε-caprolactones were used to prepare mixtures of variable enantiopurities and the mixtures were analyzed by chiral GC to ascertain their exact composition

Chemical polymerizations were carried out in 10 ml Schlenck tubes. The tubes were treated with trimethylsilyl chloride, washed with three 5 ml portions of methanol, dried at 120 °C in a oven for 12 h, flame dried and kept in a desiccator to cool down to room temperature. In a glove bag maintained under nitrogen atmosphere the lactone mixture and the catalyst (0.1 molar solution in toluene, monomer to catalyst ratio was 1:200) were transferred into the polymerization tube and capped with a rubber septum. The tubes were degassed by several vacuum purge cycles to remove the solvent in the catalyst solution and then placed in an oil bath maintained at 120 °C for 12 h. At the end of the reaction period, the crude sample was collected to estimate the conversion and the contents of the tubes were dissolved in chloroform (0.5 ml) and precipitated in methanol (30 ml) by vigorous stirring and methanol decanted. The precipitate was further washed with methanol (2 X 20 ml). The polymer was dried in a vacuum oven at 40 °C for 24 h and GPC data were recorded.

### vi) Enantiomeric Purity

The analytical separation of the two enantiomers of the racemic lactones was achieved by carefully adjusting the analysis parameters of the gas chromatograph equipped with a Cyclodex-B (J&W Scientific) chiral phase capillary column. Chiral separations were performed using the following gradient temperature program: 64 °C (2 min) to 130 °C (5 min) at 1°C followed by a 10 °C/min gradient to 180 °C (15 min). Carrier gas flow was 1.8 mL/min for analyses of lactones **1** and **3** and 2.0 mL/min for **2**. The injector and detector temperatures were maintained at 250 and 300 °C, respectively. The absolute configurations of the separated enantiomeric peaks in the racemic lactones were established by their co-injection with an authentic enantiomerically pure (S)-**1** or **2** (Fig. 2) and also from comparison of the specific rotation with that reported in literature (*4 a, 40, 41*).

The enantiomeric excess (e.e.$_p$) of the (S)-enriched hydroxy esters **4**, **5**, and **6** or polyesters obtained was calculated from the lactone conversion, $c$, and the fractions, viz. $[S/(S+R)]_S$ and $[R/(S+R)]_S$ of the (S)- and (R)-enantiomers, respectively, in the recovered (unreacted) lactone from the below equation (*28*).

$$\text{e.e.}_p = \frac{(S_p - R_p)}{(S_p + R_p)} \quad \text{Equation-1}$$

$$S_p = 0.5 - \left[(1-c)\left\{\frac{S}{S+R}\right\}_s\right]$$

$$R_p = 0.5 - \left[(1-c)\left\{\frac{R}{S+R}\right\}_s\right]$$

The enantiomeric ratios ($E$), a measure of the enantioselectivity of the lipase (Tables 1 to 4), can be related to the extent of conversion and the enantiomeric ratio and were calculated using the following equation. Where $c$ is the lactone conversion (determined by GC)/$^1$H NMR, e.e.$_s$ is the enantiomeric excess (determined by GC) of the unreacted lactone. The equation is based on the assumptions that resolution proceeds irreversibly, that the two enantiomers compete for the same active site, and that there is no product inhibition (*42*).

$$E = \frac{\ln[(1-c)(1-\text{e.e.}_s)]}{\ln[(1-c)(1+\text{e.e.}_s)]} \quad \text{Equation-2}$$

## Acknowledgements

We very much appreciate the financial support for this work from the American Cancer Society, American Lung Association, and the Division of sponsored research of the University of South Florida. We are grateful to Prof. Bill Baker for use of the digital optical polarimeter and to Prof. Mike Zaworotko for use of the DSC instrument.

## References

1. Okamoto, Y.; Nakano, T. *Chem. Rev.* **1994**, 94, 349.
2. (a) Muller, H. M.; Seebach, D. *Angew. Chem., Int. Ed. Engl.* **1993**, 32, 447.(b) Lee, S. Y. *Biotechnol. Bioeng.* **1996**, 49, 1.
3. (a) Shirahama, H.; Shiomi, M.; Sakane, M.; yasuda, H. *Macromolecules* **1996**, 29, 4821. (b) Yasuda, H.; Aludin, M.; Kitamura, N.; Tanabe, M.; Sirahama, H. *Macromolecules* **1999**, 32, 6047. (c) Spassky, N.; Wismiewski, M.; Pluta, C.; Leborgne, A. *Macromol. Chem. Phys.* **1996**, 197, 2627. (d) Radano, C.P.;Baker, G.L.; Smith, M.R. *J. Amer. Chem. Soc.* **2000**, 122, 1552.
4. (a) Pirkle, W. H.; Adams, P. E. *J. Org. Chem.* **1979**, 44, 2169. (b) Brown, H .C.; Kulkarni, S. V.; Racherla, U. S. *J. Org. Chem.* **1994**, 59, 365.
5. Pool, R. Science 1989, 245, 1187.
6. (a) Overberger, C. G.; Kaye, H. *J. Am. Chem. Soc.* **1967**, 89, 5640. (b) Matsu moto, K.; Tsutsumi, S.; Ihori, T.; Ohta, H. *J. Amer. Chem. Soc.* **1990**, 112, 9614. (c) Katoh, O.; Sugai, T.; Ohta, H. *Tetrahedron : Asymmetry* **1994**, 5, 1935. (d) Bolm, C.; Schlingloff, G.; Weickhardt, K. *Angew. Chem., Int. Ed. Engl.* **1994**, 33, 1848.(e) Gusso, H.; Baccin, C.; Pinna, F.; Strukul, G. *Organometallics***1994**, 13, 3442.
7. Hmamouchi, M.; Prud'homme, R. *J. Polym. Sci. Polym. Chem. Ed.* **1991**, 29, 1281.
8. Carriere, F.; Eisenbach, C.; Schulz, G. V. *Makromol. Chem.* **1981**, *182*, 325.
9. D'hondt, C.; Lenz, R. W. *J. Polym. Sci., Polym. Chem. Ed.* **1978**, *16*, 261.
10. Grenier, D.; Prud'homme, R. E. *J. Polym. Sci., Polym. Chem. Ed.* **1981**, *19*, 1781.
11. (a) Zhong, Z; Dijkstra, P. J.; Feijen, J. *J. Am. Chem. Soc.* **2003**, *125*, 11291. (b) Zhong, Z; Dijkstra, P. J.; Feijen, J. *Angew. Chem., International Ed.* **2002**, *41*, 4510. (c) Chamberlain, B. M.; Cheng, M.; Moore, D. R.; Ovitt, T. M.; Lobk ovsky, E. B.; Coates, G. W. *J. Am. Chem. Soc.* **2001**, *123*, 3229. (d) Ovitt, T. M.; Coates, G. W. *J. Am. Chem. Soc.* **1999**, *121*, 4072. (e) Nobuyoshi, N.; Ryohei, I.; Matsujiro, A.; Keigo, A. *J. Am. Chem. Soc.* **2002**, *124*, 5938. (f) Cheng, M.; Attygalle, A. B.; Lobkovsky, E. B.; Coates, G. W. *J. Am. Chem. Soc.* **1999**, *121*, 11583.
12. Cheng, M.; Ovit, T. M.; Hustad, P. D.; Coates, G. W. *Polym. Preprint* **1999**, 40 (1), 542.
13. Hmamouchi, M.; Prud'homme, R. E. *J. Polym. Sci., Polym. Chem. Ed.* **1988**, 26, 1593.

14. (a) Parmar, V. S.; Bisht, K. S.; Singh, A.; Jha, A. Proc. Indian Acad. Sci., Chem. Sci. **1996**, *108*, 75. (b) Jha, A.; Bisht, K. S.; Parmar, V. S. Proc. Ind. Acad. Sci., Chem. Sci. **1994**, *106*, 1191. (c) Carr, J. A.; Bisht, K. S. Tetrahedron **2003**, *59*, 7713. (d) Carr, J. A.; Al-Azemi, T. F.; Long, T. E.; Shim, J. Y.; Coates, C. M.; Turos, E.; Bisht, K. S.. *Tetrahedron.* **2003**, *59*, 9147. (e) Wong, C.-H. *Science* **1989**, 244, 1145.
15. Drauz, K.; Wladmann, H. (Eds) *Enzyme Catalysis in Organic Synthesis- A comprehensive handbook.* VCH Verlag: Weinhein, 1995.
16. Gutman, A. L.; Zuobi, K.; Bravdo, T. *J. Org. Chem.* **1990**, *55*, 3546.
17. Gutman, A. L.; Zuobi, K.; Boltansky, A. *Tetrahedron Lett.* **1987**, *28*, 3861.
18. Gutman, A. L.; Oren, D.; Boltanski, T.; Bravdo, T. *Tetrahedron Lett.* **1987**, *28*, 5367.
19. Stewart, J. D. *Curr. Org. Chem.* **1998**, 2, 211.
20. Thiel, F. *Chem. Rev.* **1995**, 95, 2203.
21. (a) Blanco, L.; Guibe-Jamplel, E.; Rousseau, G. *Tetraderon Lett.* **1988**, 29, 1915. (b) Fouque, E.; Rousseau, G. *Synthesis* **1989**, 661. (c) Fellous, R.; Lizzani -Cuvelier, L.; Loiseau, M. H.; Sassy, E. *Tetrahedron : Asymmetry* **1994**, 5, 343.
22. (a) Gross, R. A.; Kumar, A.; Kalra, B. *Chem. Rev.* **2001**, 101, 2097. (b) Ko ba yashi, S.; Uyama, H.; Kimura, S. *Chem. Rev.* **2001**, 101, 3793. (c) Bisht, K. S.; Al-Azemi, T. F. *ACS Symposium Series- Biocatalysis in Polymer Science* **2003**, *840*, 156.
23. (a) Jones, J. B. *Tetrahedron* **1986**, 42, 3351.(b) Klibanov, A. M. *Acc. Chem. Res.* **1990**, 23, 114. (c) Santeniello, E.; Ferraboschi, P.; Grisenti, P.; Manzocchi, A. *Chem. Rev.* **1992**, 92, 1071. (d) Seoane, G. *Curr. Org. Chem.* **2000**, 4, 283.
24. (a) Uyama, H.; Kobayashi, S. Chem. Lett. 1993, 1149. (b) Uyama, H.; Suda, S.; Kikuchi, H.; Kobayashi, S. *Chem. Lett.* **1997**, 1109. (c) MacDonald, R. T.; Pulapura, S. K.; Svirkin, Y. Y.; Gross, R. A.; Kaplan, D. L.; Joseph, A.; Swift, G.; Wolk, S. *Macromolecules* **1995**, 28, 73. (d) Henderson, L. A.; Svirkin, Y. Y.; Gross, R. A.; Kaplan, D. L.; Swift, G. *Macromolecules* **1996**, 29, 7759. (e) Uyama, H.; Takeya, K.; Kobayashi, S. *Acta polym.* **1996**, 47, 357. (f) Cordova, A.; Iversen, T.; Hulk, K. *Polymer* **1999**, 40, 6709. (g) Bisht, K. S.; Henderson, L. A.; Gross, R. A.; Kaplan, D. L.; Swift G. *Macromolecules* **1997**, 30, 2705. 25. (a) Al-Azemi, T. F.; Bisht, K. S.. *J. Polym. Sci., Polym. Chem.* **2002**, *40*, 1267. (b) Al-Azemi, T. F.; Harmon, J. P.; Bisht, K. S.. *Biomacromolecules* **2000**, *1*, 493. (c) Bisht, K. S.; Svirkin, Y. Y.; Henderson, L. A.; Gross, R. A.; Kaplan, D. L.; Swift, G. *Macromolecules* **1997**, 30, 7735. (d) Al-Azemi, T. F.; Bisht, K. S. *Macromolecules* **1999**, 32, 6536. (e) Kobayashi, S.; Kikuchi, S.; Uyama, H. *Macromol. Rapid. Commun.* **1997**, 18, 575. (f) Matsumura, S.; Tsukada, K.; Toshima, K. *Macromolecules* **1997**, 30, 3122.

26. Wallace, J. S.; Morrow, C. J. *J. Polymn. Sci., Polym. Chem. Ed.* **1989**, 27, 2553.
27. Svirkin, Y. Y.; Xu, J.; Gross, R. A.; Kaplan, D. L.; Swift, G. *Macromolecules* **1996**, 29, 4591.
28. Xie, W.; Li, J.; Chen, D.; Wang, P.G. *Macromolecules* **1997**, 30, 6997.
29. Kikuchi, H.; Uyama, H.; Kobayashi, S. *Macromolecules* **2000**, 33, 8971.
30. (a) Shioji, K.; Matsuo, A.; Okuma, K.; Nakamura, K.; Ohno, A. *Tetrahedron Lett.* **2000**, 41, 8799.
31. Kondaveti, L.; Al-Azemi, T. F.; Bisht, K. S. *Tetrahedron: Asymmetry* **2002**, 13, 129.
32. Al-Azemi, T. F.; Kondaveti, L.; Bisht, K. S. *Macromolecules* **2002**, 35, 3380.
33. (a) Bisht, K. S.; Al-Azemi, T. F.; Kondaveti, L.; *Polymer Preprints (American Chemical Division, Division of Polymer Chemistry)* **2002**, 43, 704. (b) Al-Azemi, T.F.; Kondaveti, L.; Bisht, K. S. *Polymer Preprints (American Chemical Division, Division of Polymer Chemistry)* **2003**, 44, 597.
34. (a) Arroyo, M.; Sinisterra, J. V. *J. Org. Chem.* **1994**, 59, 4410. (b)Holmberg, E.; Hult, K. *Biotechnology Lett.* **1991**, 13, 323.
35. (a) Talasky, G. *Angew. Chem., Int. Ed. Engl.* **1971**, 10, 548. (b) Keinan, E.; Hafeli, E. K., Seth, K. K.; Lamed, R. *J. Am. Chem. Soc.* **1986**, 108, 162. (c) Pham, U. T.; Philips, R. S.; Ljungdahl, L. G. *J. Am. Chem. Soc.* **1989**, 111, 1935.
36. Uppenberg, J.; Hensen, M.T.; Patkar, S.; Jones, T. A. *Structure* **1994**, 2, 293.
37. Haffner, F.; Norin, T. Chem. Pharm. Bull. 1999, 47, 591.
38. Uppenberg, J.; Ohrner, N.; Norin, M.; Hult, K.; Kleywegt, G. J.; Patkar, S.; Waggen, V.; Anthonsen, T.; Jones, T.A. *Biochemistry* **1995**, 34, 16838.
39. Parida, S.; Dordick, J. S. *J. Org. Chem.* **1993**, 58, 3238.
40. Stewart, J. D.; Reed, K. W.; Martinez, C. A.; Zhu, J.; Chen, G.; Kayser, M. M. *J. Am. Chem. Soc.* **1998**, 120, 354.
41. Tashner, M. J.; Black, D.J.; Chen, Q.-J. *Tetrahedron : Asymmetry* **1993**, 4, 1387. 42. Chen, C. S.; Fujimoto, Y.; Giradaukas, G.; Sih, C. J. *J. Am. Chem.. Soc.* **1982**, 104, 7284.

Chapter 27

# Enzymatic Ring Opening Polymerization of ε-Caprolactone in Supercritical $CO_2$

Takahiko Nakaoki[1], Makoto Kitoh[1], and Richard A. Gross[2]

[1] Department of Materials Chemistry, Faculty of Science and Technology, Ryukoku University, Seta, Otsu, Shiga 520-2194, Japan
[2] NSF Center for Biocatalysis and Bioprocessing of Macromolecules, Department of Chemistry and Chemical Engineering, Polytechnic University, 6 Metrotech Center, Brooklyn, NY 11201

Ring opening polymerizations of ε-caprolactone (ε-CL) catalyzed by immobilized lipases were investigated in super critical carbon dioxide ($scCO_2$). The polymerization using Novozyme-435 (immobilized lipase B from *Candida Antarctica*) at 70 °C and 10 MPa in $scCO_2$ provided more than 70 % monomer conversion within 90 min, which is as fast as the reaction in toluene. It was suggested that the reactivity concerns with the dielectric constant of $scCO_2$, which is close to that of toluene. When Novozyme-435 was used for polymerization of ε-CL after incubating over 140 °C in $scCO_2$, the reduced monomer conversion was observed. This showed that Novozyme-435 keeps enzymatic activity in $scCO_2$ below 140 °C without deactivation. The temperature completely losing the enzymatic activity was estimated as 152 °C by extrapolation. This temperature is a little higher than that in toluene. The polymerization of ε-CL in $scCO_2$ was studied as functions of temperature and pressure. When the temperature was varied in the range between 20 °C and 100 °C under the pressure of 10 MPa, in which the Novozyme-435 wasn't deactivated, the apparent reaction rate was the largest at 80 °C.

Even in non-super critical condition below 31 °C, the reaction was promoted, although it was very slow. When the pressure was changed between 1 and 18 MPa under the same condition at 70 °C, the apparent reaction rate kept constant for all pressures. Even in non-super critical state below 7.4 MPa, the reaction was promoted. These results indicate that the super critical state is not an essential factor to promote the polymerization but it depends on only the reaction temperature.

Enzymatic polymerization in organic media has been attracted for not only academic but also industrial interests over the past few years. Investigation on ring opening polymerization of lactones was performed by Kobayashi et. al. (1) Also, Knani et. al. studied ε-caprolactone (ε-CL) ring opening polymerization catalyzed by porcine pancreatic lipase (PPL) and initiated by methanol (2). After these pioneering experiments, investigations on macrolactones with larger ring size, activity of enzyme, immobilized enzyme, and so on have been carried out by many researchers (3-15).

Recently Gross et. al. reported that polymerization of ε-CL catalyzed by Novozyme-435 (immobilized lipase B from Candida Antarctica) was improved by reaction in toluene because of higher log $P$ (16). In situ $^1$H NMR measurement in toluene-$d_8$ was employed to study the effect of reaction temperature on propagation kinetics and average molecular weight. The polymerization conducted at 90 °C gave the most rapid polymerization rate.

The use of super critical fluid as nonaquous solvent for enzyme-catalyzed polymerization has been receiving increased attention as a new polymerization system. In many cases, super critical carbon dioxide (scCO$_2$) is the most popular supercritical fluid, because CO$_2$ has an easily accessible critical point with a critical temperature ($T_c$) of 31 °C and pressure ($P_c$) of 7.4 MPa. In addition, CO$_2$ has apparent advantages over other supercritical fluids, such as low cost, nonflammable, and nontoxicity. However, the low solubility for many reactants in scCO$_2$ is a negative character (16). While scCO$_2$ is a good solvent for most nonpolar molecules with low molecular weight, it acts as a poor solvent for most polymers under comparatively mild condition. In case of poly(methylacrylate) having the molecular weight $10^5$ g/mol, the soluble condition of CO$_2$ requires 200 MPa and 100 °C (17). Since enzymes are generally insoluble in organic media, the reactions were carried out through enzyme suspension and heterogeneous immobilized enzyme systems. Okahata et. al. reported that a lipid-coated enzyme soluble in organic media acts as an efficient catalyst (*18*). They applied this enzyme to the triglyceride synthesis in $scCO_2$, and obtained better reactivity than organic media such as benzene.

In this investigation, we compared the enzymatic polymerization of ε-CL in $scCO_2$ with that in organic media. The reactivity of Novozyme-435 was also reported as functions of reaction temperature and pressure in $scCO_2$.

## Experimental

**Samples.** ε-CL and toluene-$d_8$ were purchased from Wako Co. Ltd. Novozyme-435, which is immobilized *Candida Antarctica Lipase* B (CALB) on acrylic resin, was provided from Novozyme Japan. Accurel and QDE 2-3-4 immobilized CALB on polypropylene were gifts from Novo Nordisk. The characterization of these immobilizaed lipases is listed in Ref. 19.

**Polymerization of ε-CL in $scCO_2$.** The apparatus was specially designed to investigate reactions in $scCO_2$ in a reactor with a reaction cell volume of 1.0 ml. 0.2 mL of ε-CL and 20 mg of Novozyme-435 were introduced within the reactor. After sealing, pressure was added by pumping liquid $CO_2$ to the desired final pressure. The reactor was thermostated in silicone oil bath. After reacted for given time, the reactions were terminated by adding an excess of chloroform and removing the enzymes by filtration. The solution was pored into methanol and the white precipitation was obtained. The precipitate was isolated by filtration and dried in a vacuum oven. Then the monomer conversion was estimated by weight. When the reactions were carried out by using Accurel and QDE 2-3-4, the same amount of CALB on matrix as Novozyme-435 which was calculated by protein content of each sample was used for the reaction.

**Instrumental metods.** $^1H$ NMR measurements were performed on Bruker Biospin DPX 400 NMR spectrometer. The chemical shift was referenced relative to tetramethylsilane (TMS) at 0 ppm. In situ measurement was carried out as follows. Novozyme-435 (12 mg) and toluene-$d_8$ (0.6 ml) were put into NMR tube and they were incubated at 70 °C. Then ε-CL was added by syringe. The measurements were carried out after every 5-7 minutes. After every measurement, the tube was taken out, shaken well, and put in the NMR probe. This procedure was quickly performed to avoid fluctuation in the reaction temperature.

Infrared spectra were performed on a JASCO FT-IR 660 plus spectrometer equipped with a DTGS detector. Resolution and accumulation time were 2 $cm^{-1}$ and 256, respectively. Water content was measured by Kyoto Electronics MKC-510N. Novozyme-435 after treating in vaccuo for 24 h included water of 2.0 wt%.

## Results and Discussion

### Enzymatic Activity in scCO$_2$ in Comparison with Toluene

Recently lots of investigations have been reported for enzymatic reaction in scCO$_2$ instead of organic media. First the activity of Novozyme-435 for polymerization of ε-CL in scCO$_2$ was compared with that in toluene. In order to compare it in the same condition, the ratio of Novozyme-435, ε-CL, and reaction media was fixed to 0.1/1/5 (w/v/v), and the reaction temperature was adopted 70 °C. The monomer conversion was plotted against reaction temperature as shown in Figure 1. As described in previous paper, the signals of -OCH$_2$- for monomer and polymer provide the different chemical shift (7). So we can estimate the monomer conversion taking a ratio of integral intensity of monomer and polymer. The result unexpectedly provided almost the same reactivity between scCO$_2$ and toluene. One of the significant results is that Novozyme-435 kept the enzymatic activity in scCO$_2$. Secondary the activity in scCO$_2$ is as high as in toluene. So far, some investigation on esterification in scCO$_2$ of low molecular weight compounds have been carried out, and the reactivity was reported to be higher than that in organic solvent (20, 21). However our experimental result showed that the monomer conversion depending on time was comparable. A possible explanation is low solubility of polycaprolactone (PCL). With increasing reaction time, molecular chains having larger molecular weight precipitated, and as a result the propagation rate would be reduced.

Gross et. al. revealed that the organic media associated with larger log $P$ tends to provide the higher monomer conversion (7). At this time, there is no report on log $P$ for scCO$_2$ so that the another parameter should be used for comparison. Catoni et. al. investigated the esterification by *Pseudomonas sp.* immobilized on an ACR-silica gel as functions of dielectric constant, Hildebrand solubility parameter, and log $P$ (20). They found the linear relationship between monomer conversion and log $P$. With respect to the dielectric constant, the smaller values less than 5 such as chloroform, toluene, benzene and so on. gave higher monomer conversion, although there is no linear relationship between dielectric constant and monomer conversion. The dielectric constant of scCO$_2$ at 70 °C and 10 MPa takes 1.13 (22), whereas that of toluene is 2.44. Both values take less than 5. Therefore, one of the reasons to promote the polymerization in scCO$_2$ is considered to take low dielectric constant of scCO$_2$. Another reason to enhance the polymerization is perhaps due to the molecular modification of ε-CL under the condition of high pressure. Since ε-CL takes seven membered cycle, the ring is considerably strained. In Figure 2 shows infrared spectra of ε-CL in liquid state and in scCO$_2$ at 50 °C and 10 MPa. These spectra reflect the local vibrational modes because there is no molecular symmetry like solid state. As a result, the number of frequencies was almost the same, but the peak intensity and wavenumber were different between them.

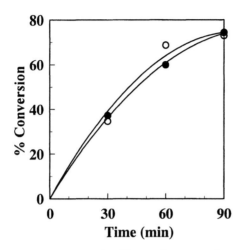

*Figure 1. Monomer conversion of ε-CL as a function of reaction time in toluene ( • ) and scCO₂•( • ). All reaction conditions are the same except reaction media used.*

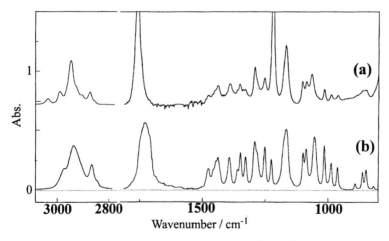

*Figure 2. Infrared spectra of ε-CL in $scCO_2$ at 50 °C and 10 MPa (a) and in liquid state (b).*

This indicates that ε-CL takes different molecular modification in $scCO_2$. Perhaps the high pressure allowed the strained structure open with ease.

### Enzymatic Activity Incubated at High Temperature in $scCO_2$

Next the enzymatic activity was investigated for Novozyme-435 incubated over 100 °C in $scCO_2$. As we reported previously, the enzymatic activity of Novozyme-435 in xylene began to lose activity over 110 °C and then it was completely lost at 138 °C (12). The same experimental procedure was applied to the system in $scCO_2$. After incubating the Novozyme-435 in $scCO_2$ for 2 h at given temperature, in situ $^1H$ NMR spectra were measured for polymerization of ε-CL in toluene-$d_8$. The result was shown in Figure 3. Since this type of measurements include experimental error, the measurements were repeated more than twice and error bars were added in the Figure. For Novozyme-435 incubated at high temperature, the monomer conversion was reduced with incubated temperature. In order to make clear the degree of deactivation, we adopted the relative monomer conversion normalized by the monomer conversion obtained from non-incubated Novozyme-435. In Figure 4 plotted the relative monomer conversion at reaction time of 30, 60, and 90 minutes against incubated temperature. Even for the enzyme incubated at 140 °C, the relative monomer conversion kept more than 90 %. This indicates that the enzyme wasn't deactivated at this temperature. However, the reactivity decreased rapidly above 140 °C. The temperature extrapolated to zero relative conversion, which corresponds to the temperature that the enzymatic activity was completely lost, was evaluated as 152 °C. Previously we reported 138 °C for the deactivation temperature of Novozyme-435 in xylene (19). Both the onset temperature to deactivate and the extrapolated temperature to zero reactivity in $scCO_2$ became slightly higher. This indicates that the deactivation receives little influence from high pressure, but the temperature is the main factor for deactivation.

### Matrix Effect for Enzymatic Activity in $scCO_2$

It has been well-known that the enzymatic activity is enhanced by immobilization on matrix and it depends on hydrophobicity of matrix. We investigated the seven different matricies for ring opening polymerization of ε-CL, and found that Accurel and QDE 2-3-4 immobilized on polypropylene provided higher activity than Novozyme-435 (19). The enhancement factors for QDE 2-3-4 and Accurel were 3.1 and 2.1, respectively. Therefore we applied these immobilized lipases to the polymerization in $scCO_2$. The reaction was conducted at 70 °C and 10 MPa in $scCO_2$ and then the monomer conversion was estimated in $CDCL_3$ solution by $^1H$ NMR. Figure 5 shows the monomer conversion versus reaction time. The conversion after 60 min was about 70 % for Novozyme-435, whereas the conversions for QDE2-3-4 and Accurel

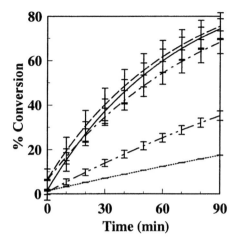

*Figure 3. Monomer conversion as a function of reaction temperature for a series of polymerization using Novozyme-435 incubated at different temperature.*

—— : non-incubated,--- : 130 °C; ---- : 140 °C; --- : 145 °C; ...... : 150 °C.

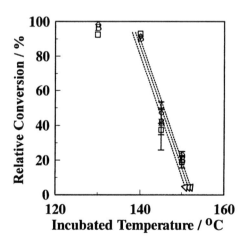

*Figure 4. Relative monomer conversion depending on incubated temperature. The relative monomer conversion was normalized by the conversion of non-incubated Novozyme-435. The relative monomer conversions were evaluated at the reaction time of 30, 60, and 90 min. • : 30 min, • : 60 min, • : 90 min.*

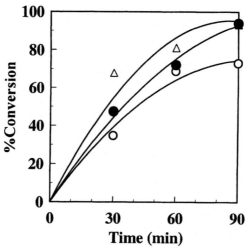

*Figure 5. Monomer conversion depending on reaction time for polymerization using different immobilized CALB in $scCO_2$ at 70 °C and 10 MPa.* ● : Accurel, ● : QDE2-3-4, and ● : Novozyme-435.

provided higher monomer conversion. This trend was the same as the reaction in toluene. When the time reached to 50 % conversion was compared among these samples, QDE 2-3-4 and Accurel were 1.6 and 1.2 times faster than Novozyme-435. The enhancement factors are not so high in comparison with the reaction in toluene. The poor solubility of PCL in $scCO_2$ would contribute to reduce the polymerization rate. Since there is no significant influence for the molecular structure of enzyme and surface structure of matrix in $scCO_2$, the enzymatic activity would be kept at high pressure.

## Enzymatic Activity in $scCO_2$ under Different Conditions

The enzymatic activity strongly depends on temperature. In case of reaction using Novozyme-435 in toluene, the polymerization rate of ε-CL was the fastest at 90 °C (7). Since Novozyme-435 can be used below 140 °C without deactivation in $scCO_2$ as described above section, the reactivity was investigated as a function of temperature below 100 °C. The initial slope of the percent monomer conversion versus time plots was used to calculate the apparent rate constant ($k_{app}$). Figure 6 shows $k_{app}$ depending on reaction temperature. The reaction provided the fastest rate at 80 °C, which is almost the same as the reaction in toluene. Even for the reaction condition at 20 °C lower than critical temperature, the enzyme catalyzed the polymerization of ε-CL although the reaction rate was very slow. This indicates that super critical state is not required for polymerization of ε-CL catalyzed by Novozyme-435. The dielectric constants at 10 MPa take between 1.54 and 1.12 at 10 °C and 80 °C, respectively. This range is not so wide that the reactivity wouldn't be affected. Rather the reaction temperature would be a dominant factor to enhance the polymerization.

Next subject is to make clear the dependence of pressure. Figure 7 shows the values of $k_{app}$ depending on pressure. In these experiments, the reaction temperature was fixed at 70 °C. As can be seen in the Figure, $k_{app}$ provided almost constant value independent of pressure. Even for the pressure below 7.4 MPa, which corresponds to the critical point of $CO_2$, the reaction rate was almost the same as that observed over 10 MPa. This result also supported that super critical state isn't required for the polymerization of ε-CL. Since the dielecric constants at 2 and 18 MPa take between 1.02 and 1.36 at 70 °C, respectively (22), the small change of dielectric constant would be independent of reaction. From these experiments as functions of temperature and pressure in $scCO_2$, it is concluded that the reaction temperature is the most important factor to enhance the polymerization.

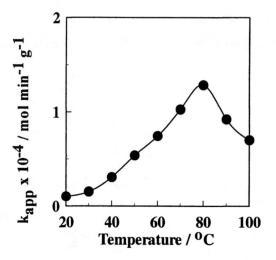

*Figure 6. Plots of apparent rate constant ($k_{app}$) depending on reaction temperature in $scCO_2$ at 10 MPa.*

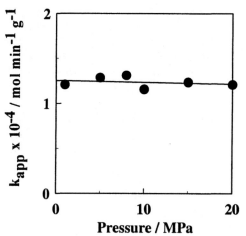

*Figure 7. Plots of apparent rate constant ($k_{app}$) depending on pressure in $scCO_2$ at 70 °C.*

## Conclusions

Enzymatic activity of immobilized CAL in scCO$_2$ was investigated by ring opening polymerization of ε-CL. When the reaction in scCO$_2$ was compared with toluene, the enzymatic activity was comparable between them. This indicates that Novozyme-435 kept enzymatic activity in scCO$_2$. The dielectric constants of scCO$_2$ and toluene are close so that the polymerization would be promoted in a similar degree. In addition, infrared spectrum of ε-CL in scCO$_2$ is different from that in liquid state, reflecting that the different molecular modification. Perhaps high pressure allowed the strained ring structure open the lactone ring.

When Novozyme-435 was incubated at 10 MPa in scCO$_2$, the enzyme started to lose activity over 140 °C. The temperature extrapolated to conversion zero was evaluated as 152 °C. This temperature corresponds that the enzymatic activity is completely lost. Compared with the reaction in organic media such as xylene, the deactivated temperature in scCO$_2$ was a little higher.

Novozyme-435 is immobilized on acrylic resin. In general, the matrix is an important factor to enhance the reactivity. Accurel and QDE 2-3-4, which are immobilized on polypropylene, provided better activity than Novozyme 435 in scCO$_2$. This result was the same trend as the reaction in toluene.

The polymerization of ε-CL in compressed and super critical CO$_2$ was studied as a function of reaction temperature. With increasing reaction temperature, the apparent reaction rate $k_{app}$ became larger, and then 80 °C was the most preferred reaction temperature for polymerization in scCO$_2$. Even for the temperature below critical temperature, the reaction was promoted. This indicates that the super critical condition is not required to promote polymerization. When the pressure was varied between 1 and 18 MPa under the constant temperature at 70 °C, the reaction rate $k_{app}$ was unchanged at any pressures. Even at 1 MPa, which is lower than critical pressure, the reaction was promoted. From these results, it was concluded that an important factor for enzymatic polymerization of ε-CL in scCO$_2$ is not the super critical condition but the reaction temperature.

## Acknowledgement

This work was financially supported by the High Tech Research Center at Ryukoku University.

## References

1. Uyama, H.; Kobayashi, S. *Chem. Lett.* **1993**, 1149.

2. Knani, D.; Gutman, A. L.; Kohn, D. *J. Polym. Sci., Part A: Polym. Chem.* **1993**, *31*, 1221.
3. MacDonald, R. T.; Puapura, S. K.; Svirkin, Y. Y.; Gross, R. A.; Kaplan, D. L.; Akkara, J.; Swift, G.; Wolk, S. *Macromolecules* **1995**, *28*, 73.
4. Svirkin, Y. Y.; Xu, J.; Gross, R. A.; Kaplan, D. L.; Swift, G. *Macromolecules* **1996**, *29*, 4591.
5. Henderson, L. A.; Svirkin, Y. Y.; Gross, R. A.; Kaplan, D. L.; Swift, G. *Macromolecules* **1996**, *29*, 7759.
6. Dong, F.; Gross, R. A. *Int. J. Biol. Macromol.* **1999**, *25*, 153.
7. Kumar, A.; Gross, R. A. *Biomacromolecules* **2000**, *1*, 133.
8. Kumar, A.; Gross, R. A. *J. Am. Chem. Soc.* **2000**, *122*, 11767.
9. Kumar, A.; Kalra, B.; Dekhterman, A.; Gross, R. A. *Macromolecules* **2000**, *33*, 6303.
10. Uyama, H.; Takeya, K.; Hoshi, N.; Kobayashi, S. *Macromolecules* **1995**, *28*, 7046.
11. Uyama, H.; Takeya, K.; Kobayashi, S. *Bull. Chem. Soc. Jpn.* **1995**, *68*, 56.
12. Uyama, H.; Kikuchi, H.; Takeya, K.; Hoshi, N.; Kobayashi, S. *Acta Polym.* **1996**, *47*, 357.
13. Namekawa, S.; Uyama, H.; Kobayashi, S. *Polymer J.* **1998**, *30*, 269.
14. Kobayashi, S.; Uyama, H.; Namekawa, S.; Hayakawa, H. *Macromolecules* **1998**, *31*, 5655.
15. Uyama, H.; Shigeru, Y.; Kobayashi, S. *Polymer J.* **1999**, *31*, 380.
16. Randolph, T. W. *Trends Biotechnol.* **1990**, *8*, 78.
17. Rindfleisch, F.; DiNoia, T.; McHigh, M. A. *Polym. Mater. Sci. Eng.* **1996**, *74*, 178.
18. Mori, T.; Kobayashi, A.; Okahata, Y. *Chemistry Letters* **1998**, 921.
19. Gross, R. A.; Nakaoki, T.; Kumar, A.; Kalra, B.; Kirk, O.; Chritensen, M. *Polymer preprints* **2002**, *43*, 892.
20. Cernia, E.; Palocci, C.; Gasparrini, F.; Misiti, D.; Fagnano, N. *J. Mol. Cat.* **1994**, *89*, L11.
21. Catoni, E.; Cernia, E.; Palocci, C. *J. Mol. Cat. A* **1996**, *105*, 79.
22. Moriyoshi, T.; Kita, T.; Uosaki, Y. *Ber. Bunsenges. Phys. Chem.* **1993**, *97*, 589.

Chapter 28

# Synergies between Lipase and Chemical Polymerization Catalyst

Bhanu Kalra[1], Irene Lai[2], and Richard A. Gross[2,*]

[1]Diagnostic System Laboratories, Incorporated, 445 Medical Center Boulevard, Webster, TX 77598
[2]NSF I/UCRC for Biocatalysis and Bioprocessing of Macromolecules, Othmer Department of Chemical and Biological Sciences and Engineering, Polytechnic University, 6 Metrotech Center, Brooklyn, NY 11201
*Corresponding author: http://chem.poly.edu/gross

The copolymerization of L-lactide with ω-pentadecalactone using a combination of a biocatalyst (lipase) and an organometallic system was investigated. The potential for the catalysts to work together, providing synergy and avoiding deactivation of either the chemical catalyst/enzyme by combination of the systems has been studied. This enables the formation of copolymers from two monomers, which could not be copolymerized either by enzyme or by a chemical catalyst alone. We have shown that PDL can be copolymerized with lactide using Novozyme-435 and a chemical catalyst (i.e., tin octanoate, tin triflate, zinc powder, aluminum catalysts) to obtain block copolymers. Furthermore, DMAP has been used in combination with Novozyme-435 to produce PDL-*co*-lactide.

Prior to application of biocatalysis to the synthesis of poly-ω-pentadecalactone (poly(PDL)) little attention had been paid to the physical properties and potential applications of this long repeating unit polyester, owing to the low molecular weight of the polymer obtained by classical polymerization methods. Recent lipase-catalysed ring-opening polymerization of ω-pentadecalactone has yielded high molecular weight PPDL,[1] whose solid-state characterization was reported elsewhere.[2] PPDL is a crystalline polymer that melts close to 100°C and has a glass transition below room temperature. The physical properties of the polyester, which contains 14 methylene units per ester group, are somehow intermediate between those of poly-ε-caprolactone (PCL) and of polyethylene (PE). The interest towards PPDL is mainly associated with its good mechanical properties and to the presence along the polymer chain of hydrolyzable ester linkages. These characteristics suggest consideration of the polyester as a biodegradable material for diversified purposes, including biomedical applications.

Polylactides, PLAs, and their random or block copolymers, have been used for applications as drug delivery systems and medical sutures.[3,4] Early studies using poly(L)-lactide, [L]-PLA, showed that it was readily resorbed by living tissues.[5] However, the high crystallinity and low hydrophilicity of [L]-PLA decreased its degradation rate and resulted in poorer soft tissue compatibility.[5] An important strategy to fine-tune PLA physical and biological properties is by copolymerization of lactide with other monomers such as other lactones and epoxides.[6] A biodegradable suture, commercialized under the trademark name of Vicryl, is one example of random copolymers of lactide with a comonomer (in this case glycolide) that gave bioresorbable fibers with useful physical and biological properties. Also, block copolymers of ε-caprolactone (ε-CL) and lactides (LA) results in a material that benefits from both the excellent permeability of PCL and the relatively more rapid degradation rate of PLA.[7,8] Furthermore, block copolymers are well known to be useful as emulsifiers for the compatabilization of the corresponding homopolymer blends.[8] That strategy is useful to improve the phase morphology, the interfacial adhesion, and consequently, the ultimate mechanical properties of immiscible polymer blends.[9,10] Block copolymerization of ε-CL and (D,L)-LA was reported by Feng and Song using bimetallic (Al, Zn) μ-oxo alkoxides as initiators.[11] Diblock copolymers based on poly(ε-caprolactone) and poly(L,L or D,L) lactide by aluminium alkoxides has also been reported.[12] More recently, the controlled polymerization of DL-lactide and ε-CL was reported with structurally well-defined alkoxo-bridged di and triyttrium(III) complexes.[13]

Lactide is efficiently polymerized by chemical catalysts based on tin(II) octonoate,[14,15] zinc powder or zinc lactate,[16,17] and aluminum based catalysts.[18,19] However, thus far, the synthesis of polylactide (PLA) has proved difficult using lipase catalysts. In contrast, lipase-catalyzed polymerization of macrolactones proceeds efficiently[20-25] but corresponding polymerizations of macrolactones with traditional organometallic and anionic catalysts are relatively slow and inefficient.

Since the copolymerization of ε-CL and LA resulted in biomaterials with interesting properties (see above), it occurred to us that copolymerizations of LA with lactones of larger ring size would be a useful area of study. Substitution of ε-CL with similar copolymer contents of macrolactones would give more hydrophobic aliphatic repeat units. Furthermore, the thermal transitions of PCL and the homopolymer from ω-pentadecalactone (PDL) are very different. PCL and poly(PDL) have peak melting temperatures of 58-63 and 97°C, respectively, and glass transition temperatures of -66 to -60 and -27°C, respectively.[2] Thus, one motivation for the work described herein was to identify catalysts and polymerization conditions that could provide high molecular weight lactide/PDL copolymers in a variety of compositions and sequence distributions. Once available, the relative merits of lactide/PDL versus lactide/CL biomaterials would be studied.

Since no single catalyst is known that can efficiently polymerize both PDL and lactide, this paper probes whether $Sn(Oct)_2$ and immobilized Lipase B from *Candida antarctica* (Novozyme-435) can function in the presence of each other. Of interest were the following: i) does $Sn(Oct)_2$ inhibit the activity of Novozyme-435?, ii) does Novozyme-435 interfere in any way with the catalytic activity of $Sn(Oct)_2$? iii) is there any advantage to having both $Sn(Oct)_2$ and Novozyme-435 together in a one-pot reaction? These studies led to an effective method for the copolymerizations of [L]-LA and PDL. Furthermore, this work creates a basis for the further exploration of organometallic and lipase-catalyst combinations where special synergies between the catalysts may be realized.

# Experimental

### Material and Methods

ω-Pentadecalactone, tin octanoate ($Sn(Oct)_2$), tin triflate, isobutyl aluminum, zinc powder, dimethyl aminopyridine and toluene were purchased from Aldrich Chemical Company, Inc. Novozyme-435 (specified activity 7 000 PLU/g) was a gift from Novozymes. L-lactide was obtained from Johnson and Johnson Toluene was dried over calcium hydride and distilled under nitrogen atmosphere. Stock solution of tin triflate, isobutyl aluminum and aluminum porphyrin was prepared in toluene and used for polymerization. Coulomat A and Coulomat C were purchased from EMScience Co. Novozyme-435, ω-Pentadecalactone and L-lactide were dried in a vacuum dessicator (0.1 mm Hg, 25 °C, 24h).

Preparation of the Initiator: The initiator solution of $Sn(Oct)_2$ was prepared in dry toluene (129.3mg/3mL of toluene).

**Instrumentation**

Proton ($^1H$) and carbon ($^{13}C$) NMR spectra were recorded on a Bruker spectrometer Model DPX300 at 300 and 75.13 MHz, respectively. The chemical shifts in parts per million (ppm) for $^1H$ and $^{13}C$ NMR spectra were referenced relative to tetramethylsilane (TMS) and chloroform as an internal reference at 0.00 and 77.00 ppm, respectively. NMR spectra of poly(PDL-co-49 mol% L-LA) (-O=C-$CH_2$-$CH_2$\{-$CH_2$-$CH_2$-\}$_5$-$CH_2$-$CH_2$-O)-(CO-CH($CH_3$)-O-CO-CH($CH_3$)-) (Product 2, Table 1, $M_n$=51.3×10$^3$ g/mol, PDI [$M_w$/$M_n$] 1.59) was as follows: $^1$H-NMR (CDCl$_3$): δ 5.1 (q, CH, L), 4.35 (q, CHOH, L), 4.08-4.02 (t, O$CH_2$, P), 3.64 (t, HO$CH_2$($CH_2$)$_{12}$), 2.28 (m, CO$CH_2$, P), 1.68-1.52 (m, $CH_3$, L and $CH_2$, P), 1.40 & 1.20 (m, remaining $CH_2$, PDL) ppm, where P=PDL, L=lactide. $^{13}$C-NMR spectra were recorded to determine the relative fractions of diad repeat unit sequences. The spectra were recorded of 8.0% w/w polymer in d-chloroform at 28 °C. The parameters of $^{13}$C-NMR experiments were as follows: pulse width 60 degrees, 18 000 data points, relaxation delay 5.0 seconds, and 14 000-18 000 transients. $^{13}$C-NMR (d-chloroform) δ 175.0 (COOH, P), 174.00 (OCO$CH_2$, P*-P), 169.64 (OCO$CH_2$, P*-L), 70.1 (OCOCH, L*-P), 69.11 (OCOCH, L*-L), 68.22 (OCO$CH_2$, P*-L), 64.60, 64.3 (O$CH_2$, P*-P), 34.40 (OCO$CH_2$, P*-P), 34.20 (OCO$CH_2$, P*-L), 29.50-29.20, 28.60 ($CH_2$), 26.10, 25.90, 25.60, 24.90 (other $CH_2$), 16.80($CH_3$) ppm. Here, P and L are abbreviations for repeat units formed by the ring-opening of PDL and lactide, respectively.

Molecular weights were determined by gel permeation chromatography (GPC) using a Waters HPLC system equipped with model 510 pump, Waters model 717 autosampler, model 410 refractive index detector, and model T-50/T-60 detector from Viscotek Corporation with 500, $10^3$, $10^4$ and $10^5$ Å ultrastyragel columns in series. Trisec GPC software version 3 was used for calculations. Chloroform was used as the eluent at a flow rate of 1.0 mL per minute. Sample concentrations of 0.2 % wt/vol and injection volumes of 100 μL were used. Molecular weights were determined based on a conventional calibration curve generated by narrow molecular weight polystyrene standards obtained from Aldrich chemical company.

Reaction initial water contents (wt-% water) were measured by using an Aqua star C 3000 titrator with Coulomat A and Coulomat C from EMscience. The water (w/w) in reaction mixtures was determined by stirring 53 mg Novozyme-435, 1.68 g toluene and 0.53 g of PDL in coulomat A, in a closed septum container that is part of the instrument, and titrating it against coulomat C. The total water content (wt/wt) in the reactions was ~ 0.8-1.3 %.

Differential scanning calorimeter (DSC) and Thermal Gravimetric Analysis (TGA) were performed using a DSC 2920 differential scanning calorimeter, and a High Resolution TGA 2950 Thermogravimetric Analyzer, commercially available from TA instruments Inc., equipped with a TA 2000 data station; amounts of samples used were between 5.0-10.0 mg and the heating rate of 10°C/minute and a nitrogen purge were applied.

**Procedure**

The polymerization ampules (10 mL) were treated with trimethylsilyl chloride, washed with three 5 mL portions of methanol, dried at 100 °C in an oven and flame-dried while being purged with dried argon. Monomers and Novozyme-435 were transferred into the ampule under inert nitrogen atmosphere. The ampule was degassed by several vacuum-purge cycles.
Dry toluene (2:1 vol/wt of the monomers) and 35 ul of $Sn(Oct)_2$ stock (1/100 by weight of lactide) from a stock solution (in toluene) were subsequently added via syringe under nitrogen into the reaction vial. The ampule was then sealed under nitrogen and placed in an oil bath maintained at a defined temperature for a predetermined reaction time. At the end of the reaction period, the contents of the ampule were dissolved in a minimum amount of chloroform. The chloroform solution was then filtered to remove the enzyme. The chloroform solution was concentrated, added into methanol containing 5% w/w of HCl to precipitate the polymer formed and dissolve the catalyst residue. The precipitate was then filtered, washed with methanol several times and then dried in a vacuum oven (0.1 mmHg, 50 °C, 24 h).

## Results and Discussion

Our laboratory has begun to explore the possibilities of integrating enzymatic and chemical polymerization, preferably in a one pot reaction. The motivation is to incorporate the best attributes of both enzymatic and chemical methods to obtain new material designs that were hitherto fore not available using one of these methods. In this study we describe the results of experiments aimed at the copolymerization of PDL and L-lactide. PDL is efficiently polymerized by the enzyme catalyst Novozyme 435 that consists of Lipase B from *Candida antarctica* physically immobilized on the macroporous resin Lewatit VP 0C. However, efforts to polymerize PDL by organometallic systems have largely been unsuccessful. Furthermore, catalysis by Novozyme 435 and other lipases have thus far been unsuccessful for the conversion of L-lactide to high molecular weight polymer. In contrast, L-lactide is rapidly polymerized using a variety of organometallic catalysts, the most common of which is

Sn(Oct)$_2$. Since copolymers of l-lactide and PDL would be of interest as new bioresorbable medical materials, their copolymerization was studied by using a combination of Novozyme 435 and Sn(Oct)$_2$. The reactions were conducted in various ways such as by changing the order of addition of these catalysts.

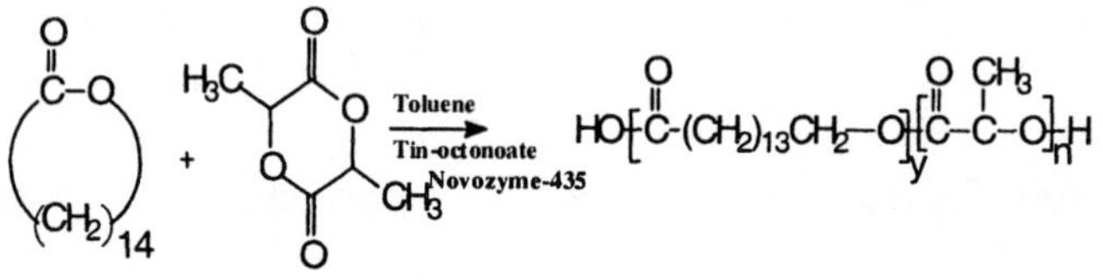

Figure 1. *Copolymerization of ω-pentadecalactone with L-lactide in toluene in the presence of Novozyme-435 and Sn(Oct)2*

The copolymerization of PDL and L-lactide, conducted by having both Sn(Oct)$_2$ and Novozyme-435 in the reaction from t=0, gave poly(PDL-*co*-52%lactide) in 55% (Rxn #1, Table 1). The $M_n$ and PDI of the product were 3140 and 1.43, respectively. In Rxn. 2, PDL was polymerized for 15 min by Novozyme-435 catalysis prior to the addition of L-lactide and Sn(Oct)$_2$ to the reaction. This change in the sequence that the reagents were added to the reaction resulted in an increase in the isolated product $M_n$ from 3140 (Rxn. 1) to 51 300. The randomness number, B for Rxns. 1 and 2, were 0.24 and 0.11, respectively. Analogous to Rxns 1 and 2, the effect of the monomer/catalyst sequence of addition was similar for Rxns 3 and 4 performed at 100 °C. Hence, Rxn 3 where both monomers and catalysts were in the reaction from t=0 gave in <5% yield with $M_n$ and PDI values of 2 630 and 1.34, respectively. In contrast, Rxn 4 performed by first reacting PDL by Novozyme-435 catalysis followed by the addition of L-lactide and Sn(Oct)$_2$ to the reaction gave a poly(PDL-*co*-50%lactide) in 49% yield with $M_n$ and PDI values of 28 800 and 1.62, respectively.

A goal of this work was to identify synergies as well as negative effects of using the two monomer/two catalyst system. Hence, a series of control experiments were performed where either one of the two monomers, one of the two catalysts, or both a monomer and catalyst were omitted from the polymerization reaction. Reaction 5 (Table 1), performed at 70 °C with the omission of PDL and by using two-catalyst system from t=0 gave little product (yield <5%). In contrast, PDL in the presence of the two-catalyst system gave poly(PDL) in 88% isolated yield with $M_n$ and PDI of 50 800 and 1.47, respectively (Rxn 6) Reaction 10 is identical to Rxn. 6, but without $Sn(Oct)_2$ addition. Comparison of the results from Rxns. 6 and 10 show that $Sn(Oct)_2$ does not inhibit but, instead, may augment Novozyme 435 activity for PDL polymerization. The copolymerization of L-lactide and PDL with only $Sn(Oct)_2$ (Rxn 7) resulted in oligomeric copolymeric products. Without PDL or Novozyme-435, the polymerization reaction of L-lactide by $Sn(Oct)_2$ at 70 °C gave <2% isolated product (Rxn 9). Similarly, L-lactide/PDL copolymerization with only Novozyme 435 (Rxn 8) resulted in no precipitated polymer. Analysis by proton NMR of the non-precipitated rxn mixture from Rxn 8 showed that 7% conversion of PDL occurred to give oligomers.

The comparative results of Rxns 6 and 10 show that $Sn(Oct)_2$ did not inhibit PDL polymerization. Comparison of Rxns 1 and 2 as well as 3 and 4 showed that high molecular weight copolymers could be formed by first polymerizing PDL with Novozyme 435 in the absence of L-lactide and $Sn(Oct)_2$. Since $Sn(Oct)_2$ did not inhibit Novozyme-435 catalyzed PDL polymerization, we believe L-lactide inhibits Novozyme-435 catalyzed PDL polymerization. In other words, the formation of low molar mass copolymers in Rxns. 1 and 3 and relatively higher molar mass copolymers in Rxns. 2 and 4 is directly related to the presence or absence of L-lactide in the reactions. Hence, the strategy of first performing Novozyme-435 catalyzed PDL polymerization and, subsequently, adding L-lactide to the reaction, is a way to circumvent L-lactide inhibition. Reactions 7 and 9 show that when Novozyme-435 is absent from the two-catalyst system, $Sn(Oct)_2$ alone at 70°C was ineffective for PDL/L-lactide copolymerization and L-lactide homopolymerization, respectively.

The copolymerization of PDL with L-lactide with Novozyme-435 and different chemical catalysts was performed where, PDL was polymerized with Novozyme-435 for 15 min. at 70 °C in toluene. Then L-lactide and chemical catalyst were added and the temperature of the reaction was changed to preffered temperatures for ring-opening polymerization of lactide with specific chemical catalysts. For eg: the preffered temperatures for $Sn-(Oct)_2$, Sn-triflate and isoButyl Al were 120 °C, whereas the preffered temperatures for zinc powder and DMAP are 140 and 70 °C, repectively. It was observed that with combination of Novozyme-435 and different chemical catalyst, a range of PDL-co-lactide copolymer with varying molecular weights and PDL to lactide compositions were obtained (Table 2).

**Table 1** Copolymerization of PDL and L-lactide in the presence of Novozyme-435 and Sn(Oct)$_2$ at 70-100 °C for 24 hrs in toluene 1:2 wt/vol of monomers

| Rxn | L-Lac /PDL mol/mol | [Sn(Oct)$_2$] /[lac] wt/wt | [Novo-435]/ [PDL] wt/wt | Temp. °C | % ppt Yield | $M_n$ | PDI | B | [P/L] |
|---|---|---|---|---|---|---|---|---|---|
| 1 | 1:1 | 0.01 | 0.1 | 70 | 55 | 3140 | 1.43 | 0.24 | 48/52 |
| 2* | 1:1 | 0.01 | 0.1 | 70 | 62 | 51300 | 1.59 | 0.11 | 47/53 |
| 3 | 1:1 | 0.01 | 0.1 | 100 | N.A | 2630 | 1.34 | N.D | - |
| 4* | 1:1 | 0.01 | 0.1 | 100 | 49 | 28800 | 1.62 | 0.04 | 50/50 |
| 5 | 1:0 | 0.01 | 0.1 | 70 | <5 | N.A | N.D | N.A | N.A |
| 6 | 0:1 | 0.01 | 0.1 | 70 | 88 | 50800 | 1.47 | N.A | N.A |
| 7 | 1:1 | 0.01 | 0.0 | 70 | 0 | <1000 | N.A | N.D | N.A |
| 8 | 1:1 | 0.0 | 0.1 | 70 | 0 | N.A | N.A | N.D | N.A |
| 9 | 1:0 | 0.01 | 0 | 70 | <2 | N.A | N.A | N.A | N.A |
| 10 | 0:1 | 0 | 0.1 | 70 | 80 | 36600 | 1.73 | N.A | N.A |

All Reactions were carried out using toluene as the solvent.

* PDL was polymerized with Novozyme-435 for 15 min at 70 °C, followed by addition of L-lactide and catalyst

$B = (f_{PDL*L}f_{PDL*PDL} + f_{PDL*L})/2F_{PDL} + (f_{L*PDL}f_{L*L} + f_{L*PDL})/2F_L$ where $f$= integral of the corresponding diad signal and $F$ is the mol fraction of the observed ω-pentadecalactone(PDL) or lactide(L) units.

**Table 2 Copolymerization of PDL and L-lactide in the presence of Novozyme-435 and different chemical catalysts for 24 hrs in toluene 1:2 wt/vol of monomers**

| Catalyst | Yields | Temp (°C) | Mn | PDI | B | P/L |
|---|---|---|---|---|---|---|
| Sn(Oct)$_2$ | 62 | 70 | 51300 | 1.59 | 0.09 | 48/52 |
| Sn(Oct)$_2$ | 88 | 120 | 31500 | 2.04 | 0.06 | 38/62 |
| Sn-Triflate | 84 | 120 | 44500 | 1.85 | 0.10 | 42/58 |
| (t-Bu)$_3$Al | 55 | 120 | 28800 | 2.04 | 0.03 | 94/06 |
| Zn Pd | 45 | 140 | 47800 | 1.91 | ND | 49/51 |
| DMAP | 89 | 70 | 26800 | 2.67 | 0.04 | 39/61 |
| DMAP-BuOH | 32 | 70 | 5900 | 1.34 | ND | 16/84 |

The thermal properties of the poly(PDL-co-[L]-LA) were analyzed by thermal gravimetric analysis (TGA) and differential scanning calorimetry (DSC). Studies were performed to determine how the thermal stability and thermal transitions were affected by the lactide composition and randomness of the copolymer. TGA thermograms of the polymers were recorded and the temperatures at which 5, 20 and 50% weight loss occurs as well as the onset decomposition temperatures are displayed in Table 3. Figure 2 shows TGA thermograms for the copolymers isolated from Rxns of PDL and lactide in the presence of Novozyme-435 and different chemical catalyst systems. Copolymers with higher lactide composition for comparable molecular weight copolymers (viz. Al catalyst and DMAP) showed lower thermal stability. Copolymers with higher molecular weight for comparable copolymer compositions (viz. Sn-triflate and DMAP) showed higher thermal stability.

The copolymer degradation onset occurs above 200 °C. PLA thermal degradation occurs about 225 °C. The PPDL thermal degradation occurs above 350 °C and is composed of a main step (I) where 90% of the initial weight is lost followed by a minor loss (II) appearing as a shoulder in the derivative curve. The two degradation events are centered at 425 °C and 475 °C, respectively.[2] Figure 3 shows the DSC thermograms from the first heating scan. Of poly(PDL-co-53 mol%Lac), and the respective homopolymers poly(PDL) and poly(LA).

**Table 3** Thermal decomposition of PDL/L-lactide copolymers synthesized by Novozyme-435/organometallic catalyst system

| Catalyst | Reaction Temp (°C) | Mn | P/L | Weight Loss Temperature (°C) 5    20%    50% | | | Decomposition Temperature (°C) (from Temp. derivative) | | |
|---|---|---|---|---|---|---|---|---|---|
| Sn-Triflate | 70 | 36700 |  | 361 | 413 | 434 |  | 355 | 434<br>479(sh) |
| Sn-Triflate | 120 | 44500 | 42/58 | 289 | 332 | 422 | 248 | 404 | 321<br>359 | 440<br>480(sh) |
| Iso-Butyl Al | 70 | 28600 | 98/02 | 333 | 408 | 432 | 251 | - | 434<br>482(sh) |
| Iso-Butyl Al | 120 | 28000 | 94/06 | 334 | 410 | 433 | 266 | - | 437<br>479(sh) |
| Al-porphyrin | 100 | 29900 |  | 197 | 335 | 422 | 143 | 321 | 434 |
| Zn-pd | 140 | 47800 | 49/51 | 161 | 281 | 422 | 130<br>240 | 397 | 289 | 434<br>476(sh) |
| DMAP | 70 | 26800 | 39/61 | 145 | 285 | 370 | 224 | 371 | 180<br>323 | 436<br>478(sh) |

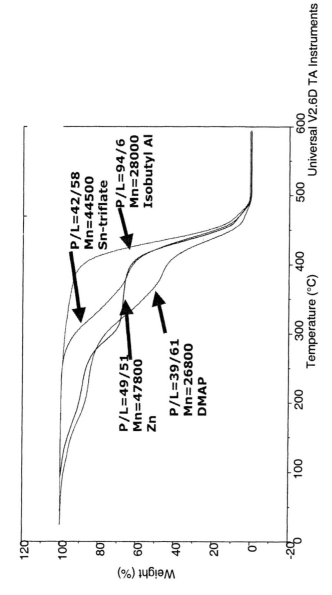

Figure 2. Thermogravimetric Analysis P(PDL-co-LA) copolymer formed by copolymerization of Lactide and PDL (1:1 mol/mol) in the presence of Novozyme-435 and different chemical catalyst systems. A scanning rate of 10 °C/min was used

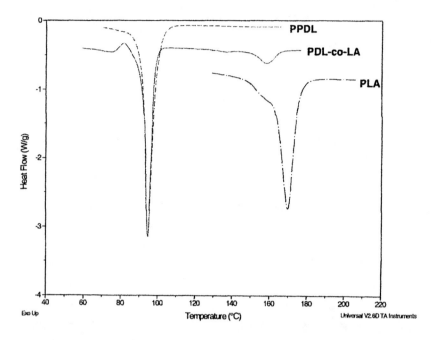

*Figure 3. DSC thermograms recorded during the second heating scan of PPDL, PLA and P(PDL-co-LA) copolymer formed by☐ copolymerization of Lactide and PDL (1:1 mol/mol) at 120 °C in the presence of Novozyme-435 and $Sn(Oct)_2$. A scanning rate of 10 °C/min was used*

## Conclusion

A new method for copolymer synthesis has been developed that involves combining a lipase with a traditional organometallic catalyst. The two catalysts work together to produce a product that would not have been formed by using only the traditional catalyst or the lipase. This concept has broad potential for being expanded to many other combinations of traditional and enzymatic catalysts. This new approach significantly expands the existing capabilities of both traditional and enzymatic systems.

## Acknowledgements

We are grateful to the members of the NSF Center for Biocatalysis and Bioprocessing of Macromolecules at the Polytechnic University for their financial support of this research. We also thank Novozymes A/S for providing us enzyme.

## References

1. Kumar, A.; Garg, K.; Gross, R. A. *Macromolecules* **2001**, *34*, 3527.
2. Focarete, M. L.; Scandola, M.; Kumar, A.; Gross, R. A. *J. Polym. Sci. Part B: Polym. Phys.* **2001**, 39, 1721.
3. Fredericks, R.J.; Melveger, A.G.; Dolegiewitz, J. *J. Polym. Sci., Polym. Phys. Ed.* **1984**, *22*, 57.
4. Eling, B.; Gogolewski, S.; Pennings, A. J. *Polymer* **1982**, *23*, 1587.
5. Bergsma, J.E.; Rozema, F.R.; Bos, R.R.M.; Boering, G.; de Bruijn, W.C.; Pennings, A.J. *Biomaterials* **1995**, *16*, 267.
6. Pitt, C.G.; Marks, T.A.; Schindler, A. *In Biodegradable Drug Delivery Systems Based on Aliphatic Polyesters: Application of Contraceptives and Narcotic Antagonists*; Baker, R., Ed.; Academic Press: New York, 1980.
7. Riess, C.; Hurttrez, C.; Bahadur, P. Block Copolymers *Encyclopedia of Polymer Science and Engineering*, 2$^{nd}$ ed.; Wiley: New York, 1985; Vol. 2, p 398.
8. Song, C.X.; Sun, H.F.; Feng, X.D. *Polym. J.* **1987**, *19*, 485.
9. Fayt, R.; Jerome, R.; Teyssie, Ph. *J. Polym. Sci., Polym. Lett. Ed.* **1981**, *19*, 79.
10. Teyssie, Ph.; Fayt, R.; Jerome, R. *Makromol. Chem., Macromol. Symp.* **1988**, *16*, 41.
11. Song, C.X.; Feng, X.D. *Macromolecules* **1984**, *17*, 2764.
12. Jacobs, C.; Dubois, Ph.; Jerome, R.; Teyssie, Ph. *Macromolecules* **1991**, *24*, 3027.
13. Chamberlan, B.M.; Jazdzewski, B.A.; Pink, M.; Hillmyer, M.A.; Tolman, W.B. *Macromolecules* **2000**, *32*, 3970.
14. Nijenhuis, A.J.; Grijpma, D.w.; Pennings, A.J. *Macromolecules* **1992**, *25*, 6419.
15. Kricheldorf, H.R.; Kreiser-Saunders, I.; Boettcher, C. *Polymer* **19 95**, *36*, 1253.
16. Chabot, F.; Vert, M.; Chapelle, S.; Granger, P. *Polymer* **1983**, *24*, 53.
17. Schwach, G.; Coudane, J.; Engel, R.; Vert, M. *Polym. Bull.* **1996**, *37*, 771.
18. Chen, X.; Gross, R.A. *Macromolecules* **1999**, *32*, 308.
19. Bhaw-Luximon, A.; Jhurry, D.; Spassky, N. *Polym. Bull.* **2000**, *44*, 31.
20. Uyama, H.; Kobayashi, S. *Chem. Lett.* **1993**, 1149.
21. Kumar, A.; Gross, R.A. *Biomacromolecules* **2000**, *1*, 133.

22. Kumar, A.; Kalra, B.; Dekhterman.; Gross, R.A. *Macromolecules* **2000**, *33*, 6303.
23. Nobes, G.A.R.; Kazlauskas, R.J.; Marchessault, R.H. *Macromolecules* **1996**, *29*, 4829.
24. Xu, J.; Gross, R.A.; Kaplan, D.L. Swift, G. *Macromolecules* **1996**, *29*, 3857.
25. Bisht, K.S.; Henderson, L.A.; Gross, R.A.; Kaplan, D.L.; Swift, G. *Macromolecules* **1997**, *30*, 2705.

# Other Examples of Biocatalysis

## Chapter 29

# Polymers from Sugars: Chemoenzymatic Synthesis and Polymerization of Vinylethylglucoside

Bhanu Kalra[1], Mania Bankova[2], and Richard A. Gross[2,*]

[1]Diagnostic System Laboratories, Inc., 445 Medical Center Boulevard, Webster, TX 77598
[2]NSF I/UCRC for Biocatalysis and Bioprocessing of Macromolecules, Othmer Department of Chemical and Biological Sciences and Engineering, Polytechnic University, 6 Metrotech Center, Brooklyn, NY 11201
*Corresponding author: http://chem.poly.edu/gross

Carbohydrates are inexpensive renewable resources that may prove valuable when incorporated into polymers. Of critical importance is to establish efficient routes to synthesize these carbohydrate monomers. To that end, the esterification of ethylglucoside with vinyl acrylate was studied without solvent and in THF. In one example, using the immobilized Lipase B from *Candida antartica* (Novozyme-435) at 35°C, 6-O'-vinylethylglucoside was obtained in 70% yield in 24 h. The monomer was homopolymerized with AIBN to yield water-soluble polymers of $M_n$ 82-90K in 80% yield. Copolymers of 6-O'-vinylethylglucoside with methacrylic acid were also synthesized and studied.

Vinyl monomers from carbohydrates have been prepared by selective enzyme-catalyzed transformations. Motivations for such work have been to: *i)* selectivity place vinyl functional groups at one site of multiple possible positions thus circumventing tedious protection-deprotection steps, *ii*) synthesize polymers from renewable carbohydrate feedstocks, and *iii*) develop a new family of functional water-soluble monomers and polymers. Since Klibanov and coworkers[1] first demonstrated selective monosaccharide acylation catalyzed by lipases, numerous reports have appeared in the literature on enzyme-catalyzed selective carbohydrate acylations and deacylations.[2,3] However, the intrinsic polarity of carbohydrate compounds causes them to be soluble in only polar solvents (e.g. dimethyl sulphoxide, DMSO, dimethylformamide, DMF, and pyridine).[4] In these media, many enzymes lose their activity due to the stripping away of critical or essential water. Consequently, in polar solvents only a small number of enzymes retain some activity that is many folds decreased relative to their use in non-polar solvents.[2,3] Alternative strategies to improve the miscibility between carbohydrates and hydrophobic organic substances have been pursued. Selected examples are as follows: (*i*) the use of solvent mixtures to achieve both satisfactory lipase activities and good solubility of the substrates,[5] (*ii*) the use of organoboronic acids to solubilize carbohydrates by complexation,[6] (*iii*) pre-adsorption of carbohydrates on silica gel,[7] (*iv*) the use of *tert*-butyl alcohol that functions as a bulky polar solvent,[8] (*v*) the prior modification of sugars by alkylation[9] and acetalization. For example, Adelhorst and coworkers[10] performed the regioselective, solvent-free esterification of simple 1-O-alkyl-glycosides using a slight molar excess of melted fatty acids. A range of 1-O-alkyl-6-O-acyl-glucopyranosides were prepared in up to 90% yield, and the process has recently undergone pilot-scale trials by Novo Nordisk. Our laboratory synthesized macromers by lipase-catalyzed ring-opening polymerizations of lactones from the hydroxyl moieties of carbohydrates.[11] In summary, Ethylglucopyranoside (EGP) was used as the multifunctional initiator and ε-caprolactone/TMC ring-opening polymerization was catalyzed by lipases.[11] Selective ring-opening from the 6-hydroxyl position was achieved.

The objective of the present work was to demonstrate efficient routes to sugar-based vinyl monomers based on enzyme selectivity, determine whether 6-O'-acryl glucoside was polymerizable and to synthesize water-soluble acrylic acid-based copolymers that also contain vinyl-sugar repeat units.

# Experimental

**Material and Methods**

*Reagents*

Vinyl acrylate and AIBN were purchased from Aldrich Chemical Company, Inc. Coulomat A and Coulomat C were purchased from EMscience. Novozyme-435 (specified activity 7,000 PLU/g) was a gift from Novozymes. Ethyl glucopyranoside (EGP) was synthesized by a procedure described in ref 10. It is a mixture of α and β anomers.

*Instrumentation*

Proton ($^1$H), DEPT-135 and carbon ($^{13}$C) NMR spectra were recorded on a DPX300 spectrometer at 300 & 75.13 MHz, respectively (Bruker Instruments, Inc.). The $^1$H and $^{13}$C NMR chemical shifts in parts per million (ppm) were referenced relative to tetramethylsilane (TMS) and chloroform as an internal reference.

The absolute molecular weights were measured by GPC-MALLS. The gel permeation chromatographic system consisted of a Waters 510 pump, a U6K model injector, and a Wyatt Optilab DSP interferometeric refractometer coupled to a Wyatt DAWN DSP multi-angle laser light-scattering photometer (Wyatt Technology, Santa Barbara, CA), a three column set consisting of Shodex OHpak KB-80M, KB806M and KB802.5M in series was used to chromatograph the samples. DMF (HPLC grade) with 10mmol LiBr was used as eluent at a flow rate of 1.0 mL/ min. The software used for data collection and processing was ASTRA (supplied by Wyatt Tech.) for GPC-MALLS analysis. Sodium acetate buffer (0.05 M, pH=7.0) was used as eluent for homopolymers of ethyl glucoside. Sodium phosphate buffer (0.1 M, pH=11.0) was used as eluent for copolymers of ethyl glucoseide acrylate with methacrylic acid [Copolymers of ethyl glucoseide acrylate with methacrylic acid were soluble in buffer (pH=12.0). However, after solubilization the pH drops to 10.2-10.4].

*Procedure*

*Novozyme-435-Catalyzed Synthesis of 6-O'-Vinylethylglucoside.* The enzyme-catalyzed 6'-O-acrylation of ethyl glucoside was carried out as follows. Ethyl glucoside (10 g, 48 mmol) and novozyme-435 (2 g) were transferred to dry round bottom flask. The flask was stoppered with a rubber septum and purged with vaccum. vinyl acrylate (11.7 mL, 98 mmol) was added via syringe. The round bottom was the placed in a constant temperature oil bath maintained at 35 °C for 24 h. The reaction was followed by TLC. After 24 h, the mixture was diluted with methanol and the enzyme was filtered off. The resultant solution was concentrated under reduced pressure. The compound was purified by column chromatography on silica gel using chloroform:methanol (98:2, v/v) as the eluent to give gummy solid in 80 % yield. $^{13}$C NMR (d$_6$-DMSO): 165.2 (C=O), 131.3 (=CH$_2$), 128.2 (=CH), 99.1(C1), 73.4 (C-3), 72.0 (C-2), 70.9 (C-4), 70.1 (C-5), 64.1 (OCH$_2$CH$_3$), 63.0 (C-6), 15.6 (OCH$_2$CH$_3$).

*Free Radical Polymerization of 6-O'-Vinylethylglucoside.* The polymerization of ethyl glucoside acrylate was carried out as follows. In a 10-mL sealed polymerization tube, a mixture containing ethyl glucoside acrylate (0.6 g, 2.3 mmol), AIBN (0.0062 g, 0.0.038 mmol) and distilled dry DMSO (1.2 mL) was maintained at 70°C for 60 h. The resulting product was precipitated in acetone. The resultant precipitated material was dried in a vacuum oven (0.1 mmHg, 50 °C, 24 h). DEPT-135 (DMSO): 94.1 (C-1), 71.6 (C-3), 71.0 (C-2), 70.6 (C-4), 69.6 (C-5), 62.3 (O$CH_2$CH$_3$), 61.3 (C-6), 18.8 (OCH$_2$$CH_3$).

## Results and Discussion

α, β-ethyl glucoside was evaluated as the substrate for Novozyme-435 catalyzed synthesis of carbohydrate based vinyl monomers because of its simple synthesis and more rapid reaction kinetics than either methyl glucopyranoside (MGP) or glucose. Unlike, α,β–methyl glucopyranoside (MGP) or its pure α and β–anomers which are solids, EGP as a mixture of α and β–anomers is a viscous oil. This allows us to carry out enzymatic synthesis in solvent free conditions.

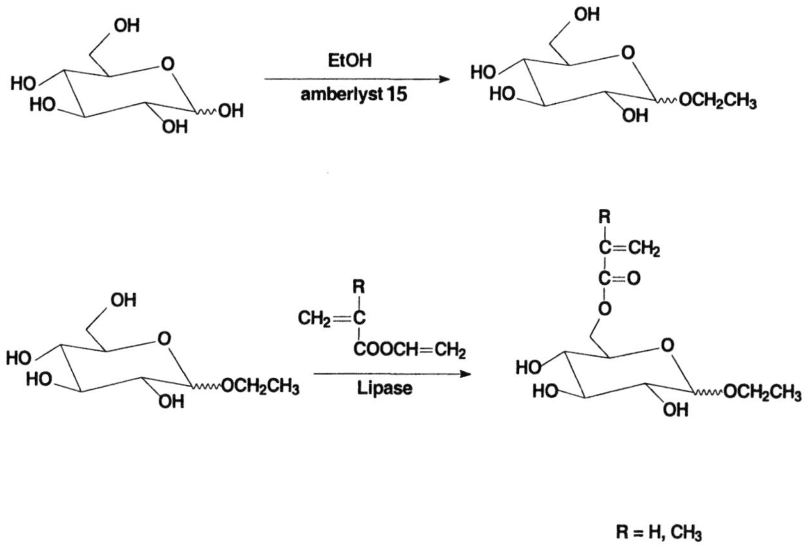

**Scheme 1.** *Chemoenzymatic Synthesis 6-O'-Vinylethylglucoside*

Novozyme-435 was to used to catalyze the synthesis of vinyl ester of EGP starting from EGP and vinyl acrylate in molar ratio of 1:2 at 35 °C in bulk conditions for 24 h. The vinyl acrylate was added in two portions. The reactions were carried out with subsequent application of vaccum. The 6-O'-vinyl ethyl glucoside was obtained as a mixture of pure α and mixture of α and β-anomers. The total yield was 80%. A comparison of the DEPT-135 spectrum of ethylglucopyranoside and the acrylated product showed that for latter, signals corresponding to EGP C-6 α and β anomers were shifted downfield by 1.1 ppm. Furthermore, signals observed at 61.1 and 61.2 ppm were not detected in product. These results indicated that the C-6 primary hydroxyl position of EGP served as site for selective acrylation. This conclusion was supported by an upfield shift for EGP C-5 by 3 ppm from 73.5 to 69.6 ppm (Figure 1).

Homopolymerization of 6-O'-vinyl ethyl glucoside using AIBN was carried out with the α–anomer. AIBN initiated free radical polymerization of 6-O'-Vinylethylglucoside in dimethyl formamide at 70 °C for 60 h did not yield any significant polymer formation. However, dimethylsulfoxide (DMSO) was found to be a suitable solvent for the above polymerization. Hence, homopolymerization and copolymerization of 6-O'-Vinylethylglucoside were performed in DMSO (Table 1). 6-O'-Vinylethylglucoside was polymerized in 82 % yields to a water soluble polymer with $M_n$ and PDI of 90000 and 1.38, respectively. Copolymerizations of 6-O'-Vinylethylglucoside with methacrylic acid in different feed compositions were performed (Table 2). A molar ratio of 6-O'-Vinylethylglucoside to methacrylic acid of 10 to 1 resulted in a water soluble polymer in 25 % yields with $M_n$ and PDI of 49000 and 1.31, respectively. As the methacrylic acid composition in copolymer was increase, the solubility of the copolymers in water was reduced, and the polymers were then soluble in basic pH (12).

**Table 1 AIBN Initiated Free Radical Polymerization of 6-O'-Vinylethylglucoside at 70 °C for 60 h in DMSO**

| ETGVA/AIBN Mol/mol | % Isolated Yields | $Mn$ | $Mw/Mn$ |
|---|---|---|---|
| 60:1 | 82 | 90000 | 1.38 |
| 20:1 | 78 | 82000 | 1.39 |

**Table 2** AIBN initiated Free Radical copolymerization of 6-O'-Vinylethylglucoside with Methacrylic Acid at 70 °C for 48 h in DMSO

| ETGVA:MA | % Isolated Yields | $Mn$ | PDI | Solubility |
|---|---|---|---|---|
| 10:1 | 25 | 49000 | 1.31 | Water |
| 1:20 | 84 | 232000 | 1.71 | Buffer (pH=12.0) |
| 1:50 | 92 | 110000 | 1.87 | Buffer (pH=12.0) |

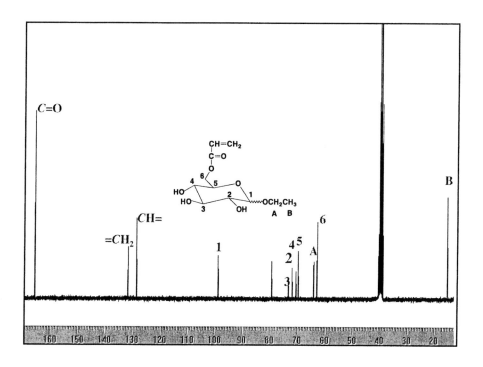

Figure 1. $^{13}$C NMR of ethyl glucoside acrylate (α−anomer)

## Conclusion

An efficient and mild route to sugar based vinyl based monomers has been demonstrated using lipase catalysis. Novozyme-435 catalyzes regioselective acylation of ethyl glucoside at 5'-position in high yields of 80% in solventless conditions in 24 h. Thus, this method overcomes previous limitations of using organic solvents, such as, DMSO, DMF, pyridine etc., long reaction time periods. Further, the vinyl based ethyl glucoside was efficiently homopolymerized and copolymerized to yield water-soluble polymers.

## Acknowledgements

We are grateful to the members of the NSF Center for Biocatalysis and Bioprocessing of Macromolecules at the Polytechnic University for their financial support of this research. We also thank Novozymes for providing us enzyme.

## References

1. Therisod, M.; Klibanov, A. M. *J. Am. Chem. Soc.* **1986**, 108, 5638.
2. Riva, S.; Secundo, F. *Chimica Oggi.* **1990**, 6, 9.
3. Drueckhammer, D. G.; Hennen, W. G.; Pederson, R. L.; Barbas, C. F. III.; Gautheron, C. M.; Krach, T.; Wong, C. H. *Synthesis* **1991**, 499.
4. Moye, C. J. *Adv. Carbohydrate Chem. Biochem.* **1972**, 27, 85.
5. Wang, Y. F.; Lalonde, J. J.; Momongan, M. *J. Am. Chem. Soc.* **1988**, 110, 7200.
6. Kitagawa, M.; Tokiwa, Y. *J. Carbohydrate Chem.* **1998**, 17, 343.
7. Ikeda, I.; Klibanov, A. M. *Biotech. & Bioeng.* **1993**, 42, 788.
8. Sharma, A.; Chattopadhyay, S. *Biotech. Lett.* **1993**, 15, 1145.
9. Oosterom, M. W.; Rantwijk, F. V.; Sheldon, R. A. *Biotech. & Bioeng.* **1996**, 49, 328.
10. Adelhorst, K.; Bjorkling, F.; Godtfredsen, S. E.; Kirk, O. *Synthesis.* **1990**, 112.
11. Bisht, K. S.; Deng, F.; Gross, R. A.; Kaplan, D. L.; Swift, G. *J. Am. Chem. Soc.* **1998**, 120, 1363.

## Chapter 30

# Enzyme-Catalyzed Condensation Reactions for Polymer Modifications

Qu-Ming Gu[1,2] and H. N. Cheng[1,*]

[1]Hercules Incorporated Research Center, 500 Hercules Road, Wilmington, DE 19808-1599
[2]Current address: National Starch and Chemical Company, 10 Finderne Avenue, Bridgewater, NJ 08807

The enzyme-catalyzed condensation is reviewed in view of its use in polymer modification reactions. The formation of esters and amides is most facile when (1) the reaction is carried out in a non-aqueous medium, and (2) the acyl donor contains a good leaving group. However, it is sometimes possible to relax these requirements. Thus, screening and selection of an appropriate enzyme and optimization of the reaction conditions can facilitate esterification and amidation reactions. Water or methanol can be physically removed through the use of vacuum or molecular sieves in order to shift the equilibrium to products and to enhance reaction yield,. For a propitious enzyme-polymer pair, the reaction can proceed even in water without a good leaving group. As illustrations, lipase-catalyzed syntheses are described for the amide of carboxymethylcellulose, substituted acrylic monomers, fatty acid esters of cationic guar, and fatty acid diesters of poly(ethylene glycol).

The use of a lipase to carry out ester synthesis is one of the earliest examples of enzyme-catalyzed reactions in organic media *(1)*. A voluminous amount of papers has been published in the literature, including several reviews and books *(2)*. In the polymer and biomaterials areas, lipases and esterases are also well known and extensively employed *(3)*. This is an active research area, and improved reactions and processes continue to be sought.

Much of the earlier work *(2-4)* entailed acids having activated leaving groups X, where X can be vinyl, trihalomethyl, 2,2,2-haloethyl, or related structures. (R and R' are organic moieties that may be part of a polymer.)

$$RCOOX \text{ (acyl donor)} + R'OH \text{ (acyl acceptor)} \rightleftharpoons RCOOR' + HOX$$

Thus, numerous papers have been published in the literature on lipase-catalyzed condensation polymerizations using activated diacids, such as vinyl esters *(3,4)*. Whereas these polymerizations are facile, they are not commercially viable due to the cost of the activated diacids. The methyl ester has been attempted (X = $CH_3$); however, the reaction tends to be sluggish. The reaction for the carboxylic acid (X = H) is even more sluggish and is only practical with the removal of water through evaporation, azeotropic distillation, or chemical drying *(2a,3)*. Most of these reactions are carried out in non-aqueous media, e.g., in bulk (neat) or in polar aprotic solvents. Recently, much progress has been made on this type of polymerizations *(5)*.

Instead of polymerization reactions, we focus in this work only on *modifications* of polymers and monomers using enzyme catalysis *(6-10)*. Selected aspects of these reactions are pointed out, especially in terms of yield improvement and process optimization. As in polymerization, the modification reactions are most facile in non-aqueous media when the acyl donor contains a good leaving group. Several cases are given herein to illustrate the scope and the applicability of these reactions.

## Results and Discussions

### 1. Fatty acid ester of hydroxyethylcellulose (HEC)

This synthesis has been reported earlier *(7,8)* and represents an optimal case of lipase-catalyzed condensation reaction.

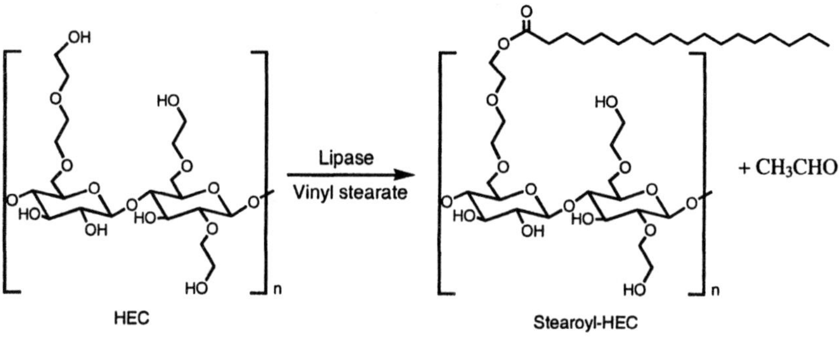

Figure 1. Lipase-catalyzed acylation of HEC

Similar to Figure 1, the palmitoyl (C-16) group has also been enzymatically grafted onto HEC. The reaction entails the incubation of HEC, vinyl palmitate, and a suitable enzyme in N,N-dimethylacetamide (DMAc) at 50°C. Several enzymes were attempted; satisfactory results were obtained with *Pseudomonas fluorescens* lipase (Amano P-30), *Pseudomonas capecia* lipase (Lipase PS), or Alcalase® immobilized alkaline protease from Novozymes A/S as a catalyst. Note that in this case a polar aprotic solvent (DMAc) is used, and one of the reactants, vinyl palmitate, has a facile leaving group (vinyl alcohol which leaves as acetaldehyde).

## 2. The amide derivative of carboxymethylcellulose (CMC)

A systematic study of the enzyme-catalyzed synthesis of CMC amides has been reported earlier *(9)* by incubating an enzyme with CMC and 1,6-hexamethylenediamine or allylamine in N,N-dimethylformamide (DMF). Although a large number of lipases and proteases were screened, the yield of the amide was still found to be 5-25% based on IR analysis.

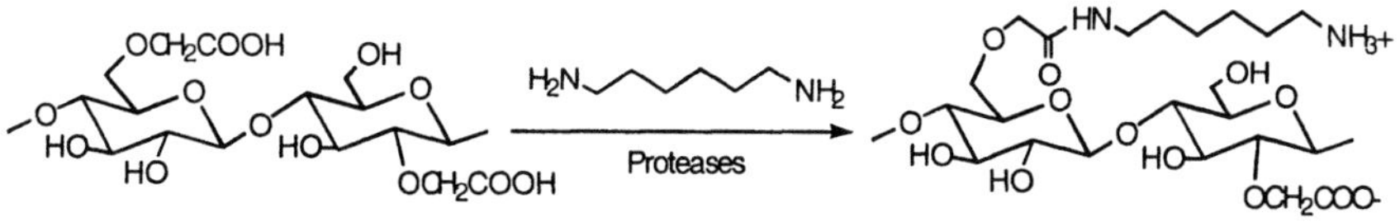

Figure 2. Hydrolase-catalyzed amidation of CMC

Note that carboxyl group of CMC has an unfavorable leaving group (OH). As a result, the reaction did not proceed too far even in a non-aqueous solvent with the help of an enzyme.

## 3. Substituted acrylic monomers

In the literature, the general chemical methodology for the synthesis of acrylic monomers is to react methyl acrylate with an alcohol or an amine that carries a desirable functionality X (Figure 3). In this way, the functionality can be incorporated into the monomer and (after polymerization) into a polymer. Enzymes can be useful catalysts for these reactions because the reaction conditions are usually mild, often give less colored products, and tend to generate less byproducts. However, it has been noted that methyl esters give low yields in such reactions *(2a)*.

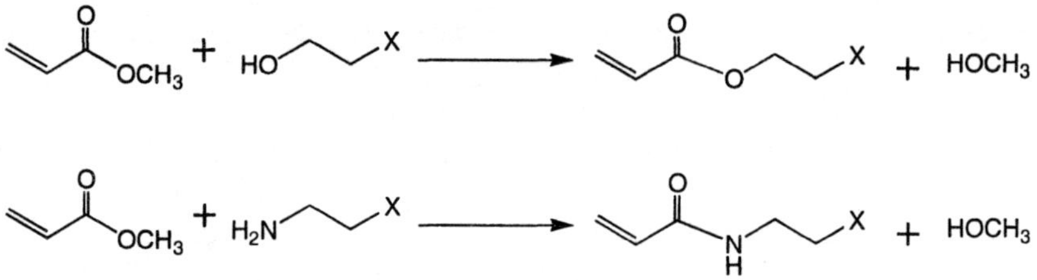

Figure 3. Synthesis of substituted acrylic monomers

The reactions shown in Figure 3 were attempted in our laboratory using Novozym® 435 lipase of Novozymes A/S as a catalyst. However, both reactions appeared to reach equilibrium at a certain point and did not proceed further, thereby leading to low yields. In this case the addition of 4A molecular sieves shifted the equilibrium toward product formation by removing methanol. Thus, after several hours of molecular sieve addition, the yield increased to 69%. In addition, the molecular sieves also eliminated the formation of byproducts due to Michael addition.

Note that in these examples the enzymatic reactions are assisted by the removal of methanol. Whereas molecular sieves are satisfactory for this purpose, the reactions can be equally enhanced through the use of vacuum (*vide infra*).

## 4. Fatty acid diester of poly(ethylene glycol)

The fatty esters of polyethers are known to be good surfactants, and many commercial products are available in the market place. Most people use chemical methods to achieve the required synthesis. Enzymatic methods can also be used; however, the reaction needs to be optimized. An example is the lipase-catalyzed synthesis of fatty acid diester of poly(ethylene glycol) (PEG).

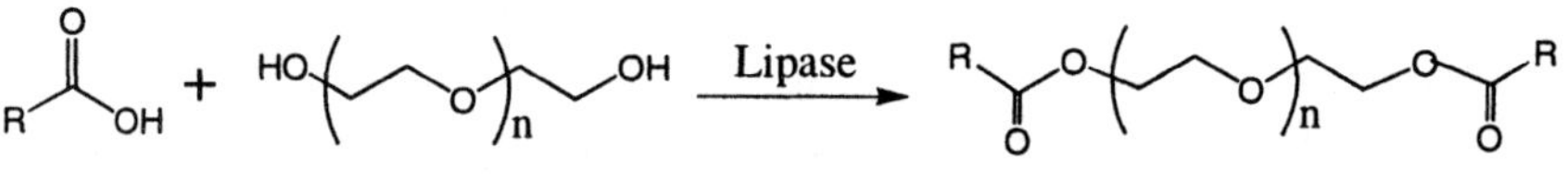

Figure 4. Lipase-catalyzed synthesis of PEG diester

To carry out this reaction, we first screened a number of commercial lipases for substrate selectivity towards the esterification of the fatty acid and poly(ethylene glycol). It was found that lipases from *Candida antarctica* and *Mucor miehei* had a strong preference for the fatty acid. The *Candida antarctica* lipase was chosen because it gave faster rates. We carried out the reaction by adding the enzyme to bulk PEG and fatty acid (at a 1.0 : 1.9 mole ratio). Although we attempted different temperatures, the yield was low, and a mixture of monoesters and diesters was obtained. We discovered that by pulling a vacuum, we could achieve high yields at 50-60°C in 8-48 hours. The final product was isolated by filtration, which also recovered the enzyme for reuse.

The same enzymatic esterification reaction was applied to the enzymatic synthesis of three PEG fatty diesters having PEG molecular weights of 2000, 8000 and 35,000. PEG 2000 fatty ester and PEG 8000 fatty ester are water insoluble, but PEG 35,000 fatty ester is water-soluble. In addition, the PEG 35,000 fatty ester has a much higher solution viscosity than the unmodified PEG in water. At 2%, the Brookfield viscosity of the PEG 35,000 fatty ester is around 340 cps at 30 RPM while the corresponding value for unmodified PEG is less than 3 cps at 30 RPM. Thus, this compound may also be used as a thickener.

## 5. Fatty acid ester of cationic guar

The enzymatic synthesis of this material was reported earlier using vinyl stearate in a reaction similar to Figure 1 *(7)*. The yield was high, more than 90%. Interestingly, this reaction could also proceed, albeit at lower yields, using palmitic acid and cationic guar in an aqueous buffer. The yield could be improved by adding a small amount of DMF to water *(10)*.

The idea originated from our observation that the low-shear solution viscosity of 1% cationic guar increased threefolds when a catalytic amount of *Novolipase* was added. As we investigated this reaction, we realized that cationic guar contained 1-2% fatty acids, which became covalently grafted onto the guar molecules. In contrast, underivatized guar gave no reaction under the same reaction conditions *(10)*.

This synthesis entails the use of unmodified fatty acid and is done in water. Thus, this represents a normally unfavorable case. There is no convenient leaving group, and water is present in abundance. The reason for this reaction to be possible is the interaction between the enzyme and the cationic charge on guar, which favors the ester formation even in the presence of water.

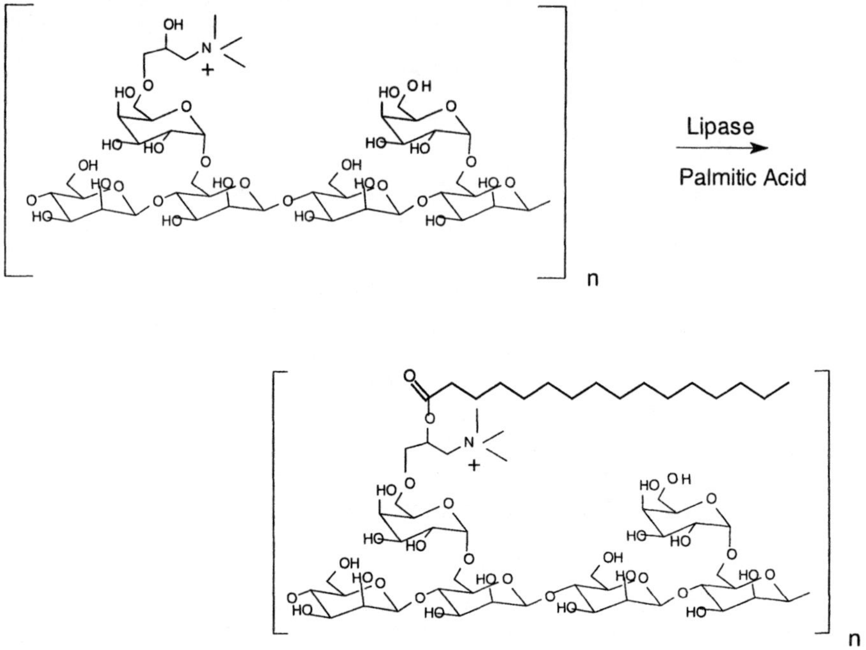

Figure 5. Lipase-catalyzed esterification of cationic guar

## Experimental

CMC, HEC, cationic guar, and fatty acids were all products of Hercules Incorporated. The other chemicals used were from commercial sources (Sigma-Aldrich). The enzymes came variously from Novozymes A/S, Amano Enzyme USA Co. Ltd., Enzyme Development Corp. (EDC), and Sigma-Aldrich.

**Fatty acid ester of hydroxyethylcellulose (HEC).** A slurry containing 40% HEC, 40% vinyl palmitate, and 10% *Pseudomonas fluorescens* lipase (P-30, Amano) in N,N-dimethylacetamide was incubated at 50°C for 1-2 days. The resulting material was treated with acetone followed by washing with isopropanol. The modified HEC was obtained as a white solid. The grafting of palmitoyl group was confirmed by IR analysis (1750 cm$^{-1}$).

**The amide derivative of carboxymethylcellulose (CMC).** The details of this reaction were reported earlier *(9)*. Basically CMC and the amine were dissolved in N,N-dimethylformamide (DMG) with a suitable lipase. Twelve lipases and eight proteases were screened. It was observed that *Pseudomonas sp.* lipase, *Subtilisin* Carlsberg, Papain, and protease from *Aspergillus saitoi* produced low levels of amides when the reactions were carried out with DMF as solvent at 50% concentration. The yields of the amides were estimated to be 5-25% based on IR analysis. No amides were observed with all the enzymes when CMC was suspended in toluene or t-butanol even at elevated temperatures (40-60°C).

**Substituted acrylic monomers.** The reactants (methyl acrylate plus alcohol or amine) were added neat or in a non-aqueous solvent together with Novozym® 435 immobilized lipase from *Candida antarctica* as a catalyst. Molecular sieves (4A) were used to remove water in order to shift the reaction equilibrium to product formation, and also to eliminate side reactions due to Michael addition that was usually enhanced by the presence of water or methanol. Unreacted starting materials were removed by evaporation, and the monomer products obtained without further purification. TLC analysis indicated that the desired products had formed. The purity of the products was confirmed by NMR and IR analysis. The two monomers were successfully polymerized in a separate step.

**Fatty acid diester of poly(ethylene glycol).** The enzymatic esterification was performed in bulk at 50-60°C under vacuum and completed in 8-48 hours. The enzyme use level was 0.1-0.5% based on the substrate. The acid numbers of the ester products depended on the enzyme type, the enzyme use level and the reaction temperature. Progress of each reaction was monitored by TLC analysis (eluted with EtOAc/Hex, 1:3; detected by the phosphomolybdic acid reagent), which showed the disappearance of fatty acid ($R_f = 0.4$) and the emergence of the PEG ester ($R_f = 0$-$0.1$). The enzyme was recovered by filtration. The product was obtained as a yellowish liquid and analyzed without further purification. All product structures were confirmed by $^1H$ and $^{13}C$-NMR analysis.

In a typical example, the fatty acid and the PEG were mixed at a molar ratio of 1.9:1. Novozym 435 lipase (0.5-1 gram per kg substrate) was added. At 60°C for 20 hours under vacuum (10-20mm Hg), the acid number was determined to be 8.5-10. $^{13}C$-NMR indicated the conversion of the acid carbon (178 ppm) to the ester carbon (174 ppm), and $^1H$-NMR spectra were used to quantify the ester formation by integrating the proton signals of $-O$-$C$-$CH_2OH$ (3.3-3.6 ppm), $-O$-$C$-$CH_2OCO$-$R$ (4.0-4.2 ppm) and $CH_3$- of fatty acids and fatty esters (0.70-0.90 ppm).

**Fatty acid ester of cationic guar.** Palmitic acid and cationic guar were dissolved in water buffered at pH 6.0 at concentrations of 0.2% and 2.0%, respectively. The enzyme, *Novolipase*, was used as the biocatalyst. Under these

Table 1. Summary of the enzyme-catalyzed condensation reactions

| Product | Reactants | Leaving Group | Enzyme | Solvent | Other Aids | Yield |
|---|---|---|---|---|---|---|
| *Case 1. Good leaving group, favorable reaction medium* | | | | | | |
| HEC ester | HEC + vinyl palmitate | CH$_3$CHO | Lipase | DMAc | None | High |
| *Case 2. Poor leaving group, favorable reaction medium* | | | | | | |
| CMC amide | NH$_2$(CH$_2$)$_6$NH$_2$ + CMC | H$_2$O | Lipase and protease | DMF | None | Low |
| *Case 3. Mediocre leaving group, favorable reaction medium* | | | | | | |
| Substituted acrylate | Alcohol + Me acrylate | CH$_3$OH | Lipase | Neat or polar aprotic solvents | Mol sieves | High/Moderate |
| Substituted acrylamide | Amine + Me acrylate | CH$_3$OH | Lipase | Neat or polar aprotic solvents | Mol sieves | High/Moderate |
| PEG diester | PEG + fatty acid | H$_2$O | Lipase | Neat | Vacuum | High |
| *Case 4. Mediocre leaving group, unfavorable reaction medium* | | | | | | |
| Guar ester | Cationic guar + fatty acid | H$_2$O | Lipase | H$_2$O, or H$_2$O-DMF (80:20) | None | Moderate |

reaction conditions after 24 hrs, part of the palmitoyl group was grafted to the cationic guar. The formation of ester bond was confirmed by IR analysis. The yield, however, was only moderate.

The yield could be improved by adding a polar aprotic solvent to the solvent medium. Thus, for example, at 40°C in a mixture of water-DMF (80:20) at pH 6.0, cationic guar and palmitic acid produced the corresponding ester in the presence of Novozym 435® lipase in 6-24 hours.

## Conclusions

In this work several hydrolase-catalyzed condensation reactions have been reviewed with respect to polymer modification reactions. In general, it is true that for these enzymatic reactions a non-aqueous medium and a good leaving group for the acyl donor can provide high yields of esters and amides. However, sometimes it is possible to achieve reasonable yields by employing appropriate reaction conditions and using suitable enzymes. At least four situations, as demonstrated, have helped to enhance these reactions: 1) screening and selection of an appropriate enzyme, 2) removal of one of the products such as water or methanol, 3) favorable enzyme-substrate interactions with less water or alcohol activities, and 4) change in the polarity of the solvent medium (e.g., the addition of DMF) that is favorable for the substrates and the products. A summary of the reactions is given in Table 1.

## Acknowledgments

Thanks are due to Sadhana Mital, Gordon F. Tozer, and Arleen J. Walton for technical assistance.

## References

1. For example, (a) Therisod, M.; Klibanov, A. M. *J. Amer. Chem. Soc.* **1986**, *108*, 5638. (b) Klibanov, A. M. *CHEMTECH* **1986**, *16*, 354.
2. For example, (a) Faber, K. *Biotransformations in Organic Chemistry*, 3$^{rd}$ Ed.; Springer-Verlag: Berlin, 1997; p. 309 ff. (b) *Biocatalysis for Fine Chemicals Synthesis*; Roberts, S. M., Ed.; Wiley: Chichester, 1999; module 1:9. (c) *Engineering of/with Lipases*; Malcata, F. X., Ed.; Kluwer: Dordrecht, 1996.
3. (a) Gross, R. A.; Kumar, A.; Kalra, B. *Chem. Rev.* **2001**, *101*, 2097. (b) Kobayashi, S.; Uyama, H.; Kimura, S. *Chem. Rev.* **2001**, *101*, 3793. (c) Cheng, H. N.; Gross, R. A. *ACS Symp. Ser.* **2002**, *840*, 1.

4. Some recent examples include: (a) Linko, Y.-Y.; Seppälä, J. *CHEMTECH* **1996**, *26(8)*, 25., and references therein. (b) Wu, X.-Y.; Linko, Y.-Y.; Seppälä, J.; Leisola, M.; Linko, P. *J. Ind. Microbiol. Biotechnol.* **1998**, *20*, 328. (c) Kline, B, J.; Beckman, E.J.; Russell, A. J. *J. Amer. Chem. Soc.* **1998**, *120*, 9475. (d) Uyama, H.; Inada, K.; Kobayashi, S. *Macromol. Rapid Commun.* **1999**, *20*, 171. (e) Binns, F.; Harffey, P.; Roberts, S. M.; Taylor, A. *J. Chem. Soc., Perkin Trans.1* **1999**, 2671. (f) Park, O.-J.; Kim, D.-Y.; Dordick, J. S. *Biotechnol. Bioeng.* **2000**, *70*, 208. (g) Matsumura, S.; Harai, S.; Toshima, K. *Macromol. Chem. Phys.* **2000**, *201*, 1632. (h) Mesiano, A. J.; Beckman, E. J.; Russell, A. J. *Biotechnol. Prog.* **2000**, *16*, 64. (i) Uyama, H.; Inada, K.; Kobayashi, S. *Polym. J.* **2000**, *32*, 440. (j) Uyama, H.; Inada, K.; Kobayashi, S. *Macromol. Biosci.* **2001**, *1*, 40. (k) Takamoto, T.; Uyama, H.; Kobayashi, S. *e-Polymers* **2001**, *4*, 1. (l) Kim, D.-Y.; Dordick, J. S. *Biotechnol. Bioeng.* **2001**, *76*, 200.
5. For example, (a) Kulshrestha, A. S.; Kumar, A.; Gao, W.; Gross, R. A. *ACS Polymer Preprints* **2003**, *44(2)*, 635. (b) Mahapatro, A.; Kumar, A.; Kalra, B.; Gross, R. A. *ACS Polymer Preprints* **2003**, *44(2)*, 595. (c) Tsujimoto, T.; Uyama, H.; Kobayashi, S. *Biomacromolecules* **2001**, *2*, 29. (d) Kline, B. J.; Lele, S. S.; Lenart, P. J.; Beckman, E. J.; Russell, A. J. *Biotechnol. Bioeng.* **2000**, *67*, 424. (e) Park, O.-J.; Kim, D.-Y.; Dordick, J. S. *J. Polym. Sci., Polym. Chem. Ed.* **2000**, *70*, 208.
6. For example, (a) Yahya, A. R. M.; Anderson, W. A.; Moo-Young, M. *Enzyme Microbial Technol.* **1998**, *23*, 438. (b) Kitagawa, M.; Tokiwa, Y. *Biotechnology Lett.* **1998**, *20*, 627. (c) Cordova, A.; Hult, K.; Iversen, T. *Biotechnology Lett.* **1997**, *19*, 15. (d) Redmann, I.; Pina, M.; Guyot, B.; Blaise, P.; Farines, M.; Graille, J. *Carbohydr.Research* **1997**, *300*, 103. (e) Shibatani, S.; Kitagawa, M.; Tokiwa, Y. *Biotechnology Lett.* **1997**, *19*, 511. (f) Bruno, F. F.; Akkara, J. A.; Ayyagari, M.; Kaplan, D. L.; Gross, R.; Swift, G.; Dordick, J. S. *Macromolecules* **1995**, *28*, 8881. (g) Chen, X.; Martin, B. D.; Neubauer, T. K.; Linhardt, R. J.; Dordick, J. S.; Rethwisch, D. G. *Carbohydr. Polym.* **1995**, *28*, 15. (h) Faber, K.; Riva, S. *Synthesis* **1992**, 895. (i) Hiratake, J.; Yamamoto, K.; Yamamoto, Y.; Oda, J. *Tetrahedron Lett.* **1989**, *30*, 1555. (j) Uemura, A.; Nozaki, K..; Yamashita, J.-I.; Yasumoto, M.; *Tetrahedron Lett.* **1989**, *30*, 248. (k) Degueil-Castaing, M.; De Jeso, B.; Drouillard, S.; Maillard, B. *Tetrahedron Lett.* **1987**, *28*, 953.
7. Gu, Q.-M. *ACS Symp. Ser.* **2002**, *840*, 243.
8. Cheng, H. N.; Gu, Q.-M. *ACS Symp. Ser.* **2002**, *840*, 203.
9. Cheng, H. N.; Gu, Q.-M. *ACS Polymer Preprints* **2000**, *41(2)*, 1873.
10. Gu, Q.-M., poster presented at the 220th National ACS Meeting in Washington, DC, August 20-24, 2000.

# Epilogue: A Rhyme on Enzymes

## H. N. Cheng

Hercules Incorporated Research Center, 500 Hercules Road, Wilmington, DE 19808-1599

The quantity of enzymes is not strain'd;
They fall as generous gifts from heaven
Among us scientists: we are fairly blest
With such friendly tools for our frequent use:
They're green chemistry at its very best,
Where molecules can be cut, trimmed, and fused
In mild conditions with few byproducts,
A tribute to nature's awesome designs,
Wherewith we sharpen our skills and our art;
But enzymes are more than mere devices;
They are involved in life's inner workings:
They bear the fingerprint of God himself;
And bodily functions work like wonders
When enzymes perform their part. Therefore, we,
Though overworked and worn, consider this,
That, at this stage of knowledge, none of us
Would change our calling: we do love enzymes,
And that same love teaches us to enhance
The use of enzymes. I have spoken thus much
To compliment the merits of enzymes;
Let's not forget now the whole cell approach,
Which has played a key role since early times.

A parody on William Shakespeare's *The Merchant of Venice* (Act IV, Scene I), by H. N. Cheng

# Author Index

Amis, Eric J., 343
Bankova, Mania, 420
Bassindale, Alan R., 164
Bentley, William E., 107
Bertozzi, Carolyn R., 96
Bisht, Kirpal S., 366
Biswas, Atanu, 141
Bond, Eric B., 280
Brandstadt, Kurt F., 164, 182
Carlson, R., 292
Chakraborty, Soma, 246
Chen, Fianhong, 107
Chen, Hong, 63
Cheng, H. N., 1, 267, 427, 438
Clarson, Stephen J., 150
Coffin, David R., 119
Conboy, Claire B., 51
Cox, Tony, 37
DeAngelis, Paul L., 232
Elliott, John T., 343
Fishman, Marshall L., 119
Frey, Holger, 354
Fujikawa, Shun-ichi, 217
Gao, Wei, 327, 343
Gitsov, Ivan, 80
Gordon, Sherald H., 141
Govindarajan, Sridhar, 37
Green, Phillip R., 280
Gross, Richard A., 1, 37, 182, 246, 318, 327, 343, 393, 405, 420
Gu, Qu-Ming, 267, 427
Gustafsson, Claes, 37
Henderson, Lori A., 14
Hsieh, You-Lo, 63
Huesing, Nicola, 150

Itoh, Ryosuke, 217
Kalra, Bhanu, 318, 405, 420
Kennedy, Scott B., 343
Kitoh, Makoto, 393
Kobayashi, Shiro, 217
Kondaveti, Leelakrishna, 366
Kulshrestha, Ankur S., 327
Kumar, Ajay, 318, 327, 343
Lai, Irene, 405
Lambrych, Kevin, 80
Lane, Thomas H., 164, 182
Lawton, John W., 141
Li, Hanfen, 192
Li, Kai, 51
Lu, Peng, 80
Mahapatro, Anil, 318
McDermott, Martin K., 107
Mei, Ying, 343
Melik, David H., 280
Miller, Lisa M., 246
Minshull, Jeremy, 37
Mori, Tomonori, 217
Morii, Hidekazu, 217
Nakaoki, Takahiko, 393
Nakas, James, 80
Narasimhan, Karunakaran, 280
Ness, Jon E., 37
Neuner, Ingo T., 354
Noda, Isao, 280
Ochiai, Hirofumi, 217
Ohmae, Masashi, 217
Patwardhan, Siddharth V., 150
Payne, Gregory F., 107
Pederson, E. N., 292
Qiao, Lei, 267

Raab, Christina, 150
Riehle, Richard J., 302
Ryan, Joseph, 80
Sahoo, Bishwabhusan, 182, 246
Satkowski, Michael M., 280
Schechtman, Lee A., 280
Sessa, David J., 141
Shao, Jun, 192
Shiba, Kiyotaka, 150
Small, David A., 107
Song, Jie, 96
Srienc, F., 292
Stewart, Jon D., 366

Tanenbaum, Stuart, 80
Taylor, Peter G., 164
Teraoka, Iwao, 246
Ursu, Mihaela, 354
Wang, Peng George, 192
Wang, Yuhong, 63
Washburn, Newell R., 343
Willett, J. L., 141
Witholt, B., 292
Yi, Wen, 192
Zhang, Bo, 292
Zhang, Hesheng, 192

# Subject Index

## A

ABO blood group system, oligosaccharide structures, 196, 197
$N$-Acetylheparosan, glycosaminoglycan, 233
$N$-Acetyllactosamine structures, glycans, 195, 196
Acrylic monomers, substituted, 429–430, 433
Activity
    enzymes for polyol polyesters, 338
    synthetic protein YT320 in bioinspired mineralisation, 160
    See also Laccase modification; Nitrilase activity; Protein engineering
Adipic acid. See Polyol polyesters; Sorbitol-containing polyesters
Adsorption, immobilization mechanism, 72, 73
Alcalase (Subtilisin enzyme)
    hydrolysis of polymer bound-CPD on polyaminopolyamide-epichlorohydrin (PAE) resin, 307$f$
    PAE resin, 305, 307, 308$f$, 309$f$
    resin performance for microbial dehalogenation and, 313, 314$f$
    See also Polyaminopolyamide-epichlorohydrin (PAE) resins
Alkaline phosphatase, polyacrylic acid-grafted polysulfone, 68
Alkoxysilanes
    trypsin-catalyzed hydrolysis and condensation of, 174–177
    See also Biosilicification
Alloys, polymer, polyhydroxyalkanoates, 289–290
Amide derivative, carboxymethylcellulose, 429, 432–433
Amide formation, pectin, 274
Amino acid variations
    benefits and functions, 45–48
    substrate specificity and physical properties of proteinase K, 40$f$
    See also Protein engineering
Ammonia fiber explosion, lignocellulosic conversion, 22, 23
Ammonia recycle percolation, lignocellulosic conversion, 22, 23
Arabidopsis thaliana
    enantioselectivity of nitrilase from, 53, 54$f$
    See also Nitrilase activity
Aspergillus oryzae, β-galactosidase on cotton, 65
Atomic force microscopy, sorbitol-containing polyesters, 347, 348$f$
Aulacoseira granulata
    scanning electron microscopy image, 151$f$
    See also Biomineralisation

## B

"Bacterial coupling" technology, oligosaccharide synthesis, 200–201
Bacterial lipopolysaccharides (LPS)
    architecture, 205$f$
    biosynthesis of O antigen, 206–207
    biosynthetic pathway of GDP-fucose, 209
    *E. coli* O128 antigen biosynthesis, 207–209
    *E. coli* O128 antigen biosynthesis gene cluster, 209$f$

E. coli O128 antigen repeat unit, 208
*Escherichia coli* O86 antigen
   biosynthesis, 210–212
   $^1$H NMR spectrum of Fucα1,2GalOMe, 210*f*
   model of O-unit assembly, 207*f*
   O antigen genes in *E. coli*, 205–206
   proposed assembly of *E. coli* O86 repeat unit, 211
   structure of *E. coli* O86 antigen, 210
   structure of LPS, 204–205
Baeyer–Villiger oxidation, homochiral lactones, 368
Benzo-α-pyrene (BP)
   biotransformation with laccase/linear-dendritic complexes, 84, 89–92
   experimental, 83–84
   *See also* Laccase modification
Biobleaching, enzymes, 4
Biocatalysis
   enzyme, 2–6
   immobilization, 7, 64
   new enzyme methodologies, 6–7
   opportunities, 30, 32
   use in polymer reactions, 2*t*
   whole cell, 6
   *See also* Biosilicification
Biocatalysts
   discovery and screening, 52
   *See also* Enzymes; Laccase modification; Nitrilase activity; Protein engineering
Biocatalysts, next generation
   enzyme discovery-development process, 32–33
   examples of enzyme-catalyzed reactions, 31*t*
   high throughput screening techniques, 32
   opportunities and needs, 30, 32
   paradox, 32–33
   toolbox, 30, 32

Biocompatibility, polyesters, 344
Biodegradability, polyesters, 344
Biodegradable copolymers. *See* Nodax™
Biodegradable films
   dissolution of films, 122
   dissolution rate of pectin/poly(vinyl alcohol) (PVOH), 135, 138
   dissolution rate of pectin/starch/glycerol films, 129–130
   effect of glycerol and PVOH on storage modulus of films, 131, 133*f*
   effect of glycerol content on loss modulus of pectin/PVOH, 131, 132*f*
   effect of glycerol level and temperature on storage modulus of pectin films, 123, 125*f*
   effect of glycerol level on storage modulus for blends of pectin and high amylose starch, 124, 125*f*
   effect of PVOH content on loss modulus of films, 130, 132*f*
   elongation to break of pectin/PVOH films, 135, 136*f*
   experimental, 121–123
   film preparation, 121–122
   Fourier transform infrared (FT–IR) spectroscopy, 122
   fracture surfaces of pectin and pectin blend films, 124, 126*f*, 127
   initial modulus of pectin/PVOH films, 135, 137*f*
   materials, 121
   mechanical testing, 122
   microscopy, 122
   oxygen permeability data of pectin/starch blends, 129*t*
   oxygen permeability testing, 123
   pectin/PVOH films, 130–138
   pectins, 120
   pectin/starch films, 123–130

polysaccharides, 120
PVOH, 120–121
scanning electron microscopy of freeze-fracture surfaces, 131, 134f, 135
starch, 120
studies, 120–121
typical FT–IR spectra for pectin/starch/glycerol (P/S/G) films, 127, 128f, 129
Bioengineering, enzyme methodology, 7
Bioethanol
 active site of family 2 cellulases, 24, 25f
 activities for developing next generation plant, 21f
 ammonia fiber explosion, 23
 ammonia recycle percolation, 23
 biomass as feedstock, 19
 Biomass Refining Consortium of Applied Fundamentals and Innovations Team, 22, 24
 biomass-to-ethanol production, 26, 29–30
 biomass treatment technologies, 20
 biorefinery, 20
 dilute sulfuric acid, 23
 DNA microarray technique for identifying genes, 26, 28f
 enzyme hydrolysis of lignocellulosics, 24, 26
 market opportunity, 20
 metabolic engineering of organisms, 29–30
 mode of operation of production plant, 26, 29
 National Renewable Energy Laboratories, 20
 pressurized hot wash, 23–24
 pretreatment technologies for lignocellulosic conversion, 22–24
 process integration project, 24
 production efficiencies, 19

*Saccharomyces cerevisiae*, 29
 separate hydrolysis and fermentation, 26, 29
 simultaneous saccharification and fermentation, 26, 29
 strategy for improving cellulases for hydrolysis, 26, 27f
 U.S. Department of Energy (DOE), 20
Biological properties, polyhydroxyalkanoate (PHA) copolymers, 284–286
Biomass
 bioethanol feedstock, 19
 production of ethanol, 26, 29–30
Biomass Refining Consortium of Applied Fundamentals and Innovations Team, 22
Biomaterials
 applications, 1–2
 biomimetic approach, 108
 novel bio-related, 7–8
 *See also* Biomimetic approach to biomaterials
Biomimetic approach to biomaterials
 application potential for tyrosinase-catalyzed gelatin-chitosan gels, 111
 applications of microbial transglutaminase (mTG) catalyzed crosslinking, 116, 117f
 experiment demonstrating grafting of GFP to chitosan, 113
 functional materials, 108
 gel formation of gelatin-chitosan mixture, 110f
 grafting gelatin to chitosan, 109–111
 pH-responsive properties of protein-chitosan, 113–114
 proposed network structure of tyrosinase-catalyzed gelatin-chitosan gels, 111
 thermal behavior of gelatin and chitosan blends, 115f

three-dimensional structure of GFP, 112*f*
transglutaminase catalyzed protein crosslinking, 114–117
tyrosinase-catalyzed protein grafting to polysaccharide, 108–114

Biomineralisation
   activity of synthetic protein YT320 in in vitro, 160
   amino acid primary sequence of synthetic protein YT320, 152*f*
   biosilicification, 152
   description, 151–152
   elongated silica structures using pre-hydrolyzed tetramethoxysilane (TMOS) and YT320, 156*f*
   experimental, 153–155
   fiber-like silica structures using pre-hydrolysed TMOS and YT320, 157*f*
   germania synthesis, 155
   in vitro mineralisation of germania, 158–159
   metal alkoxides and particle formation, 159–160
   proteins facilitating silicification, 151–152
   scanning electron microscopy (SEM) image of diatom *Aulacoseira granulata*, 151*f*
   SEM analysis of silica particles, 157–158
   SEM and EDS (energy dispersive spectroscopy) spectrum for germania, 159*f*
   SEM and EDS of YT320 mediated silification using ethylene glycol modified silane (EGMS), 158*f*
   SEM images of YT320 mediated silica particles, 155*f*
   silica formation and structure control in vivo and in vitro, 153
   silica precursors for silicification reactions, 154*t*
   *See also* Biosilicification; Mineralization strategy

Biominerals. *See* Hydrogel-biomineral composites

BioPreparation™
   cotton preparation, 15, 16, 18
   discovery, 15–19
   idea origination, 15–16
   lyases, 5
   model of interpenetrating networks within cotton cell wall, 17*f*
   polymer degradation by β-elimination reaction, 17*f*, 18

Biosilicification
   acid- and base-catalyzed silanol condensation, 169
   acid-catalyzed formation of aggregated silica gel, 170*f*
   background, 166–168
   complications, 165
   enzyme-catalyzed condensation study, 171–172
   enzyme-catalyzed siloxane bond formation, 165, 166
   experimental, 168, 171
   hydrolysis and condensation control reactions of trimethylethoxysilane, 175*f*
   hydrolysis and condensation reactions during, 165, 166, 168
   inhibition study, 178–179
   polycondensation of silicic acid, 167
   protease-catalyzed condensation study, 172–174
   proteinaceous inhibition of trypsin in hydrolysis and condensation of trimethylethoxysilane, 178*t*
   protein mediated, 152
   trypsin-catalyzed hydrolysis and condensation of alkoxysilanes, 174–177

trypsin-catalyzed hydrolysis and condensation of trimethylethoxysilane, 175f
turnover numbers of trypsin-catalyzed hydrolysis of trimethylethoxysilane and condensation of trimethylsilanol, 176f
*See also* Biomineralisation
Biosynthesis
*Escherichia coli* O86 antigen, 210–212
*See also* Carbohydrate biosynthesis associated enzymes
Biotransformation
"bacterial coupling" technology, 200–201
"living factory" technology, 203
oligosaccharide synthesis, 200–203
"superbeads" and "superbug" technology, 201–202
2,2'-Bis(hydroxymethyl) butyric acid (BHB)
branching comonomer, 357
*See also* Hyperbranched aliphatic polyesters
Bone
description, 97
*See also* Hydrogel-biomineral composites

# C

Calcium, activity of transglutaminase, 114
Calcium phosphate (CP)
morphology, chemical composition and microindentation analysis, 101f
*See also* Hydrogel-biomineral composites
Candida antartica lipase B (CALB)
catalyzed acylation of coated starch nanoparticles, 252–254
coated starch nanoparticles, 250
hydroxyethylcellulose (HEC), 247
incorporation within coated starch nanospheres, 263–264
polyesterification, 319
polyesterification as function of diacid chain length, 323f, 324
*See also* Hyperbranched aliphatic polyesters; Starch nanoparticles
ε-Caprolactone (CL)
acylation of coated starch nanoparticles with, 256–257
reaction of coated starch nanoparticles with, 251
ring-opening polymerization in super critical carbon dioxide, 394–395
*See also* Hyperbranched aliphatic polyesters; Poly(ε-caprolactone) (PCL); Poly(caprolactone)s, substituted; Polylactide (PLA) copolymers; Sorbitol-containing polyesters
Carbohydrate biosynthesis associated enzymes
fucosyltransferases, 194–195
glycosidases, 194
glycosyltransferases, 194–195
Carbohydrates
biosynthesis associated enzymes, 193–195
role in biological science, 193
structures, 197, 198
*See also* Organosilicon carbohydrates; Vinylethylglucoside
Carbon-13 distortionless enhancement by polarization transfer nuclear magnetic resonance (DEPT NMR), organosilicon carbohydrates, 185, 186f
Carbon dioxide, supercritical. *See* Poly(ε-caprolactone) (PCL)
Carboxylic quantity, bound enzymes, 72, 74

Carboxymethylcellulose, amide derivative, 429, 432–433
Catalytic activity
  bound enzymes, 72, 74
  lipases, 71
Cell density, sorbitol-containing polyesters, 351, 352f
Cellobiohydrolases, hydrolysis of lignocellulosics, 24–26
Cell spreading, sorbitol-containing polyesters, 349, 351
Cellulases
  hydrolysis of lignocellulosics, 24–26
  strategy for discovery, 26, 27f
  xanthan, 271–272
Cellulose fibers
  polyethyleneimine immobilization on, 69–70
  ultra-fine, 71
  See also Immobilization of enzymes
Chemical biotechnology
  examples, 30
  toolbox, 30, 32
Chemical polymerization catalyst. See Polylactide (PLA) copolymers
Chemoenzymatic synthesis
  enantio-enriched substituted poly(caprolactones), 382–383, 388
  See also Vinylethylglucoside
Chimeric glycosaminoglycan chains, production, 241
Chitosan
  enzyme-catalyzed grafting of gelatin to, 109–111
  grafting of green fluorescent protein to, 112–114
  thermal behavior of gelatin and chitosan blends, 115f
  See also Biomimetic approach to biomaterials
3-Chloropropanediol (CPD)
  byproduct and regulatory requirements, 302–304
  formation of polymer bound-CPD, 306
  See also Polyaminopolyamide-epichlorohydrin (PAE) resins
Chondroitin (Ch)
  enzymatic polymerization, 218–219
  glycosaminoglycan, 233
  monomer designs, 220–221
  monomer synthesis, 221, 222
  possible monomer design, 220f
  synthesis of natural Ch, 223–224
  See also Glycosaminoglycans (GAGs)
Chondroitin sulfate (ChS)
  synthesis of natural ChS, 224–225
  See also Glycosaminoglycans (GAGs)
Cloning enzymes, proteinase K, 39–41
Composites. See Hydrogel-biomineral composites
Condensation
  acid- and base-catalyzed silanol, 168, 169
  amide derivative of carboxymethylcellulose (CMC), 429, 432–433
  enzyme-catalyzed study, 171–172
  fatty acid diester of poly(ethylene glycol), 430–431, 433
  fatty acid ester of cationic guar, 431, 432f, 433, 435
  fatty acid ester of hydroxyethylcellulose (HEC), 428–429, 432
  substituted acrylic monomers, 429–430, 433
  summary of enzyme-catalyzed reactions, 434t
  See also Biosilicification; Polyesterifications, lipase-catalyzed; Polymer modifications

Contact angles, sorbitol-containing polyesters, 347, 350f
Copolymerization, graft, on polymers and enzyme binding, 67–68
Copolymers. *See* Hyperbranched aliphatic polyesters; Laccase modification; Polylactide (PLA) copolymers
Cotton, enzyme immobilization on tosylated, 65
Cotton preparation
 BioPreparation™, 16, 18
 composition of fiber, 16
 mechanism of polymer degradation, 18
 model of cell wall, 17f
 *See also* BioPreparation™
Covalent binding, immobilization mechanism, 72
Crosslinking, protein, by transglutaminase catalysis, 114–117
Crystallinity
 glycerol and sorbitol terpolyesters, 338
 polyhydroxyalkanoate copolymers, 287–288
 sorbitol-containing polyesters, 347, 349
Cyclic activity, bound lipases, 77
Cytosol of yeast. *See* Polyhydroxyalkanoates (PHAs) synthesis

**D**

Degradation. *See* Biodegradable films
Dehalogenation. *See* Polyaminopolyamide-epichlorohydrin (PAE) resins
Dendritic copolymers. *See* Laccase modification
Department of Energy (DOE), biomass treatment, 20

Diastereoselectivity, nitrilase, 53, 54f
1,3-Dichloropropanol (DPC)
 byproduct and regulatory requirements, 302–304
 microbial dehalogenation, 304
 *See also* Polyaminopolyamide-epichlorohydrin (PAE) resins
Differential scanning calorimetry (DSC)
 hyperbranched poly(ε-caprolactone), 363f
 organosilicon carbohydrates, 187, 188t
 polylactide copolymers, 409, 413, 416f
Diffusion, starch nanoparticles into lipase Novozym 435 beads, 261, 263
Directed evolution, enzyme performance, 26
Dissolution
 pectin/poly(vinyl alcohol) films, 135, 138
 pectin/starch/glycerol films, 129–130
DNA microarray
 enzyme discovery, 26, 27f
 identification of glycosyl hydrolases, 26, 28f
Drugs. *See* Pharmaceutical industry

**E**

Electrospray ionization mass spectrometry (ESI MS), organosilicon carbohydrate products, 185–186, 187f, 189f
Elongation
 native and chemically modified zein, 145–146
 pectin/poly(vinyl alcohol) films, 135, 136f
Empirical biocatalyst engineering. *See* Protein engineering

Enantiomeric purity, racemic lactones, 388–389
Enantioselectivity, lipase for reaction with lactones, 381–382
Energy dispersive spectroscopy (EDS)
  protein-mediated germania, 159f
  protein-mediated silica, 158f
Engineered yeasts. *See* Polyhydroxyalkanoate (PHA) synthesis
Enzymatic polymerization. *See* Glycosaminoglycans (GAGs)
Enzyme Commission (EC) numbers
  classification, 2
  EC1 (oxidoreductases), 4
  EC2 (transferases), 5
  EC3 (hydrolases), 2–3
  EC4 (lyases), 5
  EC5 (isomerases), 5
  EC6 (ligases), 6
Enzymes
  activity for substrates, 268
  activity in supercritical $CO_2$, 398, 399f, 400f
  applications, 6
  biocatalysis, 2–6
  carbohydrate biosynthesis associated, 193–195
  catalyzed condensation study, 171–172
  discovery and improvement, 6
  examples of enzyme-catalyzed reactions, 31t
  experimental for reactions with non-substrate polymers, 275–276
  fucosyltransferases, 194–195
  glycosidases, 194
  glycosyltransferases, 194–195
  hydrolases, 2–3
  immobilization, 7, 64
  isomerases, 5
  ligases, 6
  lyases, 5
  microbial methodologies, 7
  modification of polysaccharides, 247
  oxidoreductases, 4
  pectin and β-galactosidase, 272–273
  polyol polyesters, 328, 338
  reactions of guar, 269–270
  rhyme, 438
  specificity toward substrates, 268
  transferases, 5
  use of biocatalysts in polymer reactions, 2t
  xanthan and cellulase, 271–272
  *See also* BioPreparation™; Biosilicification; Carbohydrate biosynthesis associated enzymes; Glycosaminoglycan synthase enzymes; Immobilization of enzymes; Laccase modification; Poly(caprolactone)s, substituted; Starch nanoparticles
Epichlorohydrin. *See* Polyaminopolyamide-epichlorohydrin (PAE) resins
Escherichia coli
  biosynthesis of O86 antigen, 210–212
  entrapping within crosslinked gelatin network, 116, 117f
  O128 antigen biosynthesis, 207–209
  O128 antigen biosynthesis gene cluster, 209f
  O128 antigen repeat unit, 208
  O antigen genes in, 205–206
  proposed assembly of O86 repeat unit, 211
  structure of O86 antigen, 210
Ester formation, protease, 273, 276
Esterification reaction. *See* Organosilicon carbohydrates
Ethanol. *See* Bioethanol

**F**

Fabric preparation. *See*
  BioPreparation™
Fatty acid esters
  cationic guar, 431, 432*f*, 433, 435
  diester of poly(ethylene glycol),
    430–431, 433
  hydroxyethylcellulose, 428–429,
    432
Fibers
  enzyme immobilization, 65–66
  incorporation of proteins, 70
  ultra-fine cellulose, 71
  *See also* Immobilization of
    enzymes
Films. *See* Biodegradable films
Fluorescence microscopy
  sorbitol-containing polyesters, 349,
    350*f*
Fourier transform infrared (FT–IR)
    spectroscopy
  pectin/starch films, 127, 128*f*, 129
  procedure for films, 122
  *See also* Infrared (IR) spectroscopy
Free radical polymerization,
    methacrylic acid and
    vinylethylglucoside, 424, 425*t*
Fucosyltransferases, carbohydrate
    biosynthesis, 194–195

**G**

Galactose oxidase, polymer
    modifications, 4
β-Galactosidase, pectin, 272–273,
    275–276
Gas chromatography–mass
    spectrometry (GC–MS),
    polyhydroxyalkanoate, 298*f*, 299*f*
Gelatin
  enzyme-catalyzed grafting to
    chitosan, 109–111
  thermal behavior of gelatin and
    chitosan blends, 115*f*
  *See also* Biomimetic approach to
    biomaterials
Gel permeation chromatography
    (GPC)
  number average molecular weight
    of organosilicon carbohydrates,
    186
  *See also* Molecular weight
    characterization
Gels. *See* Biomimetic approach to
  biomaterials
Germania
  in vitro mineralisation, 158–159
  synthesis, 155
  *See also* Biomineralisation
Glycans
  ABO blood group system, 196, 197
  Galα1,3Gal structure, 197, 198
  modification of exposed GlcNac
    moieties by galactosylation, 196
  *N*-acetyllactosamine structures, 195
  structures common to, 195–197
Glycerol. *See* Biodegradable films;
  Pectin; Polyol polyesters
Glycidyl methacrylate, graft on
  polyethylene hollow fibers, 69
Glycobiology
  research area, 193
  *See also* Bacterial
    lipopolysaccharides (LPS)
Glycosaminoglycans (GAGs)
  biomacromolecules, 218–219
  chondroitin (Ch), 218–219
  chondroitin sulfate (ChS), 218–219
  enzymatic polymerization, 223–
    225
  hyaluronic acid (HA), 218–219
  linear heteropolysaccharides, 218
  mechanisms of polymerization by
    hyaluronidase, 226, 227*f*
  monomer designs, 220–221
  monomer synthesis, 221, 222

*Pasteurella* GAGs and synthases, 233*t*
possible monomer designs for synthesis of HA and Ch, 220*f*
role in living systems, 218
synthesis of natural Ch, 223–224
synthesis of natural ChS, 224–225
synthesis of natural HA, 223
Glycosaminoglycan synthase enzymes
agarose gel analysis of hyaluronan (HA)-chondroitin hybrid polymers, 241*f*
agarose gel analysis of monodisperse products of synchronized, stoichiometrically controlled reactions, 240*f*
chemoenzymatic synthesis of defined oligosaccharides, 242–243
enzymes and reactors, 236
experimental, 236–237
model for controlling size of HA products, 238–239
model for monodisperse HA production, 238
production of chimeric GAG chains, 241
research aims, 244
stepwise oligosaccharide synthesis with enzyme reactors, 237
synchronized, stoichiometrically controlled polysaccharide synthesis, 236
Glycosidases
carbohydrate biosynthesis, 194
oligosaccharide synthesis, 198–200
protocol for synthesis of oligosaccharide, 199
Glycosyl hydrolases, identification, 26, 28*f*
Glycosyltransferases
biosynthesis of oligosaccharide, 199
carbohydrate biosynthesis, 194–195
glycosaminoglycan synthases, 234–235
oligosaccharide synthesis, 198–200
polymer modifications, 5
Graft copolymerization, enzyme binding, 67–68
Grafting
gelatin to chitosan, 109–111
green fluorescent protein to chitosan, 112–114
tyrosine-catalyzed protein, to polysaccharide, 108–114
*See also* Biomimetic approach to biomaterials
Green fluorescent protein
three-dimensional structure, 112*f*
tyrosinase-catalyzed grafting to chitosan, 112–114
*See also* Biomimetic approach to biomaterials
Guar
enzymatic reactions, 269–270
fatty acid ester of cationic, 431, 432*f*, 433, 435

# H

Heat tolerance, modeling proteinase K sequence activity relationship, 42–43
Hemicellulase, enzymatic reactions of guar, 269–270
Heparosan, glycosaminoglycan, 233
High throughput screening (HTS)
enzyme activity, 52
enzyme discovery, 32, 52
frequent failure, 47–48
*See also* Enzymes; Nitrilase activity
Hyaluronan (HA)
glycosaminoglycan, 233
model for controlling size of HA products, 238–239

model for monodisperse HA
    production, 238
Hyaluronic acid (HA)
    description, 218
    enzymatic polymerization, 218–219
    monomer designs, 220–221
    monomer synthesis, 221, 222
    possible monomer design, 220f
    synthesis of natural HA, 223
    See also Glycosaminoglycans (GAGs)
Hyaluronidase, polymerization mechanisms, 226, 227f
Hydrogel-biomineral composites
    analysis of mineral surface of composite, 100–102
    anionic or adhesive monomers for copolymerization with 2-hydroxyethyl methacrylate (HEMA), 99f
    assembling hydrogel scaffold containing multiple functional domains, 98f
    bone-like composites, 97
    calcium phosphate layer thickness, 103
    effects of anionic residues, 104–105
    evaluation of mineral-hydrogel interfacial adhesion, 100
    experimental, 97, 100
    features of mineralization process, 102
    HEMA polymers and copolymers, 97
    hydrogel preparation, 97
    linear heating rates, 102, 103f
    mineralization of hydrogels with urea-mediated process, 97, 100
    morphology, chemical composition, and microindentation analysis of calcium phosphate layer on pHEMA surface, 101f
    SEM–EDS method, 100
    synthetic hydroxyapatite (HA), 100
    urea-mediated mineralization strategy, 104
    X-ray diffraction method, 100
Hydrolases
    hydrolysis, 3
    polymerizations, 2–3
    polymer modifications, 3
    reaction with xanthan, 275
Hydrolysis
    biomass-to-ethanol production, 26, 29
    biotransformation of nitrile to corresponding acid, 53, 54f
    enzyme, of lignocellulosics, 24–26
    hydrolases, 3
    nitrilase activity, 60
    oxidoreductases, 4
    proposed scheme for, of hydrocinnamonitrile, 60f
    See also Biosilicification; Nitrilase activity
Hydroxyapatite (HA)
    synthetic HA, 100
    See also Hydrogel-biomineral composites
Hydroxyethylcellulose (HEC)
    fatty acid ester, 428–429, 432
    modification using enzymes, 247
    protease, 273
    See also Starch nanoparticles
2-Hydroxyethyl methacrylate (HEMA)
    anionic or adhesive monomers for copolymerization with, 98f, 99f
    hydrogel preparation, 97
    polymerization of pHEMA-based functional hydrogel network, 97
    See also Hydrogel-biomineral composites

Hyperbranched aliphatic polyesters
  biocompatibility, 355
  ε-caprolactone (ε-CL) and 2,2'-bis(hydroxymethyl) butyric acid (BHB), 357
  differential scanning calorimetry (DSC), 363f
  experimental data, 357t
  $^1$H NMR spectra, 358f
  immobilized lipase as catalyst, 357
  interaction in hyperbranched (hb) PCL-solvent system, 362
  intrinsic viscosities, 359, 361t, 362–363
  molecular weight characterization, 358–359
  monomers representing branched and linear units, 356f
  reduced viscosity as function of concentration, 361f
  scheme of combined ring-opening polymerization/polycondensation, 356f
  synthesis, 355–356
  thermal properties, 363–364
  vapor pressure osmometry (VPO), 359, 360f

**I**

Immobilization of enzymes
  adsorption mechanism, 72
  attachment of polyethylene glycol (PEG) spacer, 73
  biocatalysis, 7, 64
  carboxylic quantity and catalytic activity of bound enzyme, 72, 74
  covalent binding mechanism, 72
  cyclic activity of bound lipases, 77
  enzyme and catalytic activity, 71
  experimental, 71–72
  fibers for enzyme, 65–66
  functionalized polystyrene for enzyme, 66
  graft copolymerization on polymers and enzyme binding, 67–68
  grafted fibers for enzyme and protein binding, 68–70
  mechanisms, 72
  pH stability, 75, 76f
  polyethylene glycol spacer, 66–67
  stability upon exposure to organic solvents, 74, 75f
  surface grafting of polyelectrolyte and enzyme adsorption, 73
  thermal stability, 75, 76f
  ultra-fine cellulose fibers, 71
Infrared (IR) spectroscopy
  ε-caprolactone in supercritical $CO_2$, 396, 397f
  zein derivative, 144
  *See also* Fourier transform infrared (FT–IR) spectroscopy
Intrinsic viscosity, hyperbranched poly(ε-caprolactone), 361t, 362–363
Isomerases, polymer science, 5

**K**

Kolmogorov–Smirnov Two-Sample (K–S) Test, sorbitol-containing polyesters, 349, 351
Kymene® wet-strength resins. *See* Polyaminopolyamide-epichlorohydrin (PAE) resins

**L**

Laccase modification
  biocatalyzed oxidation of veratryl alcohol (VA), 87–89
  embedding in polymeric micelle, 81–82
  experimental, 83–84
  general procedure, 84
  instrumentation, 83–84

integrity of micellar enzyme complex and composition, 85, 87
materials, 83
mediators N-hydroxynaphthalimide (NHN) and 1-hydroxybenzotriazole (HBT), 82, 83*f*
necessity of linear-dendritic vs. linear-linear copolymer, 90–92
procedure for biotransformation of benzo-α-pyrene (BP) with laccase/linear-dendritic complexes, 84
purification of laccase from *Trametes versicolor*, 84*t*
reaction conditions for biocatalyzed oxidation of veratryl alcohol (VA), 88*t*
schematic of complex, 87*f*
schematic of linear-dendritic enzyme complex, 81*f*
size exclusion chromatography traces of polymer, complex, and enzyme, 86*f*
specific activity of complexes, 85*t*
specific activity of laccase, 87–88
usefulness of laccase complex, 89–90
UV spectra, 86*f*
UV-vis spectra of reactions with, 89*f*, 90*f*, 91*f*
L-Lactide. *See* Polylactide (PLA) copolymers
Lactones. *See* ε-Caprolactone (CL); Poly(caprolactone)s, substituted; Polylactide (PLA) copolymers
Ligases, polymer materials, 6
Light scattering, starch nanoparticles, 249, 258–261
Lignocellulosics
enzyme hydrolysis, 24–26
pretreatment technologies, 22–24
Linear-dendritic copolymers. *See* Laccase modification
Lipases
acylation of starch nanoparticles, 252
carboxylic quantity and catalytic activity of bound, 72, 74
catalytic activity, 71
cyclic activity of bound, 77
diffusion of starch nanoparticles in Novozym 435 beads, 261, 263
effects of solvents on reactivity, 320–322
enzymatic reactions of guar, 269–270
esterification reactions, 328
organosilicon carbohydrate esterification reaction, 184
pH stability, 75, 76*f*
polyesterification as function of diacid chain length, 323*f*, 324
stability on exposure to organic solvents, 74, 75*f*
thermal stability, 75, 76*f*
*See also* Immobilization of enzymes; Organosilicon carbohydrates; Poly(ε-caprolactone) (PCL); Poly(caprolactone)s, substituted; Polyesterifications, lipase-catalyzed; Polylactide (PLA) copolymers; Polymer modifications; Starch nanoparticles; Vinylethylglucoside
Lipopolysaccharides. *See* Bacterial lipopolysaccharides (LPS)
"Living factory" technology, oligosaccharide synthesis, 203
Lyases
polysaccharides, 5
*See also* BioPreparation™

## M

Maleation, coated starch nanoparticles, 251
Maleic anhydride, acylation of coated starch nanoparticles, 256
Mechanical properties, polyhydroxyalkanoate copolymers, 286–288
Mechanical testing
　initial modulus of pectin/poly(vinyl alcohol) (PVOH) films, 135, 137$f$
　pectin/PVOH films, 130–131, 132$f$, 133$f$
　pectin/starch/glycerol films, 124, 125$f$, 127
　procedure, 122
　properties of native and chemically modified zein, 144–146
　PVOH films, 130, 132$f$
Mechanism based design, protein engineering, 37–38
Mechanisms
　enzyme immobilization on fibers, 65–66
　immobilization by adsorption, 72
　immobilization by covalent binding, 72
　polyhydroxyalkanoate production by yeast, 297, 300
　polymerization by hyaluronidase, 226, 227$f$
Mediator $N$-hydroxynaphthalimide (NHN)
　oxidation of polyaromatic hydrocarbons, 82
　structure, 83$f$
　*See also* Laccase modification
Medical applications
　nanoparticles, 248
　polylactides, 406
Melt temperature
　glycerol and sorbitol terpolyesters, 338

hyperbranched poly($\varepsilon$-caprolactone), 363
polyhydroxyalkanoate copolymers, 286$f$
Methacrylic acid, copolymerization with vinylethylglucoside, 424, 425$t$
Micelles, reverse
　incorporation of starch nanoparticles, 250
　overcoming polar aprotic solvents, 247–248
Microbial biocatalysis, polymer synthesis, 6
Microbial dehalogenation
　combined Alcalase and, treatment of polyaminopolyamide-epichlorohydrin (PAE) resin, 310–311, 312$f$, 313$f$
　1,3-dichloropropanol (DCP), 304
　PAE resin, 305, 310
　resin performance for Alcalase and, 313, 314$f$
　*See also* Polyaminopolyamide-epichlorohydrin (PAE) resins
Microbial transglutaminase (mTG), protein crosslinking, 114–117
Mineralization strategy
　effects of heating rates on urea-mediated, 102, 103$f$
　urea-mediated, 104
　*See also* Biomineralisation; Hydrogel-biomineral composites
Modeling, sequence-activity. *See* Protein engineering
Modifications. *See* Laccase modification; Polymer modifications; Starch nanoparticles
Molecular weight characterization
　hyperbranched poly($\varepsilon$-caprolactone), 358–359
　poly(4-methylcaprolactone), 378
　polylactide copolymers, 408, 410–411

## N

N-acetylheparosan, glycosaminoglycan, 233
N-acetyllactosamine structures, glycans, 195, 196
Nanoparticles
  medical applications, 248
  See also Starch nanoparticles
National Renewable Energy Laboratories (NREL), biomass treatment, 20
Nitrilase activity
  biotransformation of nitrile to acid via hydrolysis, 54f
  cultivation and biocatalysis, 55
  diastereoselectivity and regioselectivity, 54f
  experimental, 55–56
  $^1$H NMR spectrum of hydrolysis of hydrocinnamonitrile, 56f
  hydrolysis, 60
  nitrile hydrolysis bioconversion of putative *Pseudomonas* gene, 58f
  NMR spectra for supernatants of *E. coli* whole cell cultures expressing *Pseudomonas* gene, 57f
  NMR spectra for supernatants of *E. coli* whole cell lysates expressing *Pseudomonas* gene, 59f
  nuclear magnetic resonance (NMR) analysis, 55–56
  organonitrile conversions to carboxylic acids, 53
  proposed scheme for hydrocinnamonitrile hydrolysis, 60f
  resolution of racemic 2-fluoroarylnitriles, 54f
  screening genomic library of *Pseudomonas* species in *E. coli*, 56–58

Nodax™
  biodegradation of polyhydroxyalkanoate (PHA) copolymers, 285f
  biological properties, 284, 286
  crystallinity, 287–288
  experimental, 284
  granules of poly(3-hydroxybutyrate-*co*-3-hydroxydecanoate) (PHBD) in *Ralstonia eutropha*, 283
  metabolic pathways to produce, 282f, 283
  PHA copolymers, 282
  polymer alloys, 289–290
  product design space, 288f
  prototype products, 289f
  thermo-mechanical properties, 286–288
  toughness of poly(lactic acid) (PLA)/PHA blends vs. blend composition, 290f
  utilization, 288, 289f
Novel materials, biomaterials, 7–8
Nuclear magnetic resonance (NMR)
  analysis during enzyme screening, 55–56
  $^{13}$C distortionless enhancement by polarization transfer (DEPT) NMR, 185, 186f
  hydrolysis of hydrocinnamonitrile by single *E. coli* clone, 56f
  hyperbranched polymers, 358
  poly(1,8-octanyladipate-co-glyceroladipate), 336f, 337f
  poly(1,8-octanyladipate-co-sorbityladipate), 333f, 334f
  poly(4-ethylcaprolactone) and poly(4-methylcaprolactone), 380–381
  poly(sorbityladipate), 330f, 331f
  polylactide copolymers, 408, 411
  supernatants of *E. coli* whole cell cultures expressing *Pseudomonas*, 57f

supernatants of *E. coli* whole cell
  lysates expressing
  *Pseudomonas*, 59f
zein derivative, 143–144
*See also* Nitrilase activity; Polyol
  polyesters

## O

1,8-Octanediol. *See* Polyol polyesters;
  Sorbitol-containing polyesters
Oligosaccharides
  bacterial coupling technology, 200–
    201
  biocatalytic synthesis, 197–203
  biosynthesis by
    glycosyltransferase, 199
  chemoenzymatic synthesis of
    defined, 242–243
  enzyme-based synthesis, 198–200
  α-Gal superbug, 203
  globotriose production with
    superbeads, 202
  large-scale production through
    coupling of engineered bacteria,
    201
  "living factory" technology, 203
  protocol for glycosidase-based
    synthesis, 199
  stepwise synthesis with enzyme
    reactors, 237
  "superbeads" or "superbug"
    technology, 201–202
  synthesis by biotransformation,
    200–203
  transferases, 5
  *See also* Bacterial
    lipopolysaccharide (LPS);
    Polysaccharides
Optically active polymers
  preparation and uses, 367–368
  *See also* Poly(caprolactone)s,
    substituted

Organic solvents, stability of bound
  and free lipases, 74, 75f
Organometallic catalysts. *See*
  Polylactide (PLA) copolymers
Organosilicon carbohydrates
  $^{13}$C distortionless enhancement by
    polarization transfer nuclear
    magnetic resonance (DEPT
    NMR), 185, 186f
  differential scanning calorimetry
    (DSC), 187, 188t
  electrospray ionization mass
    spectrometry (ESI MS) for
    characterization, 185–186,
    187f, 189f
  experimental, 183–184
  lipase-catalyzed esterification of
    diacid-endblocked siloxanes and
    α,β-ethylglucoside during
    synthesis, 185
  lipase-catalyzed esterification
    reaction, 184
  materials, 183–184
  monitoring reactions by thin layer
    chromatography (TLC),
    184
  number average molecular weight
    by gel permeation
    chromatography (GPC), 186
  physical properties, 183
  regioselectivity of esterification
    reactions, 184–185
  sweet silicones, 183
  thermal gravimetric analysis
    (TGA), 187, 188t
Oxidation, examples of enzyme
  catalysis, 31t
Oxidoreductases
  hydrolysis, 4
  polymer modifications, 4
  polymer synthesis, 4
Oxygen permeability
  pectin/starch films, 129
  test for films, 123

## P

PAE resins. *See* Polyaminopolyamide-epichlorohydrin (PAE) resins
Papain, pectin amide formation, 274
Paper makers. *See* Polyaminopolyamide-epichlorohydrin (PAE) resins
Particles. *See* Starch nanoparticles
Pasteurella
  chemoenzymatic synthesis of defined oligosaccharides, 242–243
  glycosaminoglycans (GAGs), 233
  recombinant synthases, 237
  synthases, 234
Pectin
  amide formation, 274
  β-galactosidase, 272–273, 275–276
  mechanical properties of pectin films, 123, 125f
  mechanical properties of pectin/poly(vinyl alcohol) (PVOH) films, 130–131, 132f, 133f
  mechanical properties of pectin/starch films, 124, 125f
  pectin/PVOH films, 130–138
  pectin/starch films, 123–130
  *See also* Biodegradable films
Pectin methyl esterase, enzymatic reactions of guar, 269–270
ω-Pentadecalactone (PDL)
  copolymerization with L-lactide, 410–412
  homopolymerization, 406
  polymerization with lipase, 409
  *See also* Polylactide (PLA) copolymers
pH
  grafting green fluorescent protein to chitosan, 112–114
  mineralization process, 102
  stability of free and bound lipases, 75, 76f

Pharmaceutical industry
  drug discovery phase, 32–33
  lead molecules in drug discovery, 34f
Polar aprotic solvents, reverse micelles, 247–248
Poly(acrylic acid)-grafted poly(ethylene terephthalate), trypsin immobilization, 68–69
Poly(acrylic acid)-grafted polysulfone, alkaline phosphatase, 68
Polyamide (PA), covalently bound thermolysin, 65
Polyaminopolyamide-epichlorohydrin (PAE) resins
  alcalase hydrolysis of polymer bound-(3-chloropropanediol) (PB-CPD) on, 307f
  alcalase treatment, 307, 308f, 309f
  by-products 1,3-dichloropropanol (DCP) and CPD, 302–303
  combined alcalase and microbial dehalogenation treatment, 310–311, 312f, 313f
  experimental, 304–306
  formation of PB-CPD during PAE manufacture, 306f
  manufacture, 303f
  microbial dehalogenation, 310
  microbial dehalogenation process, 304
  performance of alcalase and microbial dehalogenation treated PAE, 313, 314f
  proteases, 306–307
  third generation products, 303–304
  typical alcalase treatment, 305
  typical combined alcalase-microbial treatment, 305–306
  typical microbial treatment, 305
Polyaniline polymers, oxidoreductases, 4
Polyaromatic hydrocarbons
  oxidation reactions, 82, 84
  *See also* Laccase modification

458

Poly(ε-caprolactone) (PCL)
  apparent rate constant vs. pressure in scCO$_2$, 402f
  apparent rate constant vs. reaction temperature in scCO$_2$, 402f
  biocompatibility, 344
  enzymatic activity at high temperature in scCO$_2$, 398
  enzymatic activity in scCO$_2$ under different conditions, 401
  enzymatic activity in scCO$_2$ vs. toluene, 396, 398
  experimental for polymerization in scCO$_2$, 395
  infrared spectra of ε-CL in scCO$_2$, 397f
  matrix effect for enzymatic activity in scCO$_2$, 398, 401
  monomer conversion of ε-CL vs. reaction time in toluene and scCO$_2$, 397f
  monomer conversion vs. reaction temperature using incubated lipase, 399f
  monomer conversion vs. reaction time for polymerization using lipase in scCO$_2$, 400f
  relative monomer conversion by incubation temperature, 399f
  ring-opening polymerization in super critical carbon dioxide, 394–395
  See also Sorbitol-containing polyesters
Poly(caprolactone)s, substituted
  (±)-ε-caprolactones structures (1, 2, and 3), 369f
  chemo-enzymatic synthesis of enantio-enriched, 382–383, 388
  enantiomeric purity, 388–389
  enantioselective ring-opening of racemic lactones, 367–368
  enantioselective ring-opening polymerization of 4-ethyl-ε-caprolactone, 374, 375t
  enantioselectivity of lipase, 381–382
  enantioselectivity of lipase for resolution of racemic lactones, 372, 373f, 374
  enzymatic synthesis of (S)-poly(4-methyl-ε-caprolactone), 377–380
  experimental procedures, 385–389
  gas chromatograms of racemic lactones, 370f
  instrumental methods, 385–386
  lipase-catalyzed enantioselective ring opening of lactones 1, 2, and 3, 369
  lipase-catalyzed solvent-free kinetic resolution of substituted racemic ε-caprolactones, 368–369, 372, 374
  lipase-catalyzed solvent-less kinetic resolution of lactones 1, 2, and 3, 371t
  molecular weight characterization, 374–375, 376f
  monomer conversion vs. time for 4-ethyl-ε-caprolactone and 4-methyl-ε-caprolactone, 376f
  NMR characterization, 380–381
  optically active lactones, 367
  procedure for enzymatic polymerization, 387
  procedure for enzymatic resolution, 386–387
  procedure for synthesis of (S) enriched 4-substituted ε-caprolactone, 387–388
  reaction temperature vs. lactone conversion in ring-openings, 372, 373f
  representation of lipase active site and its binding to lactone, 382f
  ring-opening of optically active 4-ethylcaprolactone and 4-methyl-CL catalyzed by Sn(Oct)$_2$, 384t

solventless enantioselective ring-
opening polymerization by
Novozym-435 catalysis, 374–
381
Polycondensation
acid-catalyzed, of
tetraethoxysilane, 168, 170*f*
silicic acid, 166, 167
*See also* Hyperbranched aliphatic
polyesters
Polyelectrolyte, surface grafting and
enzyme adsorption, 72, 73
Polyesterifications, lipase-catalyzed
catalysis, 319
effect of monomer building block
on $DP_{avg}$, 322–325
effect of solvent, 320–322
experimental, 319–320
extent of chain growth as function
of time and substrates, 322*f*,
323*f*
general procedure for
polycondensation, 320
linear aliphatic ω-hydroxyacids,
324*f*
materials, 319–320
selectivity, 344
*See also* Polyol polyesters;
Sorbitol-containing polyesters
Polyesters
biocompatibility and
biodegradability, 344
covalently bound trypsin, 65–66
hydrolase-catalyzed
polymerizations, 2–3
microbial biocatalysis, 6
tissue engineering possibilities,
8
*See also* Hyperbranched aliphatic
polyesters; Polyol polyesters;
Sorbitol-containing polyesters
Polyethylene glycol (PEG)
covalent binding for grafting to
fibers, 72, 73
fatty acid diester, 430–431, 433

spacer in enzyme immobilizations,
66–67
*See also* Laccase modification
Polyethylene hollow fibers, glycidyl
methacrylate (GMA) graft on, 69
Polyethyleneimine (PEI),
immobilization to cellulose fibers,
69–70
Poly(ethylene terephthalate) (PET)
fibers
covalent binding of ligand proteins,
70
poly(acrylic acid) graft and trypsin
immobilization, 68–69
Polyhydroxyalkanoates (PHAs)
biodegradable copolymers, 281
biological properties, 284–286
crystallinity, 287–288
metabolic pathways for producing,
282*f*, 283
Nodax™ copolymers, 282
poly(3-hydroxybutyrate-*co*-3-
hydroxydecanoate) (PHBD),
283
thermal stability, 287
thermo-mechanical properties,
286–288
transmission electron micrograph
of *Ralstonia eutropha*
containing PHBD granules, 283*f*
utilization of Nodax™, 288, 289*f*
*See also* Nodax™
Polyhydroxyalkanoates (PHAs)
synthesis
analysis of PHA, 294
β-oxidation mechanism, 297, 300
classes of PHAs, 293
cloning procedure, 294, 295*f*
composition of medium chain
length (MCL)-PHA from
cytosol of yeast, 296–297
experimental, 293–294
expression of PHA synthesis
pathway in cytosol of
*Saccharomyces cerevisiae*, 300*f*

expression of *P. oleovorans* PHA
  polymerase in cytosol of yeast,
  296
GC–MS analysis of PHA by *S.
  cerevisiae* BY4743, 298*f*, 299*f*
PHA content and monomer
  composition by *S. cerevisiae*
  BY4743, 297*t*
PHA content and PHA monomer
  composition of polyester by *S.
  cerevisiae* BY4743, 299*t*
production of MCL-PHA, 296
strains and media, 293–294
vectors of PHA polymerase
  (phaC1) gene expression, 295*f*
Poly(2-hydroxyethyl methacrylate
  (pHEMA). *See* Hydrogel-
  biomineral composites
Poly(lactic acid) (PLA),
  biocompatibility, 344
Polylactide (PLA) copolymers
chemical catalysts, 406–407
copolymerization of ε-caprolactone
  (ε-CL) and L-lactide (LA), 407
copolymerization of ω-
  pentadecalactone (PDL) and LA
  with lipase and Sn(Oct)$_2$, 410–
  412
differential scanning calorimetry
  (DSC) of P(PDL-*co*-LA), 409,
  416*f*
experimental, 407–409
instrumentation, 408–409
integrating enzymatic and chemical
  polymerization, 409–411
material and methods, 407–408
medical applications, 406
polymerization procedure, 409
thermal decomposition of PDL/LA
  copolymers, 414*t*
thermal properties of poly(PDL-*co*-
  LA), 413
thermogravimetric analysis (TGA)
  of P(PDL-*co*-LA), 409, 415*f*
Polylactones, hydrolases, 2–3

Polymer alloys,
  polyhydroxyalkanoates (PHAs),
  289–290
Polymer degradation, hydrolysis via
  β-elimination reaction, 17*f*, 18
Polymer hydrolysis, oxidoreductases,
  4
Polymer modifications
amide derivative of
  carboxymethylcellulose, 429,
  432–433
experimental, 432–433, 435
fatty acid diester of poly(ethylene
  glycol), 430–431, 433
fatty acid ester of cationic guar,
  431, 432*f*, 433, 435
fatty acid ester of
  hydroxyethylcellulose, 428–429,
  432
hydrolases, 3
oxidoreductases, 4
substituted acrylic monomers, 429–
  430, 433
summary of enzyme-catalyzed
  condensation reactions, 434*t*
transferases, 5
*See also* Laccase modification
Polymer reactions
enzyme biocatalysis, 2–6
use of biocatalysts, 2*t*
whole-cell biocatalysis, 6
Polymer synthesis
hydrolase-catalyzed
  polymerizations, 2–3
oxidoreductases in, 4
transferases in, 5
whole-cell biocatalysis, 6
Polymerization. *See*
  Glycosaminoglycans (GAGs);
  Vinylethylglucoside
Poly(methyl methacrylate), covalently
  binding of polyethylene glycol,
  67
Polyol polyesters
activity retention of enzyme, 338

$^{13}$C NMR spectrum of poly(1,8-octanyladipate-*co*-glyceroladipate), 337*f*
$^{13}$C NMR spectrum of poly(1,8-octanyladipate-*co*-sorbityladipate), 333*f*, 334*f*
$^{13}$C NMR spectrum of poly(sorbityladipate), 331*f*
copolymerization of adipic acid, sorbitol, and 1,8-octanediol, 332
direct condensation of adipic acid and sorbitol, 329, 332
experimental, 339–341
glycerol as polyol for poly(glyceryladipate), 335, 338
$^{1}$H NMR spectrum of poly(1,8-octanyladipate-*co*-glyceroladipate), 336*f*
$^{1}$H NMR spectrum of poly(sorbityladipate), 330*f*
instrumentation, 339–340
lipase-catalyst Novozym-435 catalyzed polymerization of sorbitol and glycerol, 329
procedure for lipase-catalyzed syntheses, 340–341
role of enzyme, 338
synthesis of aliphatic polyesters with sorbitol and glycerol, 335*t*
thermal stability, melting transitions, and crystallinity, 338
*See also* Sorbitol-containing polyesters
Polyphenols, oxidoreductases, 4
Polypropylene fabrics, grafting 2-vinyl-4,4-dimethylazlactone, 69
Polysaccharides
enzymatic reactions of guar, 269–270
hydrolases, 3
lyases, 5
microbial biocatalysis, 6
modification using enzymes, 247–248

pectin and β-galactosidase, 272–273
synchronized, stoichiometrically controlled synthesis, 236
transferases, 5
xanthan and cellulase, 271–272
*See also* Bacterial lipopolysaccharides (LPS); Biodegradable films; Glycosaminoglycans (GAGs); Oligosaccharides; Starch nanoparticles
Polystyrene, functionalization for enzyme attachment, 66
Polysulfone, alkaline phosphatase on grafted, 68
Poly(vinyl alcohol) (PVOH)
dissolution rate of pectin/PVOH films, 135, 138
elongation to break pectin/PVOH films, 135, 136*f*
initial modulus of pectin/PVOH films, 135, 137*f*
mechanical properties of pectin/PVOH films, 130–131, 132*f*, 133*f*
mechanical properties of pure PVOH films, 130, 132*f*
scanning electron microscopy (SEM) of pectin/PVOH films, 131, 134*f*, 135
*See also* Biodegradable films
Pressure, rate constants for polymerization of ε-caprolactone in supercritical $CO_2$, 401, 402*f*
Pressurized hot wash, lignocellulosic conversion, 22, 23–24
Pretreatment technologies, lignocellulosic conversion, 22–24
Process integration, bioethanol, 24
Proteases
catalyzed condensation study, 172–174
enzymatic reactions of guar, 269–270

ester formation, 273, 276
esterification reactions, 328
Protein
  attachment to poly(ethylene glycol)-grafted surfaces, 67
  grafting gelatin to chitosan, 109–111
  grafting green fluorescent protein to chitosan, 112–114
  grafting to polysaccharide by tyrosinase catalysis, 108–114
  incorporation in fibers, 70
  transglutaminase catalyzed crosslinking, 114–117
  *See also* Biomineralisation
Protein engineering
  activity of designed 1st and 2nd round variants toward modified tetrapeptide, 44$f$
  advantages and disadvantages, 38
  amino acid variations and functions, 45–48
  amino acid variations to alter substrate specificity, 40$f$
  beneficial amino acid variations, 47$t$
  creation and testing of initial variants, 41
  design, synthesis and testing of improved proteinase K, 43–45
  empiricists, 38
  failure of high throughput screens, 47–48
  mechanism-based design, 37–38
  modeling proteinase K sequence activity relationship for heat tolerance, 42–43
  principal component analysis of proteinase K variant activities, 46$f$
  process using sequence activity relationships, 39$f$
  regression coefficients of variant amino acids, 43$f$

selection of initial amino acid substitutions, 39–41
sequence-activity based strategies, 38–39
sequence-activity relationships, 38
Proteinase K
  beneficial amino acid variations, 47$t$
  cloning, from *Tritirachium album*, 39–41
  design, synthesis and testing of improved, 43–45
  modeling sequence activity relationship for heat tolerance, 42–43
  principal component analysis of variant activities, 46$f$
Pseudomonas
  cultivation and biocatalysis, 55
  library screening of genomic DNA from strain, 53, 54$f$
  *See also* Nitrilase activity


## R

Racemic seven-membered lactones. *See* Poly(caprolactone)s, substituted
Ralstonia eutropha
  poly(3-hydroxybutyrate-*co*-3-hydroxydecanoate) granules in, 283
  *See also* Nodax™
Rate constant, polymerization of ε-caprolactone in supercritical $CO_2$, 401, 402$f$
Reaction temperature, effect on starch acylation, 255
Reduced viscosity, hyperbranched poly(ε-caprolactone), 361$f$
Reduction, examples of enzyme catalysis, 31$t$
Regioselectivity
  esterification reactions, 185

nitrilase, 53, 54f
Reverse micelles
　incorporation of starch
　　nanoparticles, 250
　overcoming polar aprotic solvents,
　　247–248
Ring-opening polymerization
　ε-caprolactone, 356f
　solventless enantioselective, of
　　substituted ε-caprolactones,
　　374–377
　See also Hyperbranched aliphatic
　　polyesters; Poly(caprolactone)s,
　　substituted
Roughness, sorbitol-containing
　polyesters, 347, 348f

## S

Saccharomyces cerevisiae
　bioethanol production, 29
　See also Polyhydroxyalkanoates
　　(PHAs) synthesis
Scanning electron microscopy (SEM)
　diatom *Aulacoseira granulata*, 151f
　pectin/poly(vinyl alcohol) films,
　　131, 134f, 135
　pectin/starch films, 124, 126f, 127
　procedure, 122
　protein-mediated germania, 159f
　protein-mediated silica particles,
　　155f, 156f, 157f, 158f
　See also Biodegradable films
Scouring, cleaning process, 15
Screening
　biocatalysts, 52
　genomic library of *Pseudomonas*
　　species in *E. coli*, 56–58
　high throughput analysis, 53
　See also High throughput screening
　　(HTS); Nitrilase activity
Separate hydrolysis and fermentation
　bioethanol plant, 21f
　bioethanol production, 26, 29

Sequence activity modeling
　proteinase K for heat tolerance, 42–
　　43
　protein engineering process, 38–39
　See also Protein engineering
Silanol condensation
　acid- and base-catalyzed, 168,
　　169
Silica. See Biomineralisation
Silicic acid, polycondensation, 166,
　167
Silicification
　experimental, 153–155
　proteins facilitating, 151–152
　silica precursors for, 154t
　See also Biomineralisation;
　　Biosilicification
Silicones. See Organosilicon
　carbohydrates
Siloxane bonds
　enzyme-catalyzed formation, 165,
　　166
　See also Biosilicification
Simultaneous saccharification and
　fermentation (SSF)
　bioethanol plant, 21f
　bioethanol production, 26, 29
Simultaneous saccharification-
　cofermentation (SSCF), bioethanol
　plant, 21f
Size exclusion chromatography
　(SEC). See Laccase modification
Solvents
　effects on lipase reactivity, 320–
　　322
　See also Polyesterifications, lipase-
　　catalyzed; Supercritical carbon
　　dioxide ($scCO_2$)
Sorbitol-containing polyesters
　atomic force microscopy images of
　　poly(adipic acid-*co*-octanediol-
　　*co*-sorbitol) (PAOS) and poly(ε-
　　caprolactone) (PCL), 348f
　automated sample collecting and
　　sample analysis, 352f

cell density on PCL and PAOS films, 352f
cell spreading and proliferation, 349, 351
characterization of polymer surface, 347, 349
fluorescence microscopy images of fibroblast on PCL and PAOS films, 350f
Kolmogorov–Smirnov Two-Sample (K–S) Test results, 349, 351t
lipase-catalyzed polymerizations, 344
materials and methods, 344–345
monomer feed ratios, polymer compositions and molecular weights, 345t
PAOS preparation, 346f
roughness images of PAOS and PCL, 348f
static and advancing water contact angles of PAOS and PCL films, 350f
synthesis and characterization, 345
*See also* Polyol polyesters
Spacers, enzyme immobilizations, 66–67
Specificity, enzymes towards substrates, 268
Starch nanoparticles
acylation of anionic surfactant Aerosol–OT (AOT) coated nanoparticles with ε-caprolactone (CL), 256–257
acylation of surfactant-coated, nanoparticles with maleic anhydride, 256
acylation of surfactant coated starch nanoparticles with vinyl stearate, 250–251
acylation using free lipase SP–525, 252
characterization of unmodified and modified nanoparticles by light scattering, 258–261
description, 247
diffusion of, nanoparticles into Novozym 435 beads, 261, 263
distribution of particle diameter of nanoparticles, 262f
effect of acylating agent chain length and activation on its activity, 256t
effect of reaction temperature on starch acylation, 255
effect of substrate chain length on acylation, 255–256
experimental, 250–252
formation of surfactant coated nanoparticles, 252
incorporation of *Candida antartica* lipase B (CALB) within AOT coated nanospheres, 263–264
incorporation of nanoparticles in different surfactant systems, 252t
incorporation of nanoparticles within reverse micelles, 250
incorporation of non-immobilized CALB within nanoparticles, 250
instrumental methods, 249–250
lipase-catalyzed acylation of surfactant-coated nanoparticles, 252–254
materials and methods, 248–250
modification using enzymes, 247–248
nanoparticles, 248
reaction progress with time, 257–258
*See also* Biodegradable films
Subtilisin enzyme (Alkalase). *See* Polyaminopolyamide-epichlorohydrin (PAE) resins
Sugar polymer engineering. *See* Glycosaminoglycan synthase enzymes; Vinylethylglucoside

Sulfuric acid, dilute for lignocellulosic conversion, 22, 23
"Superbeads" or "superbug" technology, oligosaccharide synthesis, 201–202
Supercritical carbon dioxide (scCO$_2$)
 enzymatic activity in, under different conditions, 401, 402$f$
 enzymatic activity in, vs. toluene, 396, 398
 enzymatic activity incubated at high temperature in, 398
 enzyme-catalyzed polymerization, 394–395
 matrix effect for enzymatic activity in, 398, 401
 polymerization of ε-caprolactone, 395
 See also Poly(ε-caprolactone) (PCL)
Surface characterization, sorbitol-containing polyesters, 347, 349
Surface roughness, sorbitol-containing polyesters, 347, 348$f$
Surfactant Aerosol–OT (AOT). See Starch nanoparticles
Sweet silicones. See Organosilicon carbohydrates

## T

Temperature
 effect of reaction, on starch acylation, 255
 enzymatic activity of Novozym-435 in supercritical CO$_2$, 398, 399$f$, 400$f$
 heating rates and gel-mineral composites, 102, 103$f$
 mineralization process, 102
 rate constants for polymerization of ε-caprolactone in supercritical CO$_2$, 401, 402$f$
Tensile properties, native and chemically modified zein, 145–146
Tetramethoxysilane (TMOS) silica precursor, 154
 See also Biomineralisation
Textile preparation. See BioPreparation™
Thermal dynamic mechanical analysis (TDMA)
 effect of glycerol level on pectin by, 123–124, 125$f$
 procedure, 122
 See also Mechanical testing
Thermal gravimetric analysis (TGA)
 organosilicon carbohydrates, 187, 188$t$
 polylactide copolymers, 409, 413, 414$t$, 415$f$
Thermal properties, hyperbranched poly(ε-caprolactone), 363–364
Thermal stability
 free and bound lipases, 75, 76$f$
 glycerol and sorbitol terpolyesters, 338
 polyhydroxyalkanoate copolymers, 287
Thermolysin, immobilization on polyamide, 65
Thermo-mechanical properties, polyhydroxyalkanoate copolymers, 286–288
Time, starch acylation reaction progress with, 257–258
Tin octanoate [Sn(Oct)$_2$]
 integrating enzymatic and chemical polymerization, 409–412
 lactide polymerization, 406–407
 See also Polylactide (PLA) copolymers
Tissue engineering, polyesters, 8
Toluene, enzymatic activity of ε-caprolactone polymerization vs. supercritical CO$_2$, 396, 398
Trametes versicolor

catalytic activity of laccase from, 82
See also Laccase modification
Transferases, polymer synthesis and modifications, 5
Transglutaminase
 activity, 114
 protein crosslinking, 114–117
 See also Biomimetic approach to biomaterials
Transglutaminases, polymer modifications, 5
Transmission electron microscopy (TEM), poly(3-hydroxybutyrate-co-3-hydroxydecanoate) in Ralstonia eutropha, 283f
Trichoderma reesei, hydrolysis of lignocellulosics, 24
Trimethylsilanol, enzyme-catalyzed condensation, 171–172
Tritirachium album, cloning proteinase K from, 39–41
Trypsin
 catalyzed condensation of trimethylsilanol, 173–174
 catalyzed hydrolysis and condensation of alkoxysilanes, 174–177
 catalyzed hydrolysis and condensation of trimethylethoxysilane, 175f
 formation of siloxane bonds, 176–177
 immobilization on grafted poly(ethylene terephthalate) fibers, 68–69
 immobilization on polyester, 65–66
 proteinaceous inhibition, 178–179
 turnover numbers of catalyzed hydrolysis of trimethylethoxysilane and condensation of trimethylsilanol, 176f
 See also Biosilicification
Tyrosinase
 activity, 108–109
 gel formation of gelatin-chitosan mixture, 110f
 grafting gelatin to chitosan, 109–111
 polymer modifications, 4
 protein grafting to polysaccharide, 108–114
 See also Biomimetic approach to biomaterials

## U

Ultraviolet spectroscopy. See Laccase modification
Urea-mediated mineralization. See Hydrogel-biomineral composites
U.S. Department of Energy (DOE), biomass treatment, 20

## V

Vapor pressure osmometry (VPO), hyperbranched poly(caprolactone), 359, 360f
Veratryl aldehyde (VA)
 activity of laccase, 87–89
 reaction conditions for biocatalyzed oxidation, 88t
 See also Laccase modification
2-Vinyl-4,4-dimethylazlactone, grafting to polypropylene fabrics, 69
Vinylethylglucoside
 chemoenzymatic synthesis, 423
 $^{13}C$ NMR spectrum of ethyl glucoside acrylate, 25, 424, 425f
 copolymerization with methacrylic acid using AIBN, 424, 425t
 experimental, 421–423
 homopolymerization using AIBN, 424
 instrumentation, 422

materials and methods, 421–423
procedure for free radical polymerization, 423
procedure for lipase-catalyzed synthesis, 422
Vinyl stearate, acylation of coated starch nanoparticles, 250–251, 253–254

## W

Wet-strength paper. *See* Polyaminopolyamide-epichlorohydrin (PAE) resins
Whole-cell biocatalysis, polymer synthesis, 6, 438

## X

Xanthan
   cellulase, 271–272
   reaction of hydrolases with, 275

## Y

Yeast. *See* Polyhydroxyalkanoate (PHA) synthesis

## Z

Zein
   description and uses, 142
   examples of potential chemical derivatives, 143*f*
   experimental, 146–147
   infrared (IR) spectrum of derivative, 144*f*
   mechanical properties of native and chemically modified, 144–146
   nuclear magnetic resonance of derivative, 143*f*
   objective of research, 142
   reaction procedure, 146
   reactions of, 142–144
   tensile properties of native and chemically modified, 145*t*